principles
of geomorphology

don j. easterbrook

professor and chairman
department of geology
western washington state college

principles
of geomorphology

McGraw-Hill book company

new york   st. louis   san francisco   london
sydney   toronto   mexico   panama

principles of geomorphology

Library of Congress Catalog Card Number 69 - 13207

18780

1234567890 HDBP 7654321069

To the memory of

J. Hoover Mackin

An inspiring teacher, brilliant
analyst, and beloved friend who
contributed so much to geomorphology

preface

The study of geomorphology involves complex analyses of physical and chemical processes and includes a voluminous literature in the professional journals, with an annual increase in published papers that is rising rapidly. Therefore, in writing a text on the subject, it is obviously not possible to be all-inclusive and remain within the confines of readability and be digested in a single quarter or semester by geomorphology students who have had only a single course in physical geology. Since this book is intended primarily for the undergraduate student rather than the professional geomorphologist, fundamentals are stressed, and this necessitates omission or mere cursory coverage of many other aspects. To those readers who may note that this or that concept or favorite subject has been passed over briefly, the author can only plead that some selectivity in choosing subject matter and depth of coverage was required. It was often a painful task to cut out or limit certain discussions in the interest of remaining within the scope of the text.

Selected references to most topics are listed at the end of each chapter for students who may wish to refer to additional reading material. In choosing the references an attempt was made to include a blend of classic and modern literature which should be readily available in most university libraries and which should not prove unduly technical for undergraduate students. The lists are not intended to be in any sense exhaustive, and the author begs indulgence of the authors of the many excellent papers which have not been included.

From among several possible arrangements of chapters, the sequence finally selected was dictated in part by the manner in which the author has taught geomorphology for a number of years. Others who treat subjects in a different sequence should, with little difficulty, find the chapters interchangeable.

In most colleges and universities in the United States, courses in geomorphology are taught at the sophomore or junior level, followed by more advanced and specialized courses in glacial geology, fluvial morphology, map interpretation, etc. However, some institutions do not offer geomorphology until the senior or graduate level, or offer only the more specialized courses. Although this book is written primarily for the former, it may also prove useful for the latter.

Expressions of gratitude are due to the late Professor J. Hoover Mackin of the University of Texas, Professor Stephen Porter of the University of Washington, and Dr. Mark Meier of the U.S. Geological Survey for reading parts of the manuscript and offering helpful suggestions. The responsibility for any shortcomings in the text lies, of course, with the author.

Don J. Easterbrook

contents

- *preface*   *vii*

- part 1   fundamentals of geomorphology   *1*
  - *chapter 1*   *basis of geomorphology*   *3*
  - *chapter 2*   *interpretation of maps and aerial photographs*   *17*

- part 2   glaciation   *31*
  - *chapter 3*   *glaciers*   *33*
  - *chapter 4*   *alpine glaciation*   *55*
  - *chapter 5*   *continental glaciation*   *75*

- part 3   fluvial morphology   *111*
  - *chapter 6*   *stream processes*   *113*
  - *chapter 7*   *origins of stream valleys and drainage patterns*   *145*
  - *chapter 8*   *evolution of fluvial landforms*   *165*
  - *chapter 9*   *rejuvenation*   *179*

- part 4   valley-side process   *197*
  - *chapter 10*   *weathering*   *199*
  - *chapter 11*   *mass movement*   *217*
  - *chapter 12*   *evolution of slopes*   *239*

- part 5   groundwater   *247*
  - *chapter 13*   *subsurface water*   *249*

- part 6   desert and eolian landforms   *267*
  - *chapter 14*   *desert landforms*   *269*
  - *chapter 15*   *eolian landforms*   *289*

- part 7   coastal morphology   *305*
  - *chapter 16*   *shorelines*   *307*

- part 8   relationships of geologic structure to topography   *337*
  - *chapter 17*   *folded sedimentary rocks*   *339*
  - *chapter 18*   *topography associated with faulting*   *361*
  - *chapter 19*   *igneous landforms*   *381*
  - *appendix*   *411*
  - *glossary*   *420*
  - *index*   *445*

1

fundamentals
of geomorphology

basis of geomorphology

introduction

Geomorphology is the study of the origin and evolution of topographic features by physical and chemical processes operating at or near the earth's surface. The name stems from the Greek terms *geo*, meaning earth, *morphe*, meaning form, and *logos*, meaning discourse. The study of geomorphology is based on the principle that all landforms can be related to a particular geologic process, or set of processes, and that the landforms thus developed may evolve with time through a sequence of forms dependent in part on the relative time a particular process has been operating. Underlying all geologic principles, and in fact all scientific principles, is the concept of uniformitarianism, which states that the natural physical and chemical processes which operate today have operated in the past and will continue to do so in the future. The concept was first succinctly stated in 1785 by James Hutton, who taught that the present is the key to the past. Study of present phenomena which can be observed in operation allows extrapolation into the past to infer the origin of landforms which are the effects of past processes.

Grand Tetons, Wyoming, a fault block range. Jenny Lake in foreground is held in by glacial deposits. *(Austin Post, University of Washington.)*

Building upon the work of his predecessors, William Morrison Davis in 1909 synthesized many earlier geomorphic ideas and molded them into a unified system for the study of landforms. He recognized that the origin and evolution of topography was dependent on geologic structure, geomorphic process, and stage of development. This concept provides the framework for interpretive geomorphology, and thus its understanding is imperative to students of geomorphology.

geologic structure

Topography in many areas is a direct reflection of the underlying geology. Included under the category of geologic structure are not only folding, faulting, and uplift of the crust, but also other factors related to the physical and chemical characteristics of the rocks. Among them are relative resistance to mechanical and chemical weathering, dip and strike, jointing, stratification, foliation, and unconformities.

Of particular importance is the reflection of rock resistance in topographic development (Fig. 1-1). Rocks differ considerably in their weathering and erosive characteristics. Some are resistant to both mechanical and chemical weathering, some to mechanical but not chemical, and vice versa, and some are not very resistant to either type of weathering. Rocks may also vary in resistance under differing climatic conditions. A good example of this is limestone, which is generally susceptible to solution in humid climates and tends to produce lowlands. However, in arid climates, where water necessary for solution is deficient, limestone beds stand up as resistant ridges.

Generally speaking, shale, limestone, marble, and some types of schist are less resistant "valley-makers" in humid climates, whereas sandstone, quartzite, conglomerate, and various igneous rocks are resistant "ridge-makers" (Fig. 1-2). Other rock types vary more widely in their characteristics, depending on specific composition, texture, and the climatic conditions where they occur.

The effect of diastrophism must also be considered in landform development. Differences in folding and faulting among various areas may affect the spacial relationships of rocks of differing resistances. Excellent examples of this occur in the Valley and Ridge province of the Appalachian Mountain region in Pennsylvania, where the outcrop of resistant ridge-making sandstones and conglomerates is controlled by folding and subsequent erosion. In areas of recent faulting, scarps may be produced along the trace of the fault, and in areas of ancient faulting, rocks of differing resistance may be juxtaposed in such a manner as to affect topography.

Most of the major regional topographic features of the earth's surface, such as mountain ranges, plateaus, plains, and ocean basins, are directly or indirectly related to crustal warping and deformation. The reason for many such major features is geologically recent uplift which exceeds rates of weathering and erosion.

geomorphic process

Each geomorphic process imparts to the landscape distinctive features and develops characteristic assemblages of landforms. Since landforms bear the stamp of the process that developed them, it becomes possible not only

Fig. 1-1 Cliffs produced by differential erosion of resistant and weak sedimentary beds, Grand Canyon, Arizona. *(Aero Service Division, Litton Industries.)*

to describe landforms, but also to infer their origin. Interpretation of the origin and evolution of landforms is the principal objective in geomorphology.

Among the processes which shape the configuration of the earth's surface are the following:

RUNNING WATER: The precipitation which falls on the earth runs off on the surface, soaks into the ground, or evaporates back into the atmosphere. That portion which runs off on the surface of the land eventually collects into streams which continuously erode the land and deposit the material elsewhere. Landscapes sculptured by fluvial erosion bear characteristic features which differ from those developed by other processes.

Fig. 1-2   Alternating ridges and valleys developed on folded sedimentary beds of differing resistance to weathering and erosion, Loveland, Colorado. *(U.S. Department of Agriculture.)*

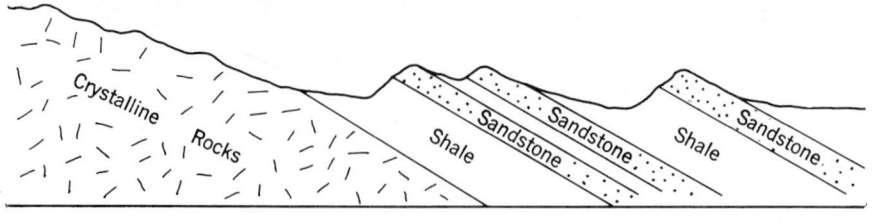

GLACIERS: Glacial ice differs from running water and other surface processes in erosive and depositional characteristics, and hence produces landforms markedly different in form. Glaciers move more slowly downslope than do streams, but are nevertheless capable of carrying large quantities of material derived by erosion from valley sides and bottoms.

GROUNDWATER: Some of the precipitation which falls from the atmosphere seeps into the ground, where it is stored until it emerges along valley sides and floors. While in contact with rock material, groundwater promotes solution and other types of chemical weathering. Removal of the weathered and dissolved material leads to the development of unique landforms, espe-

Fig. 1-3 Drainage network developing by headward erosion, McLean County, North Dakota. *(U.S. Department of Agriculture.)*

cially in areas of readily soluble rocks, such as limestone. Heating of ground-water may result in hot springs and geyser activity.

WAVES AND CURRENTS: Shorelines of oceans, seas, and large lakes are modified continuously by the abrasive action of waves beating against the shore and deposition of material by wave and current action.

WIND: Wind is a less vigorous agent of erosion and transportation of material than water, but in arid to semiarid regions, or areas having an abundant supply of loose sand, wind is locally an important topography-producing agent. The shifting sand dunes of some deserts provide an example of wind-developed topography.

Fig. 1-4 Meandering stream with cutoff meanders and oxbow lakes, Tallahatchie River, Mississippi. *(U.S. Department of Agriculture.)*

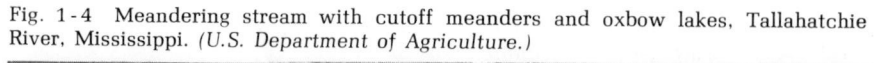

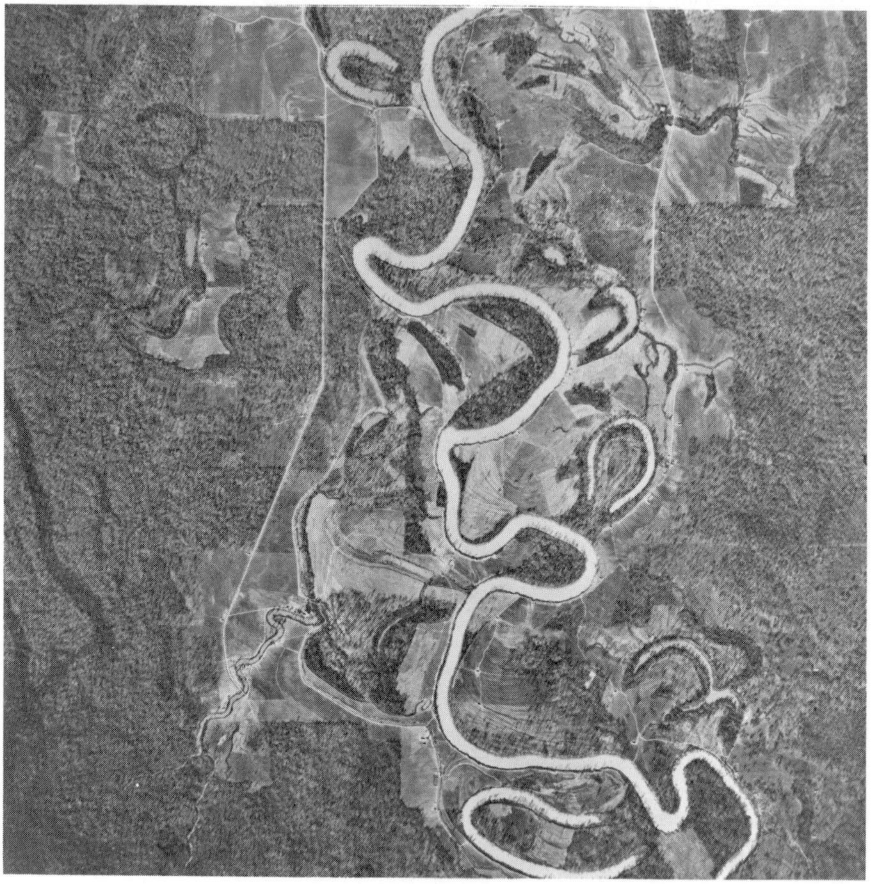

WEATHERING: Mechanical disintegration and chemical decomposition of rocks causes them to be broken up into smaller pieces. In those areas where rocks offer differing resistance to weathering, differential weathering etches out weak rock zones, and lowlands develop. In areas where mechanical weathering is dominant, the topography develops angular hillslopes, whereas in areas dominated by chemical weathering, smooth, rounded slopes are developed.

VOLCANISM: Eruption of lava on the surface produces very distinctive types of landforms, which, if not too old, are easily recognized. These include such features as volcanic cones, lava flows, and calderas.

Fig. 1-5   Modification of a mountainous region by alpine glaciation. *(Austin Post, University of Washington.)*

DIASTROPHISM: Deformation of the earth's crust by tensional and compressional forces may produce initial landforms or alter the spatial relationships of rocks of differing resistance in such a way as to cause certain landforms to develop under differential weathering. Among common topographic features produced initially by diastrophism are scarps developed along faults, uplifted mountains and plateaus, and downwarped basins.

stage of development

The topography of a region evolves through a continuous sequence of forms having distinctive characteristics at successive stages of development. This

Fig. 1-6 Commonwealth Glacier, a lobe of ice fed by ice sheet in the background. Asgard Range, Antarctica. *(U.S. Navy for the U.S. Geological Survey.)*

basic concept was put forth by Davis (1909), who recognized that the topography in any area is constantly undergoing changes because of the influence of surface processes and that, depending on the intensity of a process and how long it has been active in an area, even landforms developed by the same process will differ in detail. Davis recognized three stages, which grade imperceptibly into one another but which possess certain characteristics differing from those of other stages. The youthful stage is characterized by initial forms which progressively change as maturity is reached. The final product of long-continued erosion is the old-age stage. Although all three stages involve successive increments of time, there is no implication that the stages are of equal duration or that they can be measured in absolute terms, such as years.

Fig. 1-7   Mt. Erebus, a volcanic cone with steam rising from the summit. McMurdo Sound, Antarctica. *(U.S. Navy for the U.S. Geological Survey.)*

Because the earth's crust is dynamic in nature, a given area seldom passes completely from youth to old age without interruptions. Diastrophic or climatic changes may interrupt at any stage, and in fact such interruptions are to be expected. Thus partial cycles may be more common than completed cycles.

Two fundamental concepts, common to areas other than geomorphology and geology, i.e., *equilibrium* and *evolution*, are referred to throughout the text. Equilibrium refers to a state of balance between opposing forces, and evolution involves a gradual progressive change with time. Each of these concepts underlies the understanding of surface physical and chemical processes.

qualitative and quantitative approaches to geomorphology

The qualitative approach in geomorphology deals primarily with the understanding of phenomena which may or may not be based on numerical data. Emphasis is placed on verbalizing an idea or concept. In particular, the qualitative approach attempts to draw together relationships between cause and effect, and in so doing may rely heavily on mathematical and experimental data; the goal lies not in the numbers themselves, but rather in what they demonstrate. Much depends on reasoning and evaluation of data at each step in an investigation.

The quantitative approach in geomorphology places much emphasis on measurement and the gathering of numerical data. In this sense the approach may be intricately interwoven with the qualitative approach, and a distinction between the two may be difficult to make. For example, in studying a stream, one soon becomes aware that the size of particles being transported decreases downstream. Since particle size is easily measured, it is possible to determine quantitatively, by various statistical measures, the decrease in particle size downstream, and these data could then be used to formulate an equation showing the rate of average particle-size decrease per unit length of stream for that river. If the approach is purely quantitative, the investigation might end with the numbers and the equation which shows *how much* particle size decreases downstream for that particular river. However, such an investigation would lose much of its interest if it did not take the next step into the realm of qualitative analysis to ask the question, *why* does particle size decrease downstream? In this phase of investigation, emphasis shifts from the gathering of numerical data to the rational pursuit of cause-and-effect relationships; i.e., having observed a specific effect, the decrease in particle size downstream, the investigator turns his attention to the cause of the effect.

From the example cited, it should be obvious that the purely quantitative and qualitative approaches are somewhat different, i.e., the *how much* versus *why*, but neither approach is in an absolute sense better than the other. They are simply aimed at slightly different goals and complement each other. A qualitative statement of a concept may well have been derived from a quantitative investigation.

Since about 1940 there has been a noticeable increase in attempts to quantify various phases of geomorphology (examples are Horton, 1940; Strahler, 1950; Leopold and Wolman, 1957). As pointed out by Mackin (1963), the principal difference between qualitative and quantitative approaches is not as significant as the difference between the *rational* and *empirical* approaches. The rational approach involves critical evaluation of various lines of evidence in solving a problem. Each piece of evidence is examined for accuracy and relevance to the problem at hand, resulting in rejection of irrelevant evidence and focusing attention on evidence most critical to establishing a conclusion. Quantitative evidence may play an important role in the investigation, but each number is analyzed qualitatively for relevance to the problem before being used. The empirical method, on the other hand, places great stress on the mathematical, or quantitative, approach, minimizing relationships between cause and effect and considering all available numerical data. The end result is usually stated in the form of an equation or numerical expression which has been arrived at by plotting a large number of points on a graph and, where necessary, averaging divergent points on the graph. This method has been referred to by some as the "shotgun" approach, because there may be little in the way of evaluation of the individual points before they are placed on the graph. Perhaps the most dramatic way to visualize the point is to consider a person standing with one foot on a block of ice and the other in a bucket of boiling water—*on the average* the temperature may be comfortable, but when each is considered separately, the conclusion may not be valid. However, this approach does have advantages, and it is a very useful method in certain instances. Very often relationships which might not otherwise be envisioned show up when large amounts of numerical data are plotted. However, the point made by Mackin (1963) is that use of the empirical method alone may too often lead to erroneous conclusions because cause-and-effect relationships are not taken into account.

The approach generally followed in this text is largely rational, with quantitative data included where they are particularly meaningful in arriving at a qualitative conclusion. Lengthy mathematical calculations have, for the most part, been held to a minimum. Published papers containing mathematical treatments of various subjects are included in the references at the end of each chapter for those who wish to pursue the quantitative approach more fully.

references

Chamberlain, T. C.: 1897. The method of multiple working hypotheses, *Jour. Geology*, vol. 5, pp. 837-848.

Chorley, R. J., A. J. Dunn, and R. P. Beckinsale: 1964. The History of the Study of Landforms or the Development of Geomorphology, vol. 1, Geomorphology before Davis, John Wiley & Sons, Inc., New York.

Davis, W. M.: 1909. Geographical Essays, Ginn and Company, Boston (reprinted 1954 by Dover Publications, Inc., New York).

Geikie, Archibald: 1905. The Founders of Geology, St. Martin's Press, Inc., New York.

Gilbert, G. K.: 1877. Report on the geology of the Henry Mts., Utah, *U.S. Geog. and Geol. Survey, Rocky Mt. Region*.

———— : 1886. The inculcation of scientific method by example, with an illustration drawn from the Quaternary geology of Utah, *Am. Jour. Sci.*, 3d ser., vol. 31, pp. 284-299.

Horton, R. E.: 1945. Erosional development of streams and their drainage basins: hydrophysical approach to quantitative morphology, *Geol. Soc. America Bull.*, vol. 56, pp. 275-370.

Johnson, Douglas: 1933. Role of analysis in scientific investigation, *Geol. Soc. America Bull.*, vol. 44, pp. 461-494.

Leopold, L. B., and M. G. Wolman: 1957. River channel patterns: braided meandering and straight, *U.S. Geol. Survey Prof. Paper 282-B*.

Lyell, Charles: 1872. Principles of Geology, D. Appleton & Company, Inc., New York.

Mackin, J. H.: 1941. Study of drainage changes near Wind Gap; a study in map interpretation, *Jour. Geomorphology*, vol. 4, pp. 24-53.

———— : 1963. Rational and empirical methods of investigation in geology, in C. C. Albritton, Jr. (ed.), The Fabric of Geology, Addison-Wesley Publishing Company, Inc., Reading, Mass., pp. 135-163.

Playfair, John: 1802. Illustrations of the Huttonian Theory of the Earth, William Creech, Edinburgh.

Powell, J. W.: 1877. Report on the geographical and geological survey of the Rocky Mt. region, Government Printing Office, Washington, D.C.

Strahler, A. N.: 1950. Equilibrium theory of erosional slopes approached by frequency distribution analysis, *Am. Jour. Sci.,* vol. 248, nos. 10, 11.

Thornbury, W. D.: 1954. Principles of Geomorphology, John Wiley & Sons, Inc., New York, pp. 1 - 14.

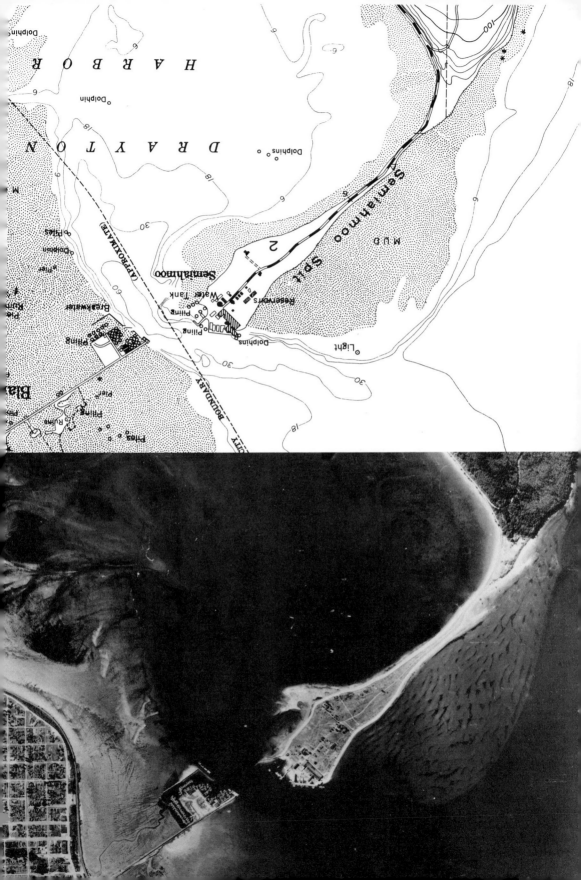

interpretation of maps
and aerial photographs

Emphasis in geomorphology centers upon the origin and evolution of land-forms, which may be interpreted on the basis of the size, shape, orientation, composition, and distribution of topographic features. These factors may be analyzed from topographic maps and aerial photographs, as well as by direct observation of the land surface. Analysis of geomorphic process, geologic structure, and stage of development, as interpreted from topographic maps and aerial photographs, is necessary to an understanding of the origin of landforms.

topographic maps

Perhaps no other single tool, with the possible exception of aerial photographs, is as useful to the geomorphologist as a topographic map. Topographic maps have the advantage of portraying three-dimensional forms on two-dimensional pieces of paper and at the same time giving accurate information as to size, shape, and elevation. Relationships of one landform to nearby asso-

Topographic map and aerial photograph of a spit being built across the mouth of a bay, Drayton Harbor, Washington. *(U.S. Geological Survey and U.S. Department of Agriculture.)*

ciated landforms may readily be shown on topographic maps. Because topographic maps are of great importance in geomorphic studies, it is imperative that the beginning student learn their use early in his career, and thoroughly understand them. Since most introductory physical geology courses cover basic map reading, these techniques are not discussed further here. However, a summary of contour maps is included in the Appendix.

In interpreting landforms from topographic maps, constant attention must be paid to the map scale, because correct interpretation of surface features involves consideration of size, as well as the other factors previously mentioned. Map scale is the ratio of distance on a map to the actual distance on the ground surface. Maps are made in a great variety of scales, depending on how much of the earth's surface is to be shown on a given size of paper and on how much detail is desired. Those which are intended to show large portions of the earth's surface are drawn at small scales, whereas maps which cover a relatively local area are made at large scales.

A map which is drawn at a scale of 1:125,000 indicates that one inch on the map equals 125,000 inches of ground distance, or any other unit of measured map distance equals 125,000 of the same units of ground distance. The scale of a map is usually stated as (1) the ratio of map distance to ground distance, (2) one inch equals a given number of miles or other unit of ground distance, or (3) a bar or graph. In any case the scale on nearly all standard maps is shown at the bottom of the map.

In order to measure distance on maps, it is necessary to measure the map distance between two points and then convert the map distance to actual distance. For example, consider two points which are 5 inches apart on a map drawn at a scale of 1:24,000. Since 1 inch on the map equals 24,000 inches of ground distance, 5 inches equals $5 \times 24,000$ inches of ground distance. To put this large number into more usable terms, it may be divided by 12 to convert it into feet, and then further divided by 5,280 to convert it into miles. A map distance of 5 inches at this scale thus represents a true distance of slightly less than 2 miles. Standard scales of maps made by the U.S. Geological Survey are shown in the table on facing page.

If a map scale is disregarded, the apparent size of a given feature on maps of differing scale is quite different, and the amount of detail shown may be significantly different. A comparison of maps of the same area at three scales, 1:250,000, 1:62,500, and 1:24,000, is shown in Fig. 2-1. As can be readily seen from these maps, much greater detail is shown on the 1:24,000 map than on either of the other two. Although detail of form becomes blurred on maps of smaller scale, such maps are still very useful because they often show relationships of landforms to one another over a larger area. Thus, in analyzing landforms in an area, some care must be taken to select the map which has

scale	conversion	1 inch equals	standard uses (quadrangle map)
1:24,000	1 inch $= \dfrac{24,000}{(12)\,(5,280)}$	0.38 mile (2,000 feet)	7½ minutes
1:62,500	1 inch $= \dfrac{62,500}{(12)\,(5,280)}$	0.99 mile	15 minutes
1:125,000	1 inch $= \dfrac{125,000}{(12)\,(5,280)}$	1.97 miles	30 minutes
1:250,000	1 inch $= \dfrac{250,000}{(12)\,(5,280)}$	3.95 miles	1 × 2 degrees
1:500,000	1 inch $= \dfrac{500,000}{(12)\,(5,280)}$	7.9 miles	

the optimum scale for the particular purpose of the investigation, and in comparing landforms shown on several maps the interpreter must be constantly aware of the map scales.

interpretation of aerial photographs

In recent years aerial photographs have come to play an increasingly important role in interpretation of landforms. It is often possible to identify features on aerial photographs which would be unnoticed during observation from the ground. Photographs show a great deal more detail than most topographic maps, but do not show elevations. An additional value of aerial photographs is that, by viewing pairs of photographs stereoscopically, the terrain may be viewed in three dimensions.

photograph scale

The scale of an aerial photograph depends on the altitude of the plane taking the photograph and the focal length of the camera. The higher the plane flies, the smaller the scale will be. In Fig. 2-2 the focal length is $f(p_1L)$, and the altitude is $H(LP)$. Since a_1p_1L and APL in Fig. 2-2 are similar triangles, image distance a_1p_1 is proportional to ground distance AP, and p_1L is proportional to PL; that is, f is porportional to H.

Fig. 2-1 Comparison of details on maps at scales of (a) 1:250,000, (b) 1:62,500, and (c) 1:24,000. (U.S. Geological Survey.)

$$\text{Photo scale} = \frac{\text{image dist.}}{\text{ground dist.}} = \frac{p_1 L}{PL} = \frac{f \text{ (focal length)}}{H \text{ (camera height)}}$$

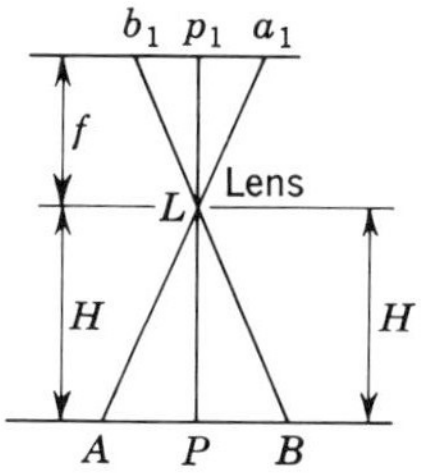

Fig. 2-2 Air-photograph scale.

If the focal length of a camera is 6 inches and altitude is 10,000 feet, the scale of the photographs will be

$$\frac{f}{H} = \frac{0.5}{10,000} = \frac{1}{20,000}$$

A certain amount of variation in scale on a photograph is introduced if the region being photographed has relief. In Fig. 2-3 several lines of sight from a camera to the surface features are shown.

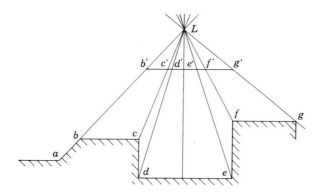

Fig. 2-3 Distortion of features on air photographs.

The line of sight from the camera L to the ground is such that the vertical cliff dc will have a horizontal displacement $d'c'$ on the photograph image, and thus will not appear as a vertical cliff. Images of points with greater elevations tend to be displaced outward from the center of the photograph. The slope ab is not vertical, but because it is parallel to the line of sight from L, the slope will appear to be vertical in the photograph image.

The horizontal distance gf is equal to cb, but because gf is at a higher elevation, the image distance $g'f'$ appears to be greater than $c'b'$. A similar type of distortion can occur if the camera is tilted from a vertical line of sight.

As with topographic maps, the scale of aerial photographs is important in determining the amount of detail shown. On small-scale photographs taken from a high altitude, some features are too small to recognize, whereas on large-scale photographs regional relationships may be missed. For reconnaissance work a scale of 1:60,000 may be satisfactory, but for detailed work a scale of something like 1:5,000 may be required.

If the camera height and focal length are not known, the scale of a photograph may be determined by measuring the distance between two prominent points and comparing it with the true ground distance as measured from a topographic map. The ratio of photograph distance to true ground distance is the scale. For example, if two buildings on an air photograph are 3 inches apart, and if the distance between buildings is determined from a topographic map to be 1 mile, then the scale of the photograph is calculated as follows:

3 inches (photograph distance) = 1 mile (actual distance)

$$\frac{1 \text{ inch (photograph distance)}}{X \text{ miles (actual distance)}} = \frac{3 \text{ inches (photograph distance)}}{1 \text{ mile (actual distance)}}$$

1 inch (photograph distance) = 1/3 mile = 1,760 feet (actual distance)

or 1 inch = (12) (1,760) = 21,120 inches and the scale is 1:21,120.

stereoscopic pairs

If two photographs of the same point are taken from slightly different positions and aligned in such a way that one eye looks at one photograph and the other eye looks at the second photograph, a three-dimensional image will be seen. Figure 2 - 4 shows two photographs of the same ridge and peak. By using a stereoscope to view the photographs, a three-dimensional image becomes visible. Relief of an area is exaggerated when viewed stereoscopically. Such exaggeration distorts the true relief, but is often very useful in bringing out details in topography. The amount of exaggeration varies with the relief of an area, the focal length of the camera, the camera height, the distance separating the photographs, and the distance between an individual's eyes.

To orient photographs correctly for stereoscopic viewing, the following procedure may be used: (1) Mark the center of each photograph and locate the corresponding point on the opposite photograph (A and A', B and B' on Fig. 2-4). (2) Connect the points. (3) Adjust the lines so that there is about 2½ inches between corresponding points. With practice this procedure need not be followed precisely. If the photographs are not properly oriented, two images will be seen, one with each eye. By moving the photographs until the two images merge into one, the proper orientation may be determined. Sometimes the relief may appear to be inverted, but this may be corrected by rotating the photographs 180 degrees.

mosaics

A number of individual air photographs may be systematically fitted together to form a mosaic, giving a composite view of a much larger area than that available in a single photograph. Since distortion occurs near the edges of

Fig. 2-4 Stereoscopic photographs of a volcanic neck and radiating dike, Shiprock, New Mexico. *(U.S. Department of Agriculture.)*

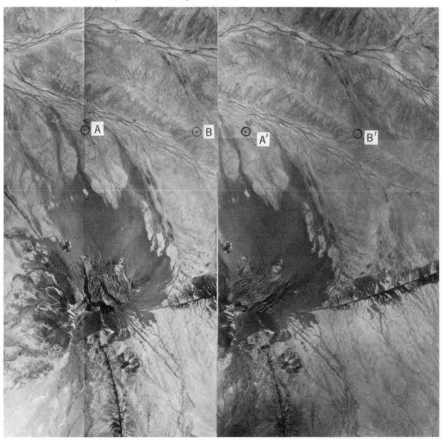

vertical photographs, a mosaic made by assemblage of a number of photographs possesses a certain amount of inaccuracies. Usually, the overall image is satisfactory, but adjacent photographs do not exactly match, and scales are distorted. Such a composite of photographs is known as an *uncontrolled mosaic*.

With modern equipment it is possible to remove many of the distortions and inaccuracies of the photographs, and a controlled mosaic may be produced, giving the viewer a photograph which appears to be a single photograph, rather than a composite of many. This is accomplished by adjusting tonal variations between individual photographs and making adjustments for distortions of scale.

Aerial photographs are available for much of the United States through a number of state and federal agencies (see the Appendix). A very useful map in determining which agency holds photography of a given area in the United States is the "Status of Aerial Photography," available without charge from the Map Information Service, U.S. Geological Survey, Washington, D.C. A similar map, "Air Photographic Coverage," showing availability of air photography in Canada, may be obtained from the Map Distribution Office, Department of Mines and Technical Surveys, Ottawa, Canada. The availability of mosaics in the United States is shown on the map of "Status of Aerial Mosaics," which may be obtained from the Map Information Service in Washington, D.C.

criteria used for interpretation

The relative amount of light reflected from a feature on a photograph is the *photographic tone*. In general, tonal differences result from differences in rock type, soil, and vegetation, allowing identification of rocks on photographs. Sandstones and conglomerates often produce light tones, whereas shales produce darker tones (Fig. 2-5).

The *degree of dissection* and *drainage texture* is related to the nature of underlying rocks, reflecting the permeability of material. Fine-grained shales have a fine-textured drainage on them as compared with the coarser drainage on sandstones or other rock types. Intense gullying and dissection are typical of fine-grained shales.

Various *patterns*, especially those consisting of straight or gently curved features, permit identification of faults (Fig. 2-6), joints (Fig. 2-5), dikes, or bedding.

Drainage patterns depend on the lithology and dip of the underlying rocks, and thus indicate the nature and structure of material. Vegetation and soil are to a large degree controlled by the underlying rock. Elements in the soil derived from weathering of bedrock influence the type of vegetation in an area. In some instances it is possible to map geologic contacts on the basis of differences in vegetation growing on different types of rock.

The *shape* of topographic forms is related to origin. Landforms such as volcanic cones, sand dunes, moraines, spits, and many others show up as obvious features on photographs. Differential erosion of dipping sedimentary beds allows tracing of individual beds and identification of geologic structures.

Examples and further discussion of the use of air photographs in identification and interpretation of landforms are incorporated in subsequent chapters.

Fig. 2-5 Air photograph of an area consisting of sandstone and shale. The sandstone has a light tone, the shale a darker tone. Parallel lines in the sandstone are joints. *(U.S. Geological Survey.)*

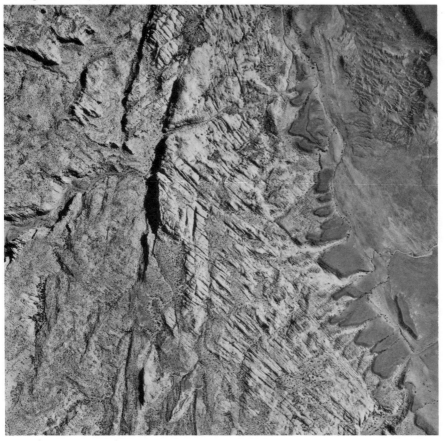

the scientific method in geomorphology

The ultimate goal in geomorphology is the understanding of surface processes and of the origin and evolution of landforms. In order to interpret the origin of a given landform, a series of steps may be followed:

1. OBSERVATION OF FACTS. The first step in any investigation is to determine as many relevant facts as possible. Among the data which might be available concerning landforms are shape, size, association with other features, composition, and orientation.

Fig. 2-6   Ridges of sedimentary beds offset by a fault. *(U.S. Geological Survey.)*

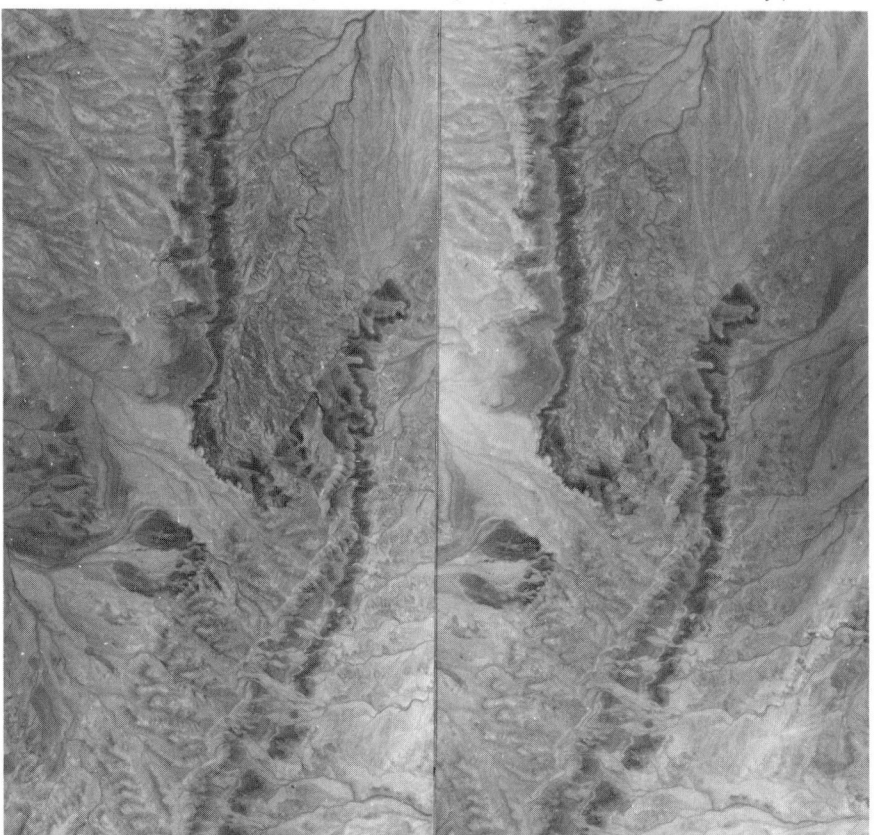

2. DEVELOPMENT OF MULTIPLE-WORKING HYPOTHESES. The data having been collected, several possible interpretations may come to mind. It is then necessary to test each interpretation against the facts and to reject hypotheses which do not fit the observed facts.

3. TESTING HYPOTHESES. After rejecting those hypotheses which are contrary to the observed facts, the next step is to seek out new information pertinent to the support or rejection of the remaining hypotheses and critically test each one.

Ideally, only one hypothesis will be consistent with all the data available. An important step in arriving at the final interpretation is the testing of multiple hypotheses against the available evidence. This procedure invariably leads to recognition of needs for certain critical evidence, and sharpens the focus upon the problem at hand.

Figure 2-7 may be used to illustrate this approach to problem solving in geomorphology. Barren River presently flows northward, joining Green River near the town of Woodbury, where they continue northward. West of Woodbury the broad valley of Black Swamp Branch presents an anomalous situation, originating where Muddy Creek makes a right-angle bend and flowing 180 degrees away from the lower portion of Muddy Creek. The problem is, what is the origin of the valley of Black Swamp Branch? Observation of the immediate area leads to the formulation of several hypotheses of origin: erosion by (1) Black Swamp Branch, (2) Muddy Creek, (3) Barren River, or (4) Green River.

Testing each hypothesis against the available evidence leads to the elimination of several possibilities. Although the valley of Black Swamp Branch is as wide as that of Barren River, there is a great difference in the size of the two and Black Swamp Branch flows only intermittently, as indicated by the dotted pattern. Thus it appears that Black Swamp Branch is not responsible for development of the valley, having occupied it only as a result of inheritance, after it had been abandoned by some previous occupant.

Muddy Creek is in such a position that it could have occupied the valley of Black Swamp Branch, and the width of the upper portion of the creek suggests that the stream could have developed the valley width now observed; however, if the lower portion of Muddy Creek flowed down Black Swamp Branch as well as its present lower channel, it would have had to develop two channels simultaneously or intermittently, a most unlikely circumstance.

Barren River is certainly large enough and close enough to have made the valley of Black Swamp Branch. However, in order to do so, it would have had to flow northward from Sproul Bend through Black Swamp Branch to Green River about a mile north of Woodbury. This, of course, would leave an aban-

doned valley in the present course of the Barren River between Sproul Bend and Woodbury, and would mean that the Barren would have had to erode two valleys at the same time, a very unlikely situation. A few attempts to place the Barren in both valleys at the same time by making bends in the river soon lead to the conclusion that the Barren is not by itself responsible for both valleys.

Could Green River have done the job? Ignoring for the moment the present streams in the various valleys, Green River could at one time have made a big bend, flowing southward from Woodbury to Sproul Bend, then turning northward through Black Swamp Branch, joining its present course again about a mile north of Woodbury. If the flat area north of Woodbury had been breached by erosion on the outside of bends in the former stream course, the river could have been diverted to its present position by a meander cutoff. If this is true, however, Barren River should have flowed *down* the abandoned channel through Black Swamp Branch, rather than up the old channel from Sproul Bend to Woodbury. Attention must therefore be focused on the elevation at Woodbury, Sproul Bend, and Black Swamp Branch to determine if, perhaps, the Barren did at one time flow through Black Swamp Branch but later changed its channel into its present course. Unfortunately, the contour interval, 20 feet, does not allow precise elevations to be determined directly from the map, but there is indirect evidence available. The 400-foot contour does not cross either Green River or Barren River anywhere on the map, indicating that the slopes of these rivers are very gentle, less than 20 feet of vertical drop in approximately 20 miles, and that there is only a few feet of difference in elevation between the channels in question. During a flood, water of the Barren very likely occupies both channels, and when floodwaters subside, the shorter, steeper channel between Sproul Bend and Woodbury would have an advantage over the longer route from Sproul Bend through Black Swamp Branch.

The point to be made from the example is that the method of analysis used in solving such a problem is a powerful tool in interpretation of many geomorphic problems. Critical evaluation of observations and testing of hypotheses at every step in an investigation are necessary factors in a rational approach.

references

Chamberlain, T. C.: 1890. The method of multiple working hypotheses, reprinted in *Science,* vol. 148, pp. 754 - 759.

Johnson, Douglas: 1933. Role of analysis in scientific investigation, *Geol. Soc. America Bull.,* vol. 44, pp. 461 - 494.

Mackin, J. H.: 1941. Study of drainage changes near Wind Gap; a study in map interpretation, *Jour. Geomorphology,* vol. 4, pp. 24-53.

Miller, V. C., and C. F. Miller: 1961. Photogeology, McGraw-Hill Book Company, New York.

Ray, R. G.: 1960. Aerial photographs in geological interpretation and mapping, *U.S. Geol. Survey Prof. Paper 373.*

$\left(2\right)$

glaciation

glaciers

Much of the rugged alpine scenery typical of mountainous terrains owes its topographic form to glacial processes. In many mountain ranges where only minor glaciers exist today, the effects of glaciers which have long since melted away have left their impression upon the landscape. Present-day glaciers are restricted mostly to high elevations or high latitudes, but in the not-too-distant geologic past, large portions of the earth's surface were covered with extensive continental ice sheets, and mountainous regions were subjected to intense alpine glaciation.

origin of glaciers

Glaciers are masses of ice and granular snow formed by compaction and recrystallization of snow, lying largely or wholly on land and showing evidence of past or present movement (Flint, 1957).

Fresh snow falling on the surface of a glacier is very porous and has a very low density, usually ranging between about 0.07 and 0.18 gram per cubic

Contorted medial moraines on the Hayes Glacier, Alaska. *(Austin Post, University of Washington.)*

centimeter (Shumskii, 1964). The lowest recorded density of snow is 0.004 gram per cubic centimeter, and the highest is approximately 0.5 gram per cubic centimeter. Snow crystals, which belong to the hexagonal crystal system, often exhibit delicate hexagonal, prismatic, or star-shaped forms. With time, points of the lacy snow crystals round off and approach nearly spherical or hexagonal shapes as a result of partial melting and recrystallization. Granular snow, which has lost the characteristic structure of snow because of partial consolidation and recrystallization, is known as *firn*, or *névé*. It may make up substantial portions of a glacier near the surface during certain parts of the year. The density of firn is typically between 0.4 and 0.8. With further compaction and recrystallization, the firn becomes more dense, eventually transforming into ice having a density of about 0.92 gram per cubic centimeter. Since ice is 8 percent greater in volume than the same weight of water, ice never approaches a density of 1.00, that of water, unless it is charged with dust or rock fragments.

The time required for conversion of firn into ice varies from glacier to glacier, depending on temperature, rate of accumulation of snow, and other factors. In certain temperate glaciers it may take only a year or so, but in very cold areas, as in the polar regions, it sometimes may require more than 100 years.

Low temperatures alone are not adequate to develop glaciers since precipitation is also required. Some arctic and antarctic areas do not support significant glaciers, because, even though the climate is cold enough, there is not enough snowfall, and conditions necessary for the conversion of snow into ice are not attained frequently enough.

Firn in glaciers often exhibits a banded appearance, representing stratification related to annual snowfall. However, as the firn is converted to ice and movement takes place, deformation may obscure the stratification. Much of the banding observed in glacial ice is related to movement along shearing planes.

types of glaciers

Small glaciers differ in form and other characteristics from large ice masses, such as those of Greenland and the Antarctic. On the basis of size, shape, and mode of occurrence, glaciers may be classified as (1) alpine, (2) piedmont, or (3) continental.

Alpine glaciers are restricted to mountainous regions and consist of (1) valley glaciers which flow in long sinuous arms from areas of high snow accumulation (Fig. 3-1) and (2) small cirque glaciers restricted to the source of accumulation.

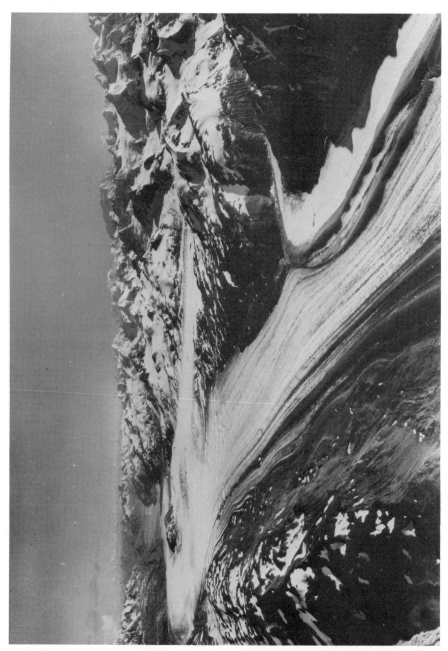

Fig. 3-1   Alpine valley glaciers in Alaska. _ (Austin Post, University of Washington.)

Piedmont glaciers form when valley glaciers emerge from the confines of mountain valleys, coalesce, and spread laterally into broad lobes beyond the mountain front. A good example of this kind of glacier is the Malaspina Glacier near Yakutat, Alaska (Fig. 3-2). The ice emerges from several narrow valleys in the Saint Elias Mountains and spreads out in a broad lobe about 30 miles across and 25 miles long.

Continental glaciers are not confined to valleys, but, like the Greenland and Antarctic ice sheets, spread out over wide areas (Fig. 3-3). The Greenland ice sheet occupies about 666,000 square miles, and the Antarctic ice sheet about 5,100,000 square miles. These glaciers exceed 10,000 feet in thickness in places, enough to bury entire mountain ranges and depress the earth's crust beneath their weight.

Fig. 3-2 Malaspina Glacier, Alaska, an example of a piedmont glacier. (Austin Post, University of Washington.)

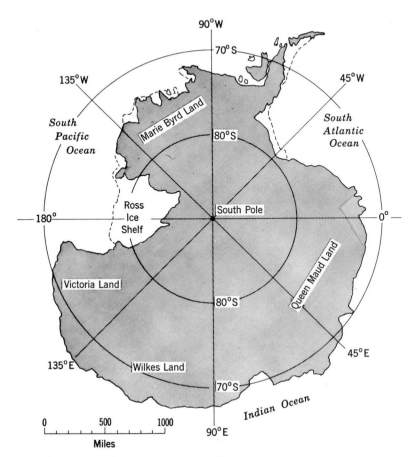

Fig. 3-3 Continental glaciers of the Antarctic.

glacial movement

evidence of movement

That alpine glaciers move continuously downvalley has long been recognized. Large boulders, blocks of ice, and other debris on a glacier become displaced with time. Rock types foreign to a local region may be traced to the source area of a glacier.

Velocities of glaciers vary considerably with ice thickness, temperature, and slope. Very cold glaciers in Antarctica may move only a few feet per year, while in an extreme case in Alaska, a glacier moved as much as 200 feet in one day. Precise measurement of surface velocities of glaciers, made with surveying instruments, proves surface velocities to be greatest near the middle of the glacier and to decrease toward the sides. Vertical drill pipes which have become bent with glacial movement illustrate that velocity of ice changes with depth. Not all parts of a glacier move at the same rate; upstream and downstream velocities may differ.

mechanism of movement

Basal and Marginal Slip. Slip occurs when ice slides along the land surface beneath a glacier. According to Sharp (1954), slip may account for as much as 90 percent of the movement of thin ice resting on steep slopes. However, it is probably responsible for only a small fraction of the movement of thick

Fig. 3-4 Antarctic polar ice plateau and Robert Scott Glacier. (U.S. Navy for U.S. Geological Survey.)

temperate glaciers on gentle slopes. Measurements of marginal slip vary from 0 to 30 percent of total movement in some glaciers, and in a few instances may account for nearly 100 percent. Measurements in drill pipes and subglacial tunnels in North American and European glaciers suggest that basal slip may account for between 0 and 90 percent of the recorded surface movement. Impressive evidence of mechanical sliding along the base and margins of glaciers may be found in glaciated regions where deep parallel grooves and striations are cut in bedrock (Fig. 3-5).

Shearing. Portions of ice in a glacier may move over other ice by fracturing along distinct shear planes. Especially near their termini, many glaciers have shear planes similar to thrust faults. The amount of movement along shear

Fig. 3-5 Glacial grooves cut in bedrock, San Juan Island, Washington.

planes relative to total movement of a glacier is still a point of controversy. Shearing may be of local importance, but few glaciologists consider it significant relative to plastic flowage. In order for ice to fracture and move along shear planes, the ice must be thin enough so that it may behave as a brittle rather than as a plastic substance. For this reason, movement by shearing is probably most likely to occur near the termini of glaciers, and at shallow depths.

Ice may behave as an elastic, plastic, or brittle material under stress. It may respond to stresses by (1) elastic deformation, in which the original shape is resumed after stress is released, (2) plastic deformation, in which the ice does not return to its original shape after stress is released, and (3) fracturing, in which the ice is brittle and breaks when stresses are great enough. The principal factors that determine which of these responses to stress takes place are the amount of stress exerted, the type of stress, the rate at which the stress is applied (i.e., rapidly or over a long period of time), and the temperature. Generally speaking, a strong force applied over a short period of time results in fracturing, whereas a smaller force exerted over a longer period of time may result in plastic deformation.

Plastic Flowage. Ice subjected to stress over long periods of time yields by plastic flowage. Where alpine glaciers are confined by valley sides, they conform to the shape of the valleys, but where they are unconfined, they spread out laterally, as does the Malaspina Glacier. In the laboratory, ice subjected to stress becomes deformed while still in the solid state.

Plastic flowage of ice takes place by *intergranular* and *intragranular shifting*. Intergranular shifting by rotation and sliding between individual ice crystals has been demonstrated in firn, but since many glaciers exhibit strong preferred crystallographic orientation of ice crystals, this mechanism is considered a relatively minor contributor to glacial flow (Sharp, 1954). Intragranular shifting by movement along glide planes within the crystals themselves, accompanied by continuous recrystallization, is thought to play a dominant role. Because of the crystal lattice of ice, gliding is easiest along planes parallel to the base of the crystals. Ice belongs to the hexagonal crystal system. If stress is applied to a hexagonal prism of ice, the response may vary, depending on the direction of application of the force. This response may be visualized by imagining a deck of cards upon which force is applied in two different directions. If the force is applied parallel to the flat surfaces of the cards, they slide past one another, deforming the original rectangular shape of the deck. In ice, a similar sliding takes place along glide planes parallel to the basal plane of the prism. If force is applied to a deck of cards at right angles to the flat surfaces, the cards will only bend. If force is applied to an ice crystal at right angles to the basal plane, the ice will bend until its strength is exceeded, at which point it will fracture.

Pressure within a glacier increases with depth. When pressure increases sufficiently, ice at the melting temperature melts; when pressure decreases sufficiently, ice at the same temperature refreezes. Melting and refreezing facilitate intergranular movement, crystal growth, and transfer of material. Melting at grain boundaries reduces friction between the grains and facilitates movement of the grains relative to each other.

In order for plastic flow to take place in a glacier, the internal shearing stress must exceed the elastic limit of ice. For a uniform layer of ice resting on an inclined surface the elastic limit λ is equal to the thickness h of the ice times the density δ of the ice times the sine of the angle of slope:

$$\lambda = h\delta \sin \alpha$$

This equation indicates that plastic flow increases with ice thickness and slope. Observations suggest that perceptible flow of ice takes place when the stress reaches approximately 10 times the elastic limit (Shumskii, 1964), so that a uniform layer of ice resting on a 10° slope would be expected to undergo perceptible plastic flow under an ice thickness of slightly over 200 feet. However, plastic flow in a glacier is a vastly complicated process because of irregularities in the subglacial floor, changes in the slope of the ice surface, differences in ice thickness from place to place, and a host of other factors. It is therefore not surprising that surface velocities measured on glacier surfaces are often jerky and irregular. Although plastic flow is not the only mechanism of glacier movement, it is considered to be the primary mechanism for most glaciers.

economy of a glacier

Accumulation is the process by which a glacier is constantly replenished. A glacier is nourished by precipitation in the form of snow and rain, by avalanches of snow from the valley sides above the glacier, and by wind-drifted snow. In mountainous regions, most snowfall is caused by the rising and cooling of moist air as it passes over the mountains. Moisture is as important as cold climate in determining the annual amount of snowfall. Thus the rate of accumulation of snow and ice tends to be great in mountainous areas adjacent to oceanic regions which supply moisture-laden storm systems.

Ablation is the process by which ice and snow are lost from a glacier. Ablation may occur by melting, sublimation, or calving. Of these, melting is by far the most important in most glaciers. The heat which causes the melting comes from solar radiation, warm air, and rain. Dark-colored rock debris on a glacier often becomes warmed more rapidly than the snow and ice because it absorbs radiation better. If the rock debris is thin enough to be warmed through, it melts its way into the underlying ice and forms a hole. However, if the material is too thick, it acts as an insulator and protects the underlying ice, and as the adjacent ice melts down, a rock table results (Fig. 3-6). Minor melting may occur as a result of hot springs and heat supplied by the geothermal gradient of the rocks beneath the ice.

Evaporation and sublimation account for a very small amount of ablation, probably less than 1 percent, primarily because of the large latent heat of vaporization (540 calories per gram). Also, cold air is limited in its capacity to absorb moisture.

Fig. 3-6   Rock table on a glacier. *(Wards Natural Science Establishment.)*

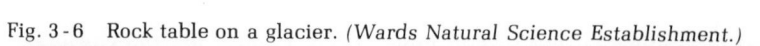

Calving of ice from the terminus of a glacier takes place where glaciers terminate in the sea or break off over cliffs. In this manner large masses of ice may become separated from the glacier without melting directly (Figs. 3-7 and 3-8).

Regimen is the relationship between the processes of accumulation and ablation. Accumulation of ice and snow exceeds the amount lost by ablation in the *zone of accumulation* at the source of a glacier (Fig. 3-9). Ice loss exceeds accumulation in the *zone of ablation* at the lower part of a glacier. This does not imply that accumulation does not occur in the zone of ablation; it refers only to the balance between the two processes. The line or zone dividing the zone of accumulation from the zone of ablation is called the *firn line*.

Fig. 3-7 Iceberg projecting 75 feet above the water in Glacier Bay, Alaska. Dark material on the iceberg is rock debris. *(G. K. Gilbert, U.S. Geological Survey.)*

In the zone of accumulation, the amount of ice passing through any given cross section equals the amount passing through a cross section upvalley *plus* the amount gained by net accumulation between the sections. Thus discharge (average rate of flow times cross-sectional area) increases from the head of a glacier to the firn line. In the zone of ablation, the amount of ice passing through a given cross section equals the amount of ice passing through a cross section upvalley minus the amount lost by net ablation in between. Thus discharge decreases from the firn line to the terminus.

The relationship between accumulation and ablation governs the position of the front of a glacier. If the relationship is balanced over a period of time, the terminus remains stationary. If accumulation exceeds ablation, the glacier has a positive economy, and the terminus advances. If ablation exceeds

Fig. 3-8   Calving of the terminus of an Alaskan glacier. *(U.S. Air Force.)*

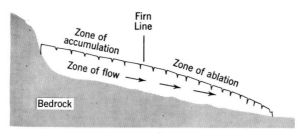

Fig. 3-9 Cross section of a glacier.

accumulation, the glacier has a negative economy, and the terminus retreats. Even when its terminus is retreating upvalley, a glacier continues to flow downvalley unless the ice is stagnant.

A spectacular advance of the terminus of the Black Rapids Glacier in Alaska occurred in 1936 and 1937. After many years of retreat, the terminus began to advance rapidly; at one stage the maximum rate of advance was about 200 feet per day. The advance was short-lived, however, and later the glacier began to retreat once again (Moffit, 1942). Similar abrupt advances have been observed in a number of glaciers now commonly referred to as *surging glaciers*. Surging glaciers are characterized by unusually rapid advances of their termini concomitant with a sudden thinning of ice in the upper portions. The cause of such glacial surges is not yet well understood.

During retreat under a negative economy, a glacier usually loses more ice by thinning or downwasting than by recession of the terminus. Thinning of the ice by ablation diminishes the rate of ice movement, and hence reduces the amount of new ice moved downvalley. As the rate of movement diminishes, the rate of shrinkage accelerates. Rate of shrinkage of a glacier is not a linear function of temperature rise. Thus accelerated shrinkage over a long period of time does not necessarily imply an accelerated warming trend.

As rate of ice flow decreases in a glacier with a protracted negative economy, enough thinning may occur to result in stagnation and, eventually, separation of the lower end of the glacier. Sharp (1951) found that 350 to 500 feet of downmelting occurred during stagnation of the lower 9 miles of the Wolf Glacier in the Saint Elias Mountains of Canada, while the terminus retreated only a few hundred yards. In this instance, a contributing factor in the stagnation was retreat of several large tributary glaciers from their junction with the Wolf Glacier, resulting in loss of nourishment of the main glacier.

From the nineteenth century until about 1945, glaciers in most parts of the world were retreating in response to a general warming of the climate (Flint, 1957). However, some glaciers, such as the Taku Glacier near Juneau, Alaska, were advancing during the same period. The terminus of the Taku

Glacier advanced about 3½ miles between 1900 and 1952. After 1945, the marked trend of negative economies of glaciers in the western United States diminished.

glacial erosion and transportation

That glaciers are active agents of erosion is well shown by features imparted to areas overridden by glaciers. Glacial erosion may take place by *abrasion*, *quarrying* (plucking), or *pushing*.

Fig. 3-10 Glacial polish and striations on bedrock, Yosemite National Park, California. *(G. K. Gilbert, U.S. Geological Survey.)*

abrasion

Rock material ranging in size from huge boulders to fine clay is embedded in ice and carried downstream by a glacier. As the glacier moves over bedrock, the embedded fragments act much like a piece of sandpaper on wood, causing abrasion of the surface over which they move. The effect of such abrasion is to wear away the rock under or against the ice. Very fine fragments embedded in a glacier polish the surface of the overridden rocks (Fig. 3 - 10), whereas large fragments gouge grooves in the bedrock. Rocks frozen in the glacier are also worn down by abrasion as they are dragged over the surface, resulting in faceted, polished, and striated pebbles (Fig. 3 - 11).

Fig. 3 - 11 Faceted, polished, and striated pebbles from glacial deposits.

Striations on flat, nearly horizontal bedrock surfaces indicate the direction of ice movement. However, local differences in direction of flow may occur. Where ice is deflected by irregularities in the topography, striations do not always indicate accurately the direction of movement of the main ice mass.

quarrying

Under certain conditions, glaciers pluck, or quarry, blocks of rock. Jointed rocks are especially vulnerable to this kind of erosion. The exact mechanism is not easy to demonstrate because the base of a glacier cannot be observed easily. However, freezing and thawing undoubtedly play an important role. Expansion caused by freezing of water in rock fractures helps to loosen blocks of material. Release of pressure in the lee of irregularities in the bedrock floor causes ice to freeze to loosened blocks, and as the ice moves, the blocks are pulled out.

Glaciers also transport a limited amount of material by pushing loose rock debris in front of the terminus, but this mechanism is much less effective on a regional scale than abrasion or quarrying.

The principal sources of the load transported by alpine glaciers are avalanches and rockfalls onto the surface of the ice. Material carried on the surface of the glacier is called *superglacial debris* (Fig. 3-12); material within the glacier is called *englacial debris*. Continental glaciers generally do not carry much superglacial debris, because not much of the land surface normally stands above the ice. Valley glaciers often carry more superglacial debris because the valley sides rise above the glacier.

glacial deposits

Glacial drift is a collective term for all rock material transported and deposited by glaciers, floating ice, and meltwater streams. The term drift is a holdover from the days when foreign rock material in Britain was believed to be related to a universal flood. Although the deposits now are known to be of glacial origin, the term persists. Glacial drift is commonly subdivided into *stratified* and *nonstratified drift*.

Nonstratified drift lacks distinct bedding and consists of poorly sorted particles. *Till*, an unstratified and poorly sorted mixture of particles ranging in size from boulders to clay (Fig. 3-13), is deposited directly from a glacier without significant reworking by the action of running water. *Lodgement till* is deposited from the base of a glacier by the plastering of rock debris over the land surface. Superglacial and englacial debris let down on the land surface by thinning and melting of a glacier are known as *ablation till*. Because it was deposited beneath the weight of an overriding glacier, lodgement till is usually more compact than ablation till. Many tills develop a fabric in which elongate pebbles tend to have a preferred orientation, with the long axes mostly parallel to the direction of ice movement. A second, less well developed, preferred orientation at right angles to the direction of movement is also found in some tills. Elongate pebbles in ablation till are randomly oriented because they were not deposited from actively moving ice.

Pebbles in till are often faceted, striated, and polished, but glaciers which override stream gravels commonly contain high percentages of rounded pebbles. Since the lithology of particles in till is related to the kind of rocks over which a glacier passes, the composition of glacial deposits often indicates the source area of a glacier.

Ice floating in marine water deposits pebbly clay and sand as *glaciomarine*

Fig. 3-12 Superglacial debris near the terminus of a glacier in Alaska. *(U.S. Air Force.)*

drift. These sediments may resemble till, but are less compact and often contain marine shells.

Stratified drift is glacial debris deposited by water or wind action, and hence is better sorted than till. Glaciofluvial sediments are composed of stratified sand and gravel (Fig. 3-14) deposited by meltwater streams. They usually consist of coarse pebbles and cobbles near the ice front, but grade into finer-grained material downstream. Glaciolacustrine sediments deposited

Fig. 3-13 Glacial till consisting of poorly sorted, unstratified pebbles, sand, and clay resting on polished and striated bedrock. (Wards Natural Science Establishment.)

in glacial lakes are mostly clay and silt, some of which may show annual layer-ing, known as *varves*. During the summer, meltwater brings sediment into glacial lakes, but during the winter, when the lake is frozen over, only very fine clays settle out. Hence an annual deposit consisting of a coarse-grained and a fine-grained layer may result. Wind-blown *glacioeolian deposits* com-monly are made up of fine silt known as *loess*, much of which is glacial silt picked up by winds blowing across outwash deposits.

Fig. 3-14   Stratified, moderately well-sorted glacial outwash deposits.

references

Chamberlain, R. T.: 1936. Glacier movement as typical rock deformation, *Jour. Geology,* vol. 44, pp. 93-104.

Demorest, Max: 1938. Ice flowage as revealed by glacial striae, *Jour. Geology,* vol. 46, pp. 700-725.

———— :1942. Glacial thinning during deglaciation, Part I, Glacial regimens and ice movement within glaciers, *Am. Jour. Sci.,* vol. 240, pp. 31-66.

———— : 1943. Ice sheets, *Geol. Soc. America Bull.,* vol. 54, pp. 363-400.

Flint, R. F.: 1957. Glacial and Pleistocene Geology, John Wiley & Sons, Inc., New York.

Johnson, D. W.: 1941. Normal ice retreat or downwasting?, *Jour. Geomorphology,* vol. 4, pp. 85-94.

Kamb, B.: 1964. Glacier geophysics, *Science,* vol. 146, no. 3642, pp. 353-365.

Meier, M. F.: 1960. Mode of flow of Saskatchewan Glacier, Alberta, Canada, *U.S. Geol. Survey Prof. Paper 351.*

———— : 1964. Ice and glaciers, in V.T. Chow (ed.), Handbook of Applied Hydrology, McGraw-Hill Book Company, New York.

Moffit, F. H.: 1942. Geology of the Gerstle River District, Alaska, with a report on the Black Rapids Glacier, *U.S. Geol. Survey Bull.* 926-B, pp. 146-157.

Nye, J. F.: 1952. The mechanics of glacier flow, *Jour. Glaciology,* vol. 2, pp. 82-93.

Sharp, R. P.: 1949. Studies of superglacial debris on valley glaciers, *Am. Jour. Sci.,* vol. 247, pp. 289-315.

———— : 1951. Glacial history of Wolf Creek, St. Elias Range, Canada, *Jour. Geology,* vol. 59, pp. 97-117.

———— : 1954. Glacial flow: a review, *Geol. Soc. America Bull.,* vol. 65, pp. 821-838.

Shumskii, P. A.: 1964. Principles of Structural Glaciology, Dover Publications, Inc., New York.

alpine glaciation

erosional landforms

cirques

Steep-walled, semicircular basins known as *cirques* occur at the upper ends of alpine glaciers. In areas where glaciation has not been sufficiently intense to form long valley glaciers, cirques may occur as isolated basins cut into the mountain slopes. The characteristic features of a cirque include a steep *headwall*, which may vary from less than 100 feet to several thousand feet high, overlooking a rock *basin*, often with a *threshold* at its downvalley edge (Fig. 4-1). The steepness of the headwall is often accentuated by a lack of talus at its base. Cirque basins are commonly occupied by lakes called *tarns*, which are held in by the threshold at the lower edge of the cirque.

The origin of cirques relates to several processes. Early in the development of a cirque, nivation, a combination of freeze and thaw processes, along with mass wasting, apparently plays an important role above the snow line. Melt-

Alpine glacier in a cirque, Price Glacier, Mt. Shuksan, Washington. *(Austin Post, University of Washington.)*

water, trickling into rock crevices beneath a bank of firn and snow, freezes and expands, breaking the rock apart. Loosened fragments then move by creep, solifluction, and rill wash. These processes deepen the depression and steepen its walls.

If enough snow, firn, and ice accumulate to cause ice flowage, a glacier forms, and glacial quarrying and abrasion then contribute to development of the cirque. Meltwater percolating into rock fractures and crevices at the *bergschrund*, a deep crevasse at the head of a glacier, freezes and promotes intense shattering of the rocks. Removal of loose blocks by quarrying steepens the headwall of the cirque and extends it headward. Since the bergschrund does not extend to the base of the glacier, all the quarrying process is not

Fig. 4-1 Cirques developed by former glaciers in the Wind River Mountains of Wyoming. *(Austin Post, University of Washington.)*

confined to the open bergschrund. Lewis (1938, 1940) suggested that some water melts its way below the bergschrund between the headwall and the glacier, where freezing pries rocks loose and quarrying removes the material from the headwall. The floor of the cirque basin is scoured by abrasion as the ice moves downvalley out of the cirque.

Cirques may form singly or may be compound, with several semicircular scallops which coalesce downvalley to form a broad basin. The heads of some glaciated valleys may be marked by cirques at several levels. Such features may form when the snowline is progressively raised from lower to higher elevations either during successive glaciations or within a single glaciation.

arêtes and horns

As cirques on opposite sides of a divide enlarge and extend headward, a narrow, sharp, serrated ridge known as an *arête* may remain between the cirque headwalls (Fig. 4-2). Continued erosion of the cirque headwalls results in the development of a *col*, a gap or sag in the ridge.

Fig. 4-2 Stereoscopic air photographs of alpine glacial erosion features near Gunnison, Colorado. *(U.S. Forest Service.)*

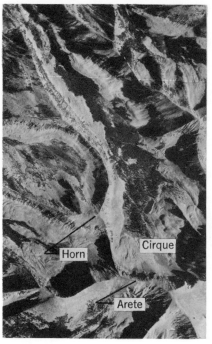 

Several cirques with their headwalls back to back may erode away all but a sharp spire-shaped peak which remains as a *horn* (Fig. 4-3). The most famous example of such a peak is the Matterhorn in Switzerland. In map view, horns often have triangular shapes due to the semicircular shape of the impinging cirques.

glacial troughs

Glaciers normally do not create new valleys because they flow down already existing stream valleys which offer the lowest course. They do, however, considerably modify stream valleys by deepening and widening them

Fig. 4-3 Horn peak formed by the headwalls of three cirques, north of Mt. Cleveland, Montana. *(Austin Post, University of Washington.)*

into *glacial troughs* (Fig. 4-4). Ice, filling the stream valley, scours and erodes the sides and floor of the valley until the V-shaped cross profile becomes U-shaped. Since stiff and massive glaciers do not bend around sharp curves as readily as streams, they trim away irregularities in the valley sides, leaving *truncated spurs.*

The long profile of a stream valley is a relatively smooth concave-upward curve with minor irregularities. The long profile of a glaciated valley, however, is likely to be quite irregular, with steps and breaks in slope. The steplike profile of some glaciated valleys may resemble a giant stairway, or *cyclopian stairs* (named after the mythical giant Cyclops). Each step may be characterized by a relatively flat *tread,* or basin, a steep *riser* at the upper end, and a

Fig. 4-4 U-shaped glacial trough and hanging tributary valley, Yosemite National Park, California. *(Aero Service Division, Litton Industries.)*

riegel, which is a rock knob between the tread and the next lower riser. Other steps may form at the junction of tributary glaciers, where increased erosion deepens the valley floor below the junction. *Paternoster* lakes, consisting of a chain of small lakes connected by streams, are common along some glaciated valleys.

Preglacial tributary streams normally join the mainstream at an accordant level. However, since ice in larger valleys is usually thicker and capable of more vigorous erosion than in tributary valleys, main valleys are often deepened by glacial erosion at a more rapid rate than side-stream valleys. Retreating glaciers may leave a sharp discordance between tributary valleys and the main valley. A side-stream valley which joins a main valley at a discordant junction is called a *hanging valley.*

The bedrock floors and sides of glaciated valleys sometimes retain glacial polish and striations for thousands of years after glaciers disappear from an area.

depositional landforms

moraines

Rock debris which accumulates along the margin of a glacier forms *lateral moraines* that appear as long dark ridges along each edge of the glacier (Fig. 4-5). The material composing the moraines is derived partly from rock falls and avalanches from the valley side and partly from erosion of the valley side by the glacier. At the junction of tributary valleys the lateral moraines on one side of each valley merge and become a *medial moraine* some distance out on the glacier itself. At each tributary junction the confluence of lateral moraines may add new medial moraines, resulting in a number of dark ribbons of debris extending down the glacier. Although they are common on present-day glaciers, medial moraines seldom remain as topographic features after a glacier retreats because meltwater and postglacial streams rework material on the valley floor. Lateral moraines, however, remain as long narrow sharp-crested ridges along the valley sides because they often lie above the normal erosive range of postglacial streams in the main valley.

Terminal, or *end, moraines* are ridges of glacial debris which form at the termini of glaciers (Fig. 4-6). They vary in size from prominent ridges a few hundred feet high to low discontinuous mounds of debris. End moraines usually have the shape of a curved arc, in map view, with the convex side pointing downvalley. Upvalley on the concave side, the end moraine is often continuous with lateral moraines along the valley sides (Fig. 4-7).

The stability of a glacier governs the position of the terminus and, hence, the size and degree of development of a terminal moraine. When a glacier achieves equilibrium between accumulation and ablation, the front of the glacier remains at one place, and movement of ice downvalley supplies a continuous load of rock debris which collects at the margin of the ice. The longer such equilibrium is maintained, the bigger the end moraine will grow. If the balance is upset, advance of the glacier terminus may destroy the end moraine; if the terminus retreats, the moraine ceases to grow. If, after the terminus of a glacier retreats upvalley from an end moraine, a new equilibrium between accumulation and ablation occurs, a *recessional moraine* will develop. However, all moraines upvalley from a terminal moraine are not necessarily recessional since a readvance of a glacier also could produce an end moraine. Moraines which extend all the way across a valley impound *morainal lakes* behind them. Drainage of the lake may be by an outlet stream or by percolation of water through the material in the moraine. Since postglacial streams may erode gaps in an end moraine, and since meltwater streams may breach a moraine while it is being formed, morainal lakes do not always form upvalley from a moraine.

Ground moraine consists of glacial till deposited as an irregular sheet be-

Fig. 4-5   Lateral moraine along the margin of the Athabaska Glacier, Alberta, Canada.

neath a glacier. After the glacier retreats from an area, the ground moraine consists of low undulating mounds with depressions.

glaciofluvial landforms

As glacial ice melts, it releases rock debris which meltwater carries away. *Outwash deposits,* consisting of gravel, sand, and silt laid down by meltwater streams, develop a relatively flat surface extending downvalley from the terminus of the glacier (Fig. 4-8). Glaciofluvial sediments confined within a valley make up a *valley train.* Most coarse material is deposited close to the ice margin, while finer material is carried farther downvalley before it is deposited. The action of running water results in a much better sorted and stratified deposit

Fig. 4-6   End moraine around the terminus of an alpine glacier in Alaska. *(Austin Post, University of Washington.)*

than that found in till. Where outwash streams terminate in lakes, alluvial fans and deltas form.

ice-contact forms

Ice-contact sediments form under conditions where a portion of the material being deposited is directly in contact with glacial ice, either upon, against, or underneath it. Melting of ice adjacent to sediments removes support and makes the sediments susceptible to slumping. Hence many ice-contact deposits are characterized by intense deformation caused by collapse.

Blocks of stagnant ice which become isolated from the main ice mass and

Fig. 4-7   Lateral and terminal moraines of an alpine glacier below Convict Lake, California. *(John Shelton.)*

are buried or surrounded by sediments form depressions, known as *kettles*, when the ice later melts (Fig. 4-9). The size of kettles varies from a few feet up to several miles in diameter, depending on the size of the buried block of ice. The depressions often fill with water and become lakes and swamps.

Eskers are long narrow sinuous ridges, ranging in height from a few feet to about 75 feet and ranging in length from a few hundred feet to several hundred miles. They are composed of moderately well stratified and well sorted sand and gravel, indicating that running water played an important role in their development. Some eskers terminate at, or grade into, fans or deltas at their downstream ends. The most common origin or eskers appears to be deposition of material by streams flowing in tunnels at the base of stagnant ice. The subglacial streams are confined by the walls of the tunnel, and when

Fig. 4-8   Braided outwash streams issuing from glaciers in Alaska. *(U.S. Air Force.)*

the glacier melts away, the deposits are left standing as a ridge. Eskers apparently form only in association with stagnant ice, since vigorous ice movement would close the tunnels in the ice and destroy the topographic form of the eskers. Some eskers may result from superglacial or englacial streams rather than from subglacial streams.

Kames are low steep-sided hills composed of ice-contact stratified drift. Some consist of material accumulated in openings or on the surface of stagnant ice. When the ice melts, the rock debris is dumped on the land surface. Other types of kames may consist of small alluvial fans or deltas built against stagnant ice along the margin of a glacier. When the supporting ice melts, the sediments collapse into an irregular mound. Short elongate ridges simi-

Fig. 4-9 Kettle forming by melting of buried ice in outwash of the Hidden Glacier, Alaska. (*G. K. Gilbert, U.S. Geological Survey.*)

Fig. 4-10 Topographic map of alpine glacial features. *(Holy Cross quadrangle, U.S. Geological Survey.)*

lar in many ways to kames will sometimes originate as *crevasse fillings.*

Kame terraces are composed of stratified sediments deposited between a glacier and an adjacent valley (Fig. 4-11). When the glacier melts away, sediments remain as terraces along the valley sides, but unlike normal stream terraces, kame terraces do not represent former valley floors which once extended all the way across the valley. Since they form in contact with the glacier, when the supporting ice disappears, portions of terraces may collapse. Blocks of ice buried in the sediments later melt, often leaving the terrace surface pitted with kettles.

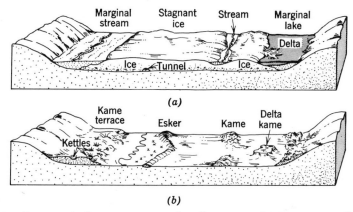

Fig. 4-11 Ice-contact deposits associated with a mass of stagnant ice. The upper diagram shows the deposits being formed, the lower diagram after the ice has disappeared. *(From A. N. Strahler, John Wiley & Sons, Inc., New York, 1960.)*

recognition of multiple alpine glaciation

Evidence of more than a single advance and retreat of former glaciers occurs in most mountainous regions which have undergone alpine glaciation. Fluctuation of climatic conditions during the Pleistocene resulted in several advances and retreats of alpine glaciers in many parts of the western United States, and similar events are recorded elsewhere in the world. Recognition of multiple stages of glaciation must be based on careful analysis of features associated with glacial deposits and landforms.

topographic criteria

The topographic position of moraines often may be useful in determining their relative ages. End moraines upvalley are younger than those downvalley because an advancing glacier would destroy any pre-existing moraines across its path. Moraines flanked on valley sides or enclosed by other moraines are younger than the enclosing moraines (Fig. 4-12), and moraines which cut across other moraines are younger (Fig. 4-13).

The degree of gullying and erosion of moraines is often indicative of age. Older moraines are more dissected by erosion and more subdued in form than younger moraines (Fig. 4-14). Older moraines often show a greater degree of destruction by streams transverse to the moraines.

Alpine glacial stages may also be distinguished from relationships between glacial features, stream terraces, faults, lava flows, ash falls, and lake shore-

Fig. 4-12   Lateral and end moraines of different ages at Lee Vining Canyon, California. They are progressively younger from the outer edge of the valley toward the center. *(Austin Post, University of Washington.)*

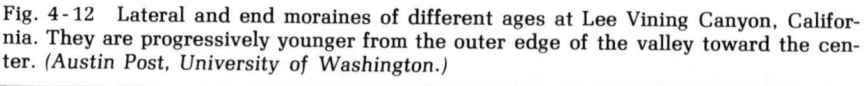

lines. Along the east side of the Sierra Nevada Range in California, Blackwelder (1931) found that shorelines of Pleistocene Mono Lake had been carved on moraines extending out from the mountain front, indicating that the moraines are older than the last rise of the lake. Blackwelder also found that some moraines had been cut by faults, some had been buried by lava flows from nearby volcanoes, and others were plastered up against older lava flows. Stream terraces related to glacial advances end upvalley at moraines deposited while the meltwater streams were building the terraces.

Fig. 4-13 Crosscutting lateral and end moraines near Walker Lake, California. The moraines in the upper left corner cut across those in the lower center of the photograph and are thus younger. *(Aero Service Division, Litton Industries.)*

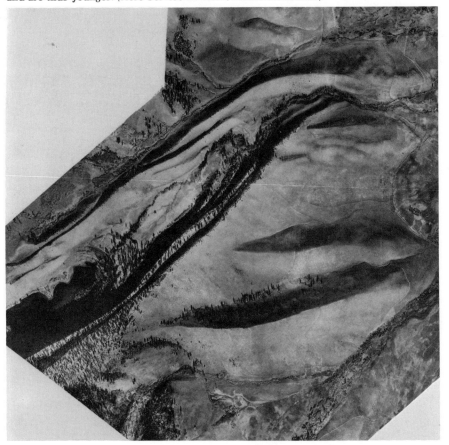

weathering

When other factors are equal, older constituents in tills or other glacial deposits are more weathered than younger constituents because they have been exposed longer. Pebbles and cobbles in relatively old glacial deposits may have weathering rinds on them, whereas younger deposits may be completely fresh. The thickness of weathering rinds on basalt pebbles varies with age, and within a local area may be used as a rough estimate of the age of moraines. Basalt pebbles in tills having similar thicknesses of weathering rinds may be correlated, those having the greatest thickness being the oldest.

Fig. 4-14 Moraines of two different glaciations, Wind River Mountains, Wyoming. *(U.S. Forest Service.)*

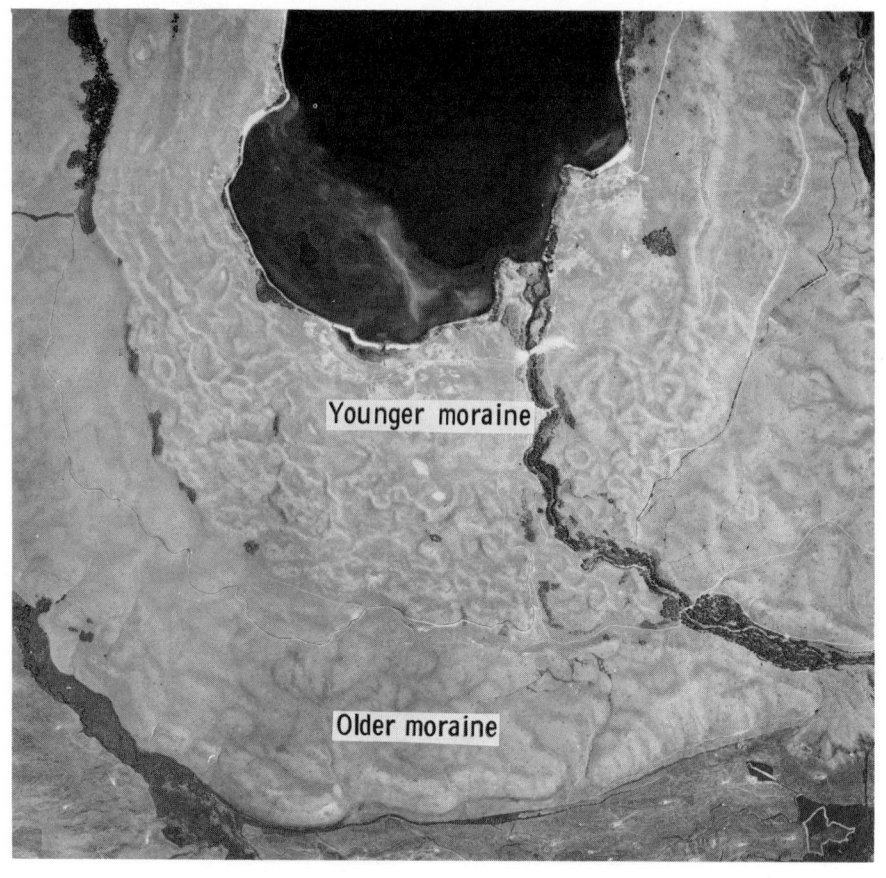

Blackwelder (1931) used "granite weathering ratios" to indicate age. Using only granitic boulders, he classified boulders on a moraine as being either (1) almost unweathered, (2) decayed but solid, or (3) rotten, and then counted the number of boulders in each category. He found that ratios of 90:10:0 between the three categories were typical of young moraines, 30:60:10 were typical of moraines of intermediate age, and 0:30:70 were typical of older moraines. However, considerable care must be exercised in using granite weathering ratios because the ratios may be affected by (1) position of boulders on a moraine, (2) knocking off of weathered material from boulders as they roll downslope, (3) local variation in dampness, and (4) other factors relating to local weathering conditions.

Degree of soil development varies for moraines of different ages, since the older a moraine is, the greater the time it has been affected by weathering. In the Rocky Mountains, Richmond (1957, 1960, 1962) used differences in soil thickness, color, structure, acidity, accumulation of clay, and other characteristics of soil profiles on moraines to determine their relative ages.

The amount of talus accumulated in cirques, amount of vegetation, and other features relating to general "freshness" differ among cirques. These features can sometimes be used in a general way to estimate relative ages of cirques in a particular region.

Degree of preservation of glacial polish and striae on bedrock can sometimes indicate age, but newly uncovered bedrock surfaces beneath glacial deposits may give false impressions of freshness and age.

stratigraphic criteria

Glacial and nonglacial deposits superimposed upon one another allow determination of relative ages. Two glacial deposits separated by nonglacial sediments could represent two separate glaciations or oscillations of a single glacier. However, buried soils and other features often help to identify true interglacial sediments between tills.

Ash on moraines may sometimes indicate the age of moraines near Pleistocene and Recent volcanoes. In the vicinity of Mono Lake, California, Blackwelder (1931) found little ash from nearby volcanoes on the youngest moraines in the area, 2 to 7 feet of ash on the next older moraines, and 6 to 20 feet of ash on still older moraines. On Mt. Rainier, Crandell and Mullineaux (1962) used lithology and thickness of ash from several Cascade volcanic eruptions of different ages to date Recent moraines and cirques.

The thickness of loess deposits on moraines or other glacial features may be used to determine relative age in much the same way as volcanic ash. A moraine which has been in existence during several episodes of loess deposition will have a greater thickness of loess, other things being equal, than a younger moraine which has been present only during the latest deposition of loess. Often weathering profiles may be preserved between loess blankets on a moraine, further strengthening relative-age interpretations. Since a given loess deposit is not likely to be of exactly the same thickness everywhere, use of loess thickness as a criterion for correlation is most valid within a relatively local area, where variations in thickness are not as likely to be present.

references

Birman, J. H.: 1964. Glacial geology across the crest of the Sierra Nevada, Calif., *Geol. Soc. America Spec. Paper 75.*

Blackwelder, Eliot: 1931. Pleistocene glaciation in the Sierra Nevada and Basin Ranges, *Geol. Soc. America Bull.,* vol. 42, pp. 865-922.

Cook, J. H.: 1946. Kame-complexes and perforation deposits, *Am. Jour. Sci.,* vol. 244, pp. 573-583.

Cotton, C. A.: 1941. The longitudinal profiles of glaciated valleys, *Jour. Geology,* vol. 49, pp. 113-128.

Crandell, D. R., and others: 1962. Pyroclastic deposits of Recent age at Mt. Rainier, Wash., *U.S. Geol. Survey Prof. Paper 450-D,* pp. D64-D68.

Fisher, J. E.: 1948. The pressure-melting point of ice and the excavation of cirques and valley steps by glaciers, *Am. Alpine Jour.,* vol. 7, pp. 62-72.

Flint, R. F.: 1928. Eskers and crevasse fillings, *Am. Jour. Sci.,* ser. 5, vol. 15, pp 410-416.

———: 1957. Glacial and Pleistocene Geology, John Wiley & Sons, Inc., New York.

Holmes, C. D.: 1947. Kames, *Am. Jour. Sci.,* vol. 245, pp. 240-249.

Johnson, Douglas: 1941. The function of meltwater in cirque formation, *Jour. Geomorphology,* vol. 4, pp. 252-262.

Lewis, W. V.: 1938. A meltwater hypothesis of cirque formation, *Geol. Mag.* (Great Britain), vol. 75, pp. 249-265.

———: 1940. The function of meltwater in cirque formation, *Geog. Rev.,* vol. 30, pp. 64-83.

Nelson, R. L.: 1954. Glacial geology of the Frying Pan River drainage, Colo., *Jour. Geology,* vol. 62, pp. 325-343.

Richmond, G. M.: 1957. Three pre-Wisconsin glacial stages in the Rocky Mts. region, *Geol. Soc. America Bull.*, vol. 68, pp. 239-262.

——— : 1960. Glaciation of the east slope of Rocky Mt. National Park, Colo., *Geol. Soc. America Bull.*, vol 71, pp. 1371-1382.

——— : 1962. Quaternary stratigraphy of the La Sal Mts., Utah, *Geol. Survey Prof. Paper* 324.

Sharp, R. P.: 1948. The constitution of valley glaciers, *Jour. Glaciology*, vol. 1, no. 4, pp. 182-189.

——— : 1960. Pleistocene glaciation in the Trinity Alps of northern Calif., *Am. Jour. Sci.*, vol. 258, pp. 305-340.

continental glaciation

Continental glaciers are observable today at only a few places in high latitudes, the best examples being the antarctic ice sheet and the Greenland ice sheet. However, during the Pleistocene, much of northern North America and Europe were covered with thick continental glaciers. The topography of the areas covered by these Pleistocene glaciers is to a large extent related to depositional and erosional processes of the ice.

Recognition of the former presence of immense glaciers dates back to the early part of the nineteenth century in Europe. Previously, various men had recognized fluctuations in existing glaciers, but in 1821 Venetz, a Swiss engineer, presented a paper in which he stated that alpine glaciers in the Alps had once been very much more extensive than at present. In 1824 Jens Esmark expressed a similar idea about the glaciers of the mountains in Norway. Several years later, in 1829, Venetz concluded that much of northern Europe, as well as the Alps, had once been covered with extensive glaciers, and a German named Bernhardi presented evidence in 1832 that glaciers from the arctic region once extended into Germany. Although Jean de Charpentier presented a paper in 1834 confirming the ideas of Venetz, Louis Agassiz, a

Antarctic polar ice plateau and outlet glaciers. Shackleton Glacier in left foreground. *(U.S. Navy for U.S. Geological Survey.)*

zoologist, was skeptical, and so in 1836 he arranged a visit with Charpentier to view the evidence in the field. Rather than refuting Venetz and Charpentier, Agassiz came away convinced that they were right, and in 1837 he suggested the former advance of enormous glaciers from the Arctic into northern Europe during a "great ice age." The idea of the *Ice Age* and evidence to support it was not the work of a single man, but it was to a large extent through the efforts of Agassiz that the concept eventually was accepted.

erosion by continental glaciers

Large portions of the Northern Hemisphere were affected by erosion of Pleistocene continental glaciers as they moved southward from high latitudes. Residual soil and the weathered rock mantle were stripped off, and the bedrock beneath scoured and abraded by the ice. Much of the Canadian and Scandinavian Shield areas today consist of scoured rock surfaces. Linear grooves oriented parallel to the direction of ice movement are common in these areas. North-south preglacial stream valleys were deepened, and less resistant rock was scoured out to form rock basins.

The amount of glacial erosion produced by Pleistocene continental glaciers varied considerably from place to place, depending on (1) differences in erodibility of material beneath the ice, (2) the character of the topography over which the ice moved, (3) thickness and rate of movement of the glacier, and (4) amount and character of fragmental material held in the ice.

finger lakes

Preglacial stream valleys oriented parallel to the direction of glacier movement are susceptible to extensive glacial erosion because the ice is thicker in the valleys than over the adjacent divide areas and there may be fewer topographic barriers to inhibit glacial movement. Such valleys may be greatly scoured and deepened.

The Finger Lakes of New York are a classic example of this feature. They consist of a group of elongate troughs, trending approximately north-south, which were covered by Pleistocene ice sheets several times. The largest lakes, Seneca and Cayuga, are about 40 miles long, and both extend to depths below sea level. Seneca Lake is more than 600 feet deep, 175 feet below sea level, and the bedrock floor lies an unknown distance beneath unconsolidated deposits on the lake floor.

Other examples of lakes occupying ice-scoured basins are numerous in northern Europe and North America, especially in Canada and the Scandinavian countries.

depositional landforms

A variety of landforms are produced by deposition of material from continental glaciers, largely as a result of the different ways in which debris may be deposited from a glacier. Material deposited from beneath the ice differs in character from that deposited along the margins or in front of a glacier, and consequently different landforms result (Fig. 5-1).

end moraines

Around the front margin of glaciers a ridge of rock debris may accumulate to form an *end moraine*. The size and degree of development of the end moraine depend on the attainment of equilibrium between the rate of ice advance and the rate of ablation. As long as equilibrium is maintained, the front of the ice remains stationary and the end moraine grows continually larger. Most of the material in moraines is deposited by dumping of rock debris from the ice at its margin. Pushing up of rock debris in front of a glacier may account for some of the material, but the amount of debris glaciers can push is limited. A large amount of debris in front of an advancing glacier produces shearing in the ice and may cause the glacier to override the material in front of it.

Most end moraines have a curved form in map view, resulting from the lobate nature of the glacier that made them. Although individual moraines may have great lateral extent, they often are interrupted by gaps where meltwater streams have flowed through them. Some gaps in moraines are produced by local oscillation of the ice front.

The topographic expression of moraines varies from low indistinct mounds to ridges more than 100 feet high. Some have a single well-defined crest; others have multiple crests recording successive stands of the glacial margin. Most moraines consist of hummocky ridges with many closed depressions caused by melting of blocks of buried ice and irregular deposition of material (Fig. 5-2).

In areas such as the north-central United States multiple glaciation and pauses in the retreat of glaciers have produced a series of moraines in concentric loops marking former ice margins. Particularly striking examples occur around the south end of Lake Michigan, where a complex system of concen-

Fig. 5-1 Depositional features on moraine, Coteau des Prairies east of Aberdeen, South Dakota. (John Shelton.)

tric end moraines exists. Some of these moraines are recessional moraines, representing pauses in glacial retreat; others represent glacial readvances.

interlobate moraines

Ice sheets having two or more lobes may build interlobate moraines between adjacent lobes (Fig. 5-3). Such moraines grade into end moraines around the front margins of each lobe. The Kettle Moraine along the west side of Lake Michigan in Wisconsin was formed between a lobe of ice occupying Lake Michigan and the Green Bay lobe to the west.

ground moraine

A *ground moraine* is a drift-covered area of low relief lacking transverse linear ridges. It consists of lodgement or ablation till deposited as a blanket over a glaciated region, and is likely to cover large areas. Large portions of the northern United States and Europe are mantled by ground moraine which forms undulating plains with gentle swells and basins.

Most ground moraine is deposited as till from the base of a glacier, but some contains scattered lenses of gravel or sand as well. Topographic relief on ground moraine may be a result of nonuniform deposition of till, or it may

Fig. 5-2 Moraine in southern Saskatchewan, pitted by kettles. *(National Air Photo Library, Surveys and Mapping Branch, Dept. of Energy, Mines, and Resources.)*

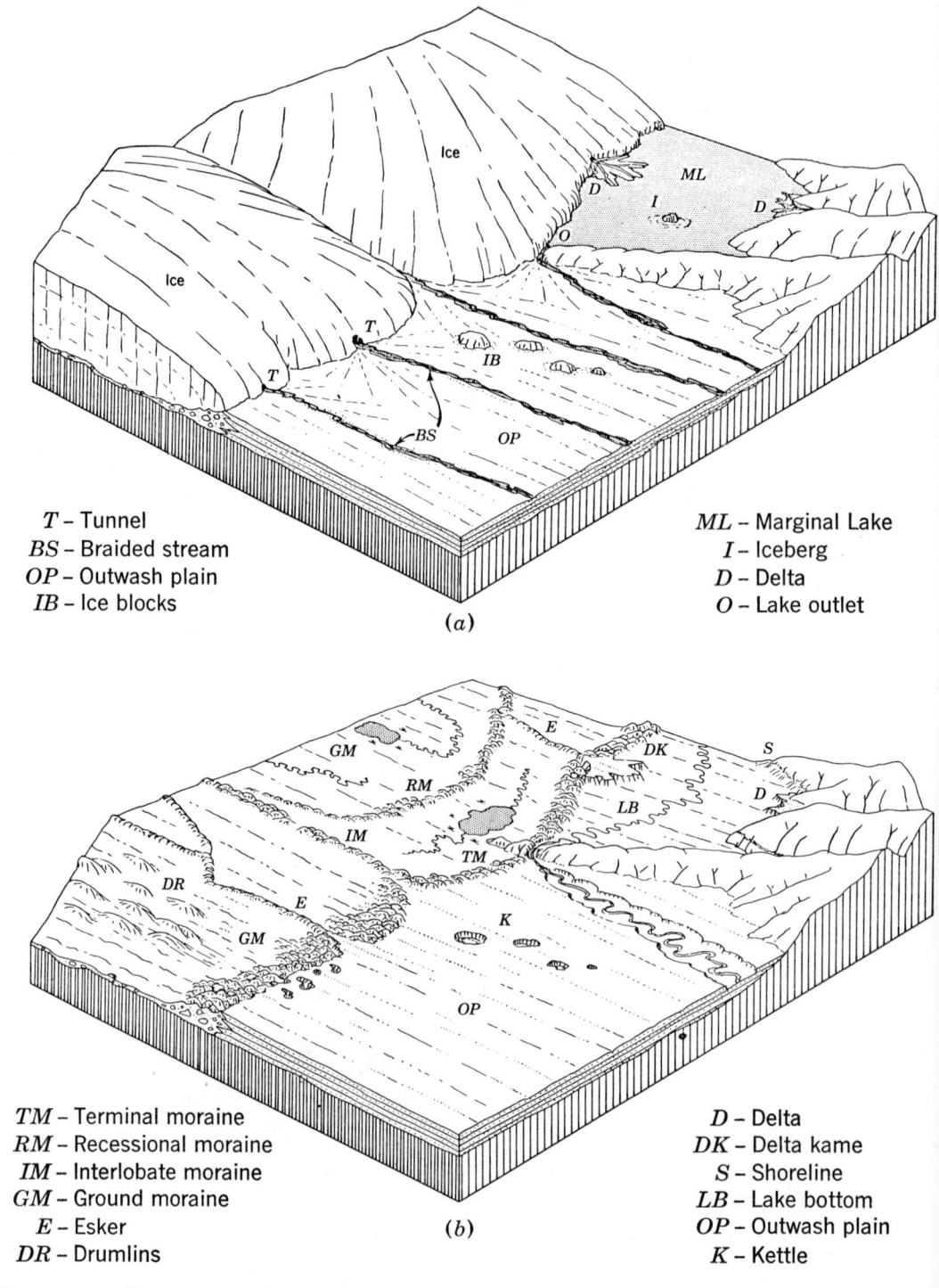

T – Tunnel	*ML* – Marginal Lake
BS – Braided stream	*I* – Iceberg
OP – Outwash plain	*D* – Delta
IB – Ice blocks	*O* – Lake outlet

(a)

TM – Terminal moraine	*D* – Delta
RM – Recessional moraine	*DK* – Delta kame
IM – Interlobate moraine	*S* – Shoreline
GM – Ground moraine	*LB* – Lake bottom
E – Esker	*OP* – Outwash plain
DR – Drumlins	*K* – Kettle

(b)

Fig. 5-3 Landforms produced by deposition near the margin of a continental glacier. *(From A. N. Strahler, John Wiley & Sons, Inc., New York, 1960.)*

reflect topography buried beneath the till mantle. Road cuts through areas of ground moraine often expose cores of preglacial hills which have been plastered with a mantle of till. In other areas, entire hills may consist of till of one glaciation. Irregularities in preglacial topography may be obliterated by deposition of a thick blanket of till to form a till plain.

drumlins

Low elliptical egg-shaped hills known as *drumlins* are common in some areas of ground moraine (Figs. 5-4 and 5-5). They typically exhibit streamlined forms resembling inverted teaspoons, with the long axes oriented parallel to the direction of glacial movement. The upstream ends of the drumlins are often steeper than the lee sides, which taper in the downstream direction. Although drumlins may vary considerably in size and shape, most are about 1/3 to 1/2 mile long, 1/4 mile wide, and less than 150 feet high (Fig. 5-5).

Fig. 5-4 Drumlin field near Rochester, New York. *(John Shelton.)*

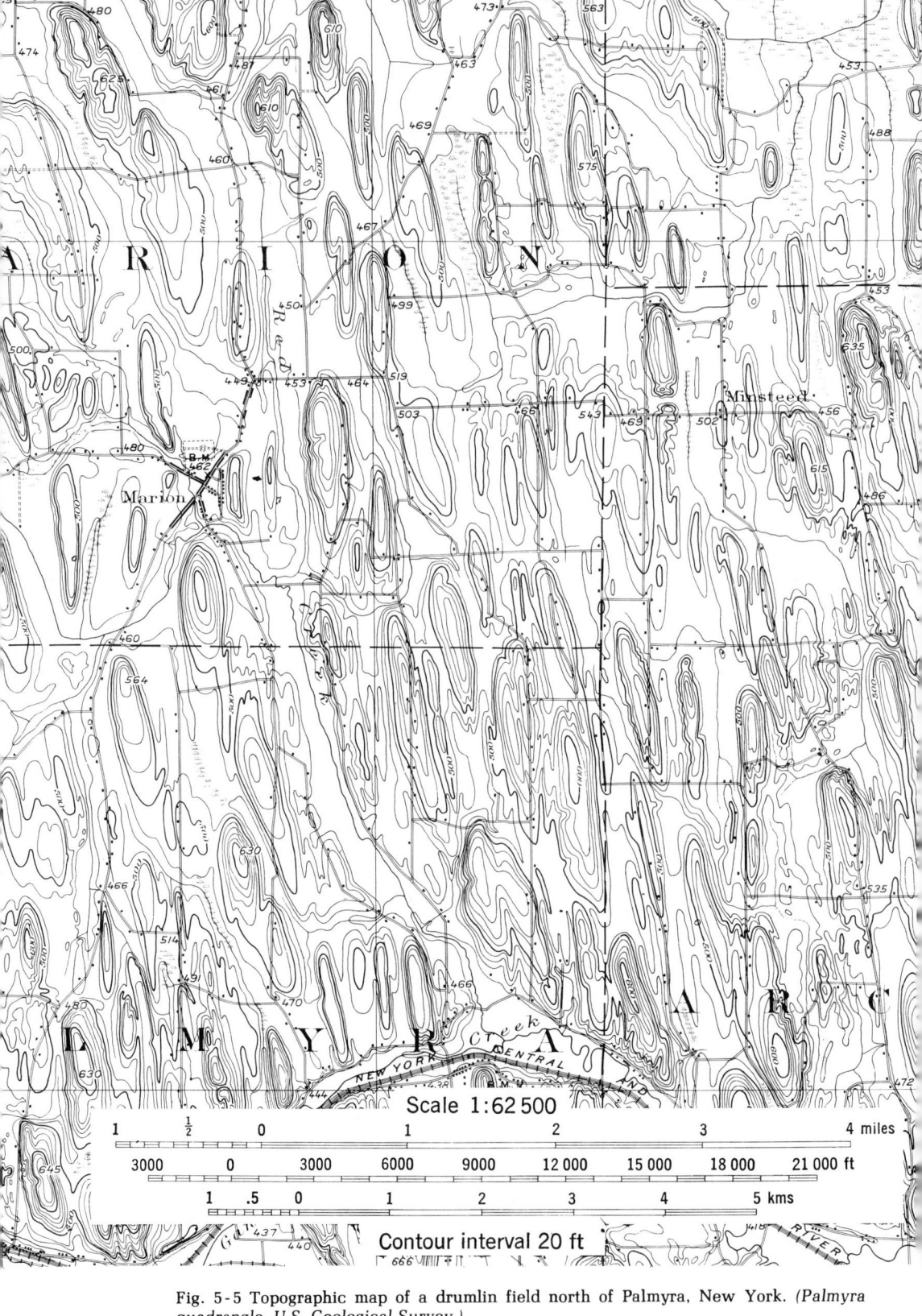

Scale 1:62 500

| 1 | ½ | 0 | | 1 | | 2 | | 3 | | 4 miles |

| 3000 | | 0 | 3000 | 6000 | 9000 | 12 000 | 15 000 | 18 000 | 21 000 ft |

| 1 | .5 | 0 | | 1 | 2 | 3 | 4 | | 5 kms |

Contour interval 20 ft

Fig. 5-5 Topographic map of a drumlin field north of Palmyra, New York. *(Palmyra quadrangle, U.S. Geological Survey.)*

Drumlins seldom appear singly, but generally occur in large numbers. Notable examples of drumlin fields in North America are those in southeastern Wisconsin, west-central New York, New England, and Nova Scotia. In Europe, drumlin fields are found in Ireland, England, and Germany.

Many drumlins consist of till similar to the ground moraine between adjacent hills, but others consist almost entirely of bedrock or sorted drift. There appears to be gradation from drumlins composed entirely of till to *rock drumlins* composed almost entirely of bedrock. Some forms, known as *craig-and-tail,* consist of bedrock knobs in the lee of which till has been deposited by an overriding glacier (Flint, 1957).

In some fields there is a striking relationship between drumlin orientation and the position of end moraines. Whereas end moraines lie transverse to the direction of glacier flow, drumlins lie parallel to the direction of movement. Hence, in areas such as southeastern Wisconsin, where the ice possessed a lobate form, the axes of drumlins diverge radically toward long loops made by end moraines.

The streamlined shape and parallel orientation of the long axes of drumlins indicate that drumlins are shaped by moving ice. However, geologists do not agree generally whether drumlins are essentially depositional or erosional in origin. According to the depositional theory of origin, drumlins are formed by deposition of concentric layers of drift at the base of the ice. Russell (1895) suggested that concentration of debris in a glacier decreases the rate of flow. Ice with less rock material may then override the debris-clogged ice and drift, shaping it into drumlins. Rock cores in some drumlins suggest that accumulation of drift may have been controlled by rock hills. Stratified sediments in the cores of some drumlins indicate reshaping of earlier deposited material, because meltwater presumably could not deposit this material under the same conditions which controlled formation of the drumlin.

The erosional theory of drumlin origin states that actively moving ice reshapes previously deposited drift or bedrock. Rock drumlins, obviously, are erosional in origin, and drumlins cored with stratified material must be at least in part erosional. However, in some areas, drumlins are composed of the very youngest till, deposited during the same glaciation that formed the drumlins, and is the same as the surrounding drift between drumlins.

Rock drumlins and drumlins consisting of till occur in the same drumlin fields, suggesting that both types, depositional and erosional, may form at the same time in the same area.

fluted topography

In some glaciated areas, long narrow ridges and straight parallel grooves give the topography a fluted appearance. Very straight elongate ridges up to several miles in length and 5 to 30 feet high occur in parts of North Dakota (Lemke, 1958) and in Alberta (Gravenor and Meneley, 1958) and other parts of Canada. The long axes of the ridges and grooves are parallel to the direction of ice movement. These unusually straight and narrow forms are gradational with drumlins and probably form in a somewhat similar manner, although some of them may be formed by squeezing up water-soaked drift into tunnels or cracks in the base of a glacier.

ice-contact forms

Sediments deposited against ice are subject to collapse when the supporting ice melts away, leaving topographic features collectively known as *ice-contact landforms*. They characteristically consist of contorted sediments deformed during collapse of the supporting ice, and are most commonly found associated with stagnant ice.

eskers

Among the common ice-contact features formed by continental glaciers are *eskers* (Fig. 5-6). They typically exhibit long winding sinuous courses, often uniting with tributary eskers and sometimes extending uphill and across divides. Most are less than one hundred feet high, with rounded crests and fairly steep sides (Fig. 5-7). Some unusually long ones are several hundred miles in length. Large numbers of eskers occur in Canada, near the latitude of Hudson Bay, in parts of Maine, and in the Scandinavian countries.

Most eskers are composed of sand and gravel deposited by streams flowing in tunnels at the base of a glacier. Since actively moving ice would tend to close tunnels in the ice, formation of eskers is apparently restricted to stagnating ice. Some eskers have been observed in the process of forming (Lewis, 1949; Meier, 1951).

kames and kame terraces

Kames and kame terraces are associated with stagnating ice of continental glaciers, as well as with alpine glaciers. Kames are common in areas where thin ice has become detached from the main glacier or where ice in the vicinity of the terminus loses its motion and downwastes by ablation, leaving hummocky landforms with numerous depressions, often referred to as *kame-and-kettle topography*. Kame terraces may be built by streams and lakes between stagnating ice and valley sides. A pitted appearance is sometimes imparted to them by numerous kettles.

Stagnant ice near the terminus of a glacier is characterized by many depressions and fissures (Fig. 5-8). Lakes formed in the depressions are floored and held in by ice for a period of time sufficiently long to allow sediment to accumulate on the lake bottom. Crevasses and meltwater channels in the ice become filled with glacial drift, and when the ice melts away, a variety of landforms result, most of which could logically be considered kames, but which may be given more specific names, depending on their origin.

Figure 5-9a illustrates the origin of one such landform, an *ice-walled lake*

Fig. 5-6 Esker, between Lac de Gras and Bathhurst Inlet, Northwest Territories, Canada. *(National Air Photo Library, Surveys and Mapping Branch, Department of Energy, Mines, and Resources.)*

plain, common in stagnant ice deposits of part of North Dakota and Saskatchewan (Gravenor and Kupsch, 1959; Clayton, 1967). Flat upland areas, underlain by horizontally laminated lake silts and clays, stand perched above the
surrounding area (Fig. 5-9b), formed by deposition of sediments in ice-walled
depressions in stagnant ice, and when the supporting ice melts away, inversion of topography takes place and the former lake floor stands as a topographic
high.

disintegration ridges and trenches

Filling of crevasses or supcrglacial channels with drift creates ridges or
trenches when the ice melts away. *Disintegration ridges* (Gravenor and Kupsch,

Fig. 5-7   Esker near Clinton, Massachusetts. *(W. C. Alden, U.S. Geological Survey.)*

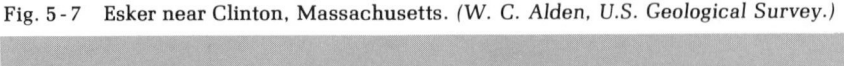

1959) are linear ridges formed by accumulation of drift in crevasses or super-glacial channels which is let down on the land surface when the support-ing ice melts. They are usually a few tens of feet high and a few hundred feet to a few miles in length. A related landform, a *circular disintegration ridge,* consists of a circular ridge a few tens of feet high and a few hundred feet in diameter, formed by the accumulation of superglacial drift in a depression in stagnant ice. As the supporting ice walls melt away, inversion of topog-raphy takes place because of the insulating effect of the drift in the bottom of the depression (Fig. 5-10), leaving a circular hill of drift with an ice core. Melting of the ice core causes collapse of the center of the hill and forma-

Fig. 5-8   Stagnant ice near the terminus of the Barnard Glacier, Alaska. Note the super-glacial lakes and ice-walled depressions. *(U.S. Geological Survey.)*

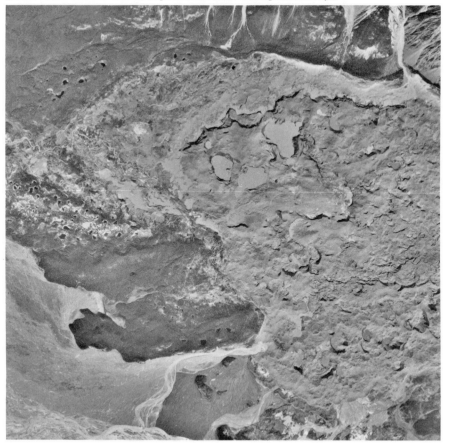

tion of a central depression, the net result being a circular ridge somewhat resembling a doughnut (Gravenor, 1955).

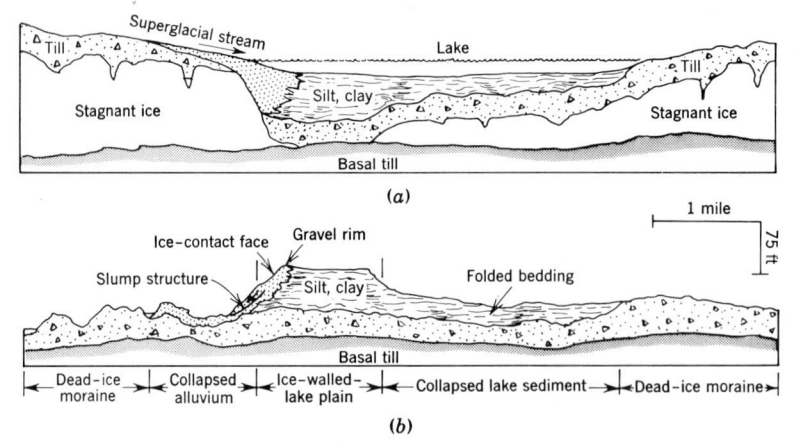

Fig. 5-9   Cross sections illustrating the development of an ice-walled lake plain. *(a)* Conditions at the time of formation; *(b)* the resulting landform. *(Lee Clayton, North Dakota Geological Survey.)*

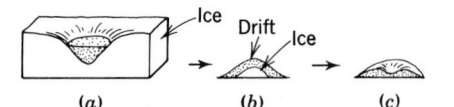

Fig. 5-10   Stages of development of a circular disintegration ridge. *(Lee Clayton, North Dakota Geological Survey.)*

Disintegration trenches are shallow linear depressions a few feet deep and several hundred feet to a mile in length (Clayton, 1962). The following series of events leads to their development (Fig. 5-11): (1) a superglacial channel or crevasse is filled with drift; (2) the supporting ice melts, and an ice-cored ridge is left by inversion of topography, due to the insulating effect of the drift; (3) the ice-cored ridge is buried by stream sediment; and (4) melting of the ice core causes collapse of the ridge, and a linear depression is produced in the overlying sediment as a result of the collapse.

outwash plains

Outwash plains are moderately flat, sloping plains beyond the margin of a glacier. They consist of stratified sand and gravel deposited by meltwater

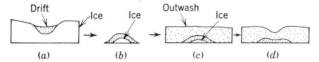

Fig. 5-11 Stages of development of a disintegration trench.
(Lee Clayton, North Dakota Geological Survey.)

streams. The slope of the plain away from the glacier results from the gradient of the meltwater streams, and may vary from a few to many tens of feet per mile, depending on the size of material being transported and the discharge of the streams.

The flatness of an outwash plain is often interrupted by numerous kettles

Fig. 5-12 Ice-walled lake plain and circular disintegration ridges, Montrail County, North Dakota. *(U.S. Department of Agriculture.)*

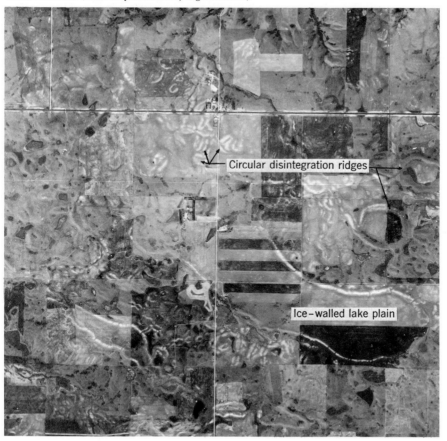

near the margin of the glacier, forming a *pitted outwash plain* (Fig. 5-15). Most large kettles in an outwash plain result from the burying of large blocks of stagnant ice detached from the terminus of a glacier.

When a glacier recedes from an area, the source of water for meltwater streams is lost, and abandoned channels are left on outwash plains. Abandoned outwash channels may also be found cutting through previously deposited moraines (Fig. 5-16).

Fig. 5-13 Disintegration trenches, east of Jamestown, North Dakota. *(U.S. Department of Agriculture.)*

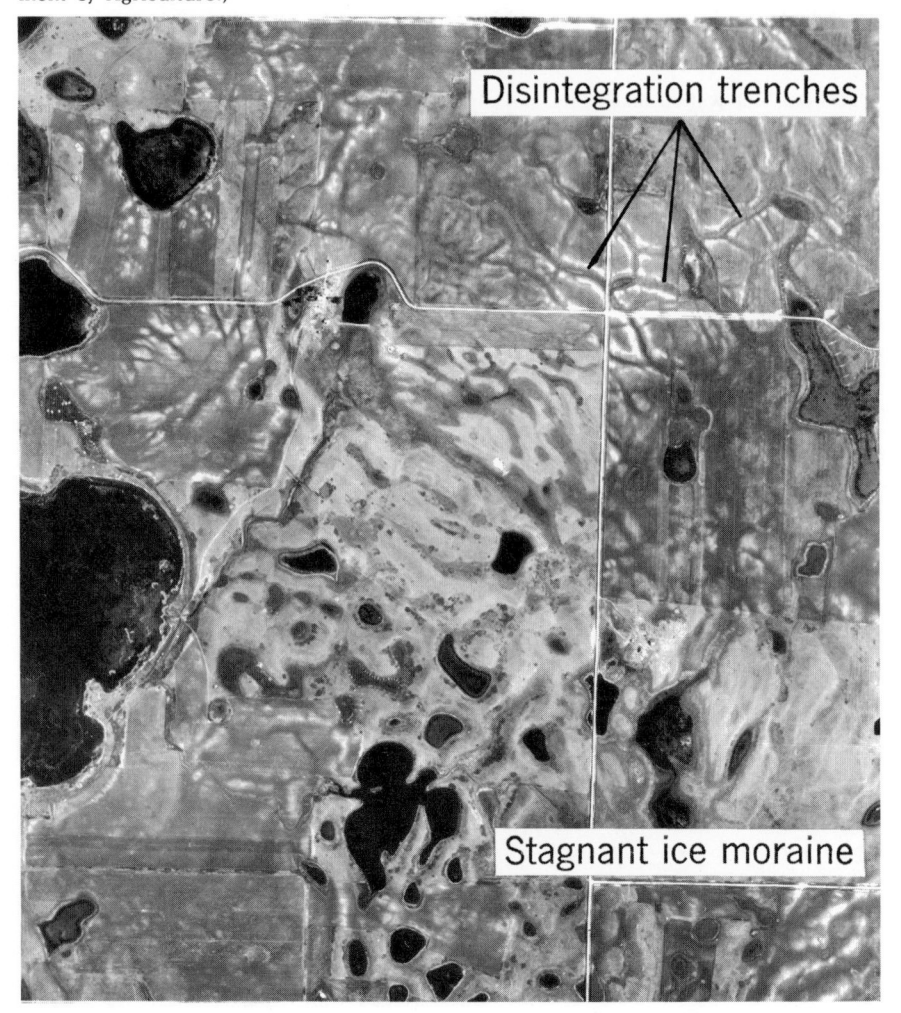

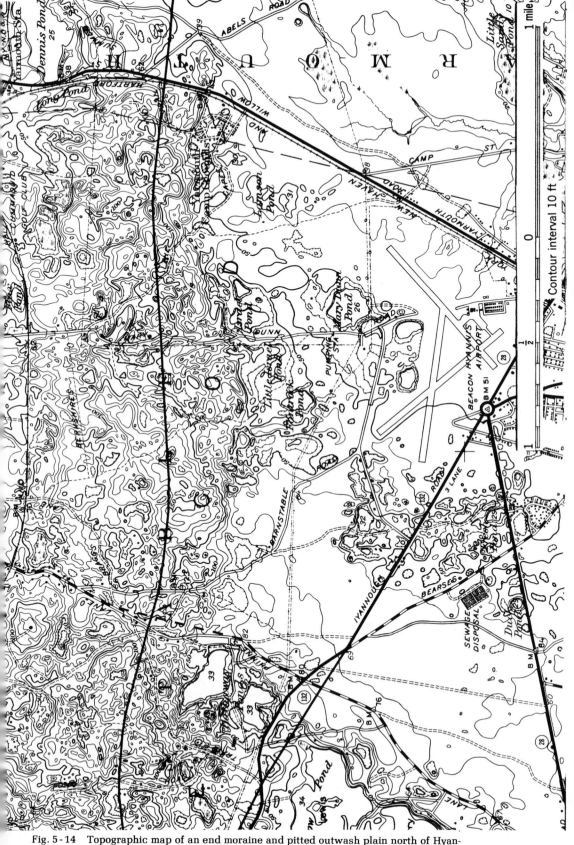

Fig. 5-14 Topographic map of an end moraine and pitted outwash plain north of Hyannis, Massachusetts. *(Hyannis quadrangle, U.S. Geological Survey.)*

criteria for recognizing glacial stages

Many criteria used to identify separate stages of alpine glaciation (pages 67-72) are useful in distinguishing stages of continental glaciation. Among them are (1) topographic position, (2) degree of gullying and erosion, (3) degree of weathering of pebbles in drift, (4) depth of soil development, and (5) stratigraphic superposition of glacial and nonglacial deposits. Other methods are discussed below.

pollen

Pollen shed by trees and other plants is carried by air currents, and accumulates in peat bogs and lake sediments. The pollen in a given layer of peat or lake clay normally reflects the type of vegetation growing nearby at the time of deposition. Since vegetation is dependent on climate, the type of trees in an area reflects, among other things, the nature of the climate in an area. In the Pacific Northwest, Hansen (1947) established a postglacial forest sequence

Fig. 5-15 Partially drowned, pitted outwash plain near Falmouth, Massachusetts. End moraine at upper right. *(John Shelton.)*

based on pollen analyses of late- and postglacial peat bogs. He found, for example, that lodgepole pine was able to withstand a fairly cool, moist, rigorous climate, and invaded a deglaciated area soon after the ice retreated. Later, during climatic warming, the lodgepole pine forests were replaced by Douglas fir and other trees adapted to the milder climate. Thus pollen analysis of nonglacial sediments between two glacial tills may allow interpretation of climate during deposition. If pollen in the intertill sediments reflects a warm climate like that of the present, the deposits represent a true interglacial period, and the two tills belong to distinctly different glaciations. If, however, the pollen suggests a cold, rigorous climate, the two tills may represent oscillation of the ice front during a single glacial episode.

radiocarbon dating

Bombardment of nitrogen atoms by neutrons in the upper atmosphere produces a radioactive isotope of carbon by knocking a proton out of the nucleus of the nitrogen atom.

Fig. 5-16   Abandoned meltwater channel in moraine, west of Devils Lake, North Dakota. *(John Shelton.)*

$$\text{Neutron} \rightarrow_7 N^{14} \xrightarrow[\rightarrow]{\rightarrow \text{proton}} {}_6 C^{14} \xrightarrow[\rightarrow]{\rightarrow \text{beta particle}} {}_7 N^{14}$$

The radioactive carbon nucleus gives off a beta particle, changing the C^{14} to nitrogen. The decomposition of radiocarbon takes place at a regular rate, decreasing the original amount of radiocarbon by one-half each 5,700 years. The time for half of a given amount of any radioactive isotope to decay is known as the *half-life*.

Both radiocarbon and normal carbon combine with oxygen to form carbon dioxide, which is taken in by living organisms. As long as an organism is alive it continues to take C^{14} into its system, maintaining a fairly constant ratio of C^{14} to C^{12} (the nonradioactive isotope). However, after an organism dies, it ceases to take in new radiocarbon, and the amount already present continuously declines as a result of radioactive disintegration. Thus the older a carbon-bearing sample is, the smaller the amount of radioactivity it has (Fig. 5-17). Samples older than about 40,000 years have so little radioactivity remaining that dating becomes difficult beyond this age. Some laboratories, however, are able to date material as old as 60,000 to 70,000 years by using special techniques.

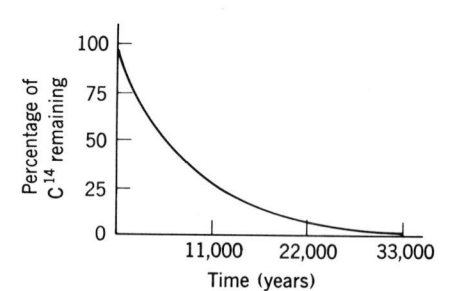

Fig. 5-17   Relationship of C^{14} in an organism to time. 5,700 years after an organism dies, half of the C^{14} has disintegrated. After about 11,400 years, only one-fourth of the original C^{14} is left.

The most suitable material for radiocarbon dating is wood, peat, or shells. Radiocarbon dating of organic material in glacial and interglacial deposits provides an important means of correlation for late Pleistocene events on a worldwide scale.

Figure 5-18 illustrates use of C^{14} dating in establishing the age of a glaciation. If wood from a forest overridden by a glacial advance yields a C^{14} date of 20,000 years, and peat from the base of a bog or top of the glacial drift gives a date of 11,000 years, the age of the glaciation is bracketed by the two dates.

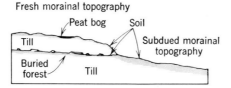

Fig. 5-18 Relationships between tills of two glaciations. The younger till has fresh morainal topography preserved, a thin soil has developed, and it lies on a buried forest. The older till has a more subdued, dissected topography and a thicker, better-developed soil and is separated from the younger till by a buried soil and forest.

potassium-argon dating

Potassium-40 is a radioactive isotope of potassium which decays to form argon-40 and calcium-40 at a constant rate, providing an exceedingly useful means of absolute age determination of potassium-bearing minerals. It may be used for dating rocks over a wide range of ages, from as young as 70,000 years to as old as 5 billion or more. The dating technique involves analysis of the relationship between potassium-40 and its daughter product, argon-40. When potassium-40 decays to argon-40, the argon is trapped in the crystal lattice of the mineral, and is retained until the mineral is melted, recrystallized, or heated sufficiently to allow the argon to diffuse through the lattice. Thus the ratio of radiogenic argon-40 to potassium-40 in a mineral is directly related to the time at which the mineral crystallized and began to accumulate argon-40.

The decomposition of potassium-40 to argon-40 takes place at a constant rate, decreasing the original amount of potassium-40 by one-half each 1.31 billion years (the half-life). To find the age of a given sample it is necessary to determine the amount of potassium-40 and argon-40 present in the sample today. Very old rocks will have more argon-40 than young rocks because there has been a longer period of accumulation of radiogenic argon from potassium-40, and the age in years may be calculated from the known rate of decay, i.e., the half-life.

Potassium-bearing minerals which are suitable for age determination include biotite, muscovite, sanidine, plagioclase, hornblende, and some pyroxene. In some cases it is feasible to date whole rock samples. The age obtained from these minerals is the age of crystallization; so it is not possible to obtain an accurate age on any of these minerals from sedimentary rocks. Glauconite is about the only sedimentary mineral which is suitable for potassium-argon dating. The principal use of potassium-argon dating in correlation and age determination of glacial stages lies in the dating of lava flows or volcanic ash associated in some way with glacial deposits. A date obtained from a lava flow or ash layer overlying glacial drift establishes the youngest limit for the age of the glaciation that deposited the drift. Conversely, a date obtained from a lava flow beneath glacial drift establishes the oldest limit for the drift; i.e., the drift cannot be older than the age of the flow.

The use of potassium-argon dating of glacial deposits is limited by (1) the lack of datable material within the glacial deposits and (2) the accuracy of dates from very young material. The first of these limits may be circumvented by dating associated material such as lava flows and ash layers. For young samples, between about 70,000 and 1 million years old, the amount of radiogenic argon in a sample is very small because of the long half-life of potassium-40. Such small amounts of argon-40 introduce difficulties in accuracy because a small error in determination of the argon or a small amount of contamination from alteration or atmospheric argon causes a large percentage of error in determination of the age of the sample. Despite these difficulties, the potassium-argon dating method is extremely useful in correlation and age determination of glacial deposits. When combined with the radiocarbon method, it is possible in many instances to obtain dates from deposits ranging in age from a few thousand to many millions of years.

paleomagnetic correlation

A recently developed method of correlation is the use of paleomagnetism in igneous rocks and some sedimentary rocks. The method is based on the facts that the earth's magnetic field changes from time to time and that certain rocks are magnetized in the earth's magnetic field at the time they are formed.

The earth's magnetic field is dipolar; i.e., it has a "north pole" and a "south pole." If the present condition of the magnetic field is taken as *normal,* a change in conditions such that the magnetic poles are reversed may be referred to as a *reversed* magnetic field. Measurements of the paleomagnetism (the magnetism imparted to the rock when it was formed) of igneous rocks and some sedimentary rocks show that in the geologic past there have been times when the earth's field was normal and other times when the field was reversed (Cox and Doell, 1960; Cox, Doell, and Dalrymple, 1965).

When volcanic rocks cool below the Curie temperature, the temperature at which minerals first become ferromagnetic, they assume a remnant magnetism which is related to the magnetic field at that time but which remains locked in unless the rock is reheated or demagnetized. Thus, if a volcanic rock cools in a normal magnetic field, it will retain normal remnant magnetism, but if it cools in a reversed magnetic field, it will retain reversed remnant magnetism. All the volcanic rocks in the world which cooled during a particular time period will have the same remnant magnetism unless disturbed.

Potassium-argon dating of volcanic rocks having normal and reversed remnant magnetism has resulted in the development of a paleomagnetic polarity time scale (Cox, Doell, and Dalrymple, 1965) which shows the polarity of magnetism in rocks formed at a given time within about the last 4 million years. During the last million years the earth's magnetic field has had normal polarity; between about 2.5 and 1 million years ago the polarity was reversed; between about 2.5 and 3.4 million years ago the polarity was normal; and between about 3.4 and 4.0 million years ago the polarity was reversed. Within each of these polarity epochs there are minor changes of short duration. By measuring the remnant magnetic polarity of a section of rocks it is possible to determine which were formed during a normal period of magnetic polarity and which were formed during a period of reversed polarity. From this information correlation may be made with rocks elsewhere in the world having the same polarity sequence, and they may be fitted into a time scale of years by comparing them with polarity epochs based on potassium-argon dates.

lithology

Differences in the composition of glacial deposits of different ages and differences in composition of glacial and nonglacial deposits are often important in interpretation of glacial stages. The composition of a glacial deposit reflects the kind of rock over which the glacier passed, and the composition of a nonglacial deposit reflects the type of rock in the drainage basin of streams.

Where the lithology of a glacial deposit differs from the lithology of local bedrock and stream alluvium, it may be possible to identify a deposit as being of glacial origin simply by analyzing its composition. In some cases it may be possible to trace certain rock types in a till back to their bedrock source. In the Puget Lowland of Washington, each glacial advance in the lowland from Canada brought large amounts of debris derived from the mountains of British Columbia. During interglacial episodes fluvial sediments were deposited from the Cascade Range to the east. Because the source rocks differ in certain respects, it frequently is possible to distinguish glacial from interglacial deposits on the basis of differences in lithology.

Occasionally, differences in direction of flow of successive glacial advances will produce glacial deposits of different composition.

reconstruction of former glacial advances

Much evidence concerning the extent and duration of former glaciations may be derived from glacial deposits and the topography developed upon them. The former maximum thickness of ice in an area may be inferred from the maximum altitude of erratics and other glacial deposits. It is sometimes possible to detect the maximum altitude of glacial scour along valley sides, which indicates the upper limit of the ice surface. The direction of movement of continental glaciers is determined by noting the direction of long axes of drumlins or grooves and striations on flat bedrock surfaces (Fig. 5-19). The

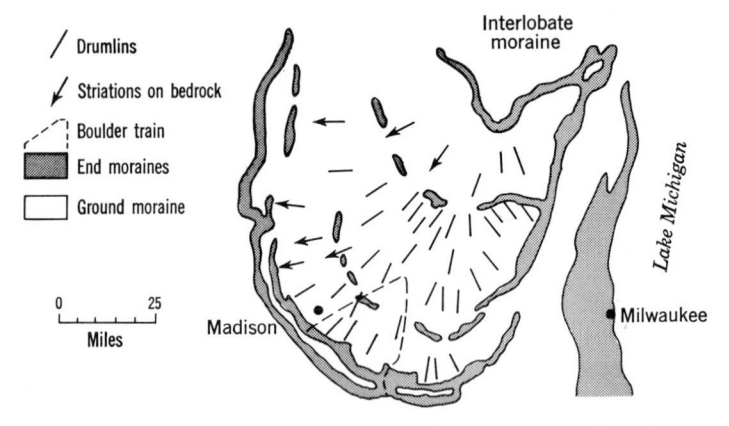

Fig. 5-19 Map of an area in Wisconsin showing relationships between end moraines, drumlins, and various indicators of the direction of ice movement. The glacier here was lobe shaped. Note the divergent trend of ice flow, approximately at right angles to the end moraines. (Modified from "Glacial Map of the United States, East of the Rocky Mountains," Geological Society of America, 1959.)

source of a former glacier is often identified by analyzing the lithology of pebbles in till and matching certain rock types with bedrock in the source area. Ice-contact features such as kames and eskers suggest that at a given place ice was stagnant, whereas drumlins and grooved bedrock indicate that ice was actively moving. Moraines mark the margins of former glaciers.

Using these and other criteria, a great deal of information about former glaciers may be determined. At least four major periods of glaciation have now been identified in North America, Europe, Asia, and various other parts of the world. Each of these stages of glaciation was followed by a retreat of the glaciers and a return to climates somewhat similar to those of today.

In North America ice accumulated to great thicknesses in Canada and flowed southward into the northern United States (Fig. 5-20), while at the same time alpine glaciers advanced in the Rocky Mountains, the Sierra Nevada, the Cascade Range, and various other places. The present topography in much of the northern United States is related to Pleistocene glaciation.

The youngest glaciation, the Wisconsin, was characterized by several advances and retreats during the general episode of glaciation. Deposits of this glaciation are widespread and many moraines occur throughout the midwestern United States (Fig. 5-21). Moraines of older glaciations emerge from beneath the Wisconsin drift cover in parts of Nebraska, Kansas, Iowa, Missouri, Illinois, and Indiana, and are exposed in minor areas farther east.

In Europe, continental ice spread radially from Scandinavia, extending southward into northern Europe. The British Isles were covered by a smaller ice cap. Portions of northern Asia, particularly Siberia and the northern plains of Russia, were glaciated. A summary of the glaciations in North America and Europe is given in the table below:

mid-continental United States	Europe
Wisconsin Glaciation — Valders, Two Creeks, Woodfordian, Farmdale, Altonian	Wurm Glaciation ... Riss Glaciation
Sangamon Interglacial	
Illinoian Glaciation	Mindel Glaciation
Yarmouth Interglacial	
Kansan Glaciation	Gunz Glaciation
Aftonian Interglaciation	
Nebraskan Glaciation	Donau Glaciation

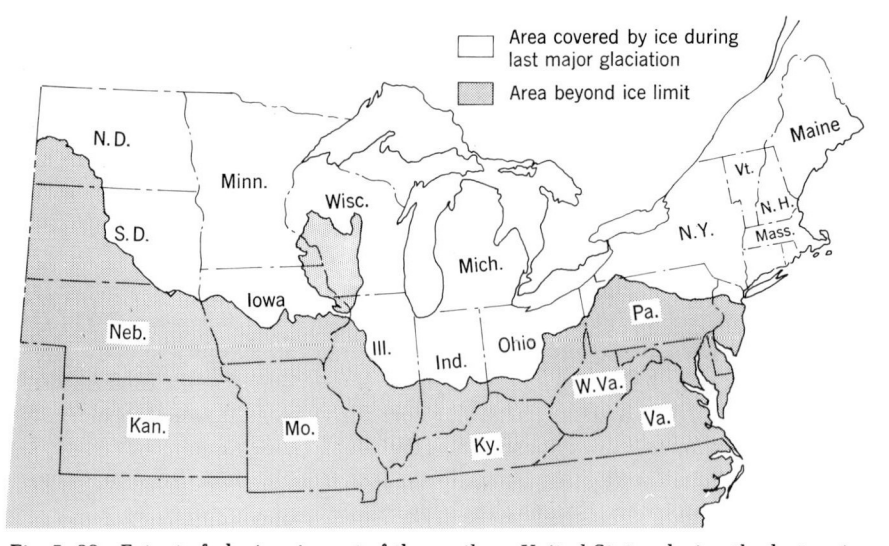

Fig. 5-20 Extent of glaciers in part of the northern United States during the last major glaciation. (Modified from "Glacial Map of the United States, East of the Rocky Mountains," Geological Society of America, 1959.)

other effects of Pleistocene glaciation

sea-level changes

Water evaporated from the sea provides the source of moisture for most of the precipitation that falls on the land. Rainfall normally is returned back to the oceans as runoff in streams, but when the moisture is in the form of snow and remains on the land as glaciers, the amount returned to the oceans is greatly reduced. The size of Pleistocene continental glaciers was so large that a substantial amount of water was tied up in them, resulting in a lowering of sea level during glacial periods. When the glaciers receded during an interglacial period, sea level rose again.

The amount of sea-level fluctuation during a glacial episode depends on how much water is tied up in glaciers. If the total volume of the Pleistocene ice sheets could be calculated, the amount of sea-level fluctuation could be determined. However, the volume of present-day ice sheets in the Antarctic and Greenland is not accurately known, and the total volume of Pleistocene ice sheets is even less precisely known. Estimates based on approximations of the volume of ice on land during a glaciation suggest that the amount of sea-level change was probably of the order of magnitude of 300 to 400 feet. The lowest stand of sea level during the late Wisconsin Glaciation apparently occurred between about 17,000 and 20,000 years ago [when sea level stood

Fig. 5-21 Moraines deposited in the northern United States during the last major glaciation. *(Modified from "Glacial Map of the United States, East of the Rocky Mountains," Geological Society of America, 1959.)*

approximately 400 feet below present sea level (Curray, 1965)]. During this time, broad areas of the present continental shelves must have been above sea level.

Other evidence for Pleistocene sea-level changes is to be found in marine terraces now above or below sea level. During some of the interglacial periods sea level may have been even higher than at present. If the present-day ice caps in the Antarctic and Greenland were to melt, sea level would rise several hundred feet. Since we do not know how much ice was in these areas during earlier interglacials, sea level could have been either somewhat higher or lower than at present. Marine terraces and beaches up to several hundred feet above sea level are found along coastlines in many parts of the world, but there is as yet no general agreement on their significance with regard to change in worldwide sea level since some of them are due to uplift of the land.

isostatic rebound

The weight of a large ice sheet on the land depresses the earth's crust beneath the glacier. Strong evidence for postglacial uplift of regions occupied by large continental glaciers may be found in the Scandinavian countries and North America (Gutenburg, 1941; Flint, 1957). The evidence includes:

1. In Scandinavia and North America the outer limit of uplift parallels the limit of the last glaciation.

2. In both regions the greatest postglacial uplift has occurred where the ice was thickest and lines connecting areas of equal uplift are concentric around the region of maximum ice thickness.

3. In both regions the rate of uplift has been approximately the same.

4. In both regions marine deposits now above sea level indicate that the areas were formerly below sea level.

5. In both regions negative-gravity anomalies are found, indicating that there are deficiencies in mass below the surface and that equilibrium has not yet been attained.

It is now generally recognized that crustal warping due to glaciation depends on (1) density of the ice, (2) thickness of the ice, (3) density of subcrustal rocks, (4) elasticity and plasticity of the crust beneath the glacier, and (5) length of time an ice sheet covers an area. The density of glacial ice is approximately 0.9, although it may be higher if a large quantity of rock debris is incorporated in it, and the density of rocks in the region where plastic transfer takes place is believed to be between about 3.0 and 3.3. Assuming that crustal adjustment to glacial loading is complete, that the average density of ice is about 0.9, and that the average density of the rocks is about 3.0, the maximum subsidence beneath a large continental glacier should be approximately one-third the thickness of the ice (Flint, 1957). Thus the thicker the ice, the greater amount of postglacial rebound is expectable. The rate of postglacial uplift is generally greatest soon after unloading occurs, possibly beginning during thinning of the ice, and is usually discontinuous. In the Baltic area the total postglacial uplift has been estimated to be between 800 and 1,800 feet in the last 10,000 years (Gutenburg, 1941). On the basis of marine features now above sea level, the Hudson Bay region is thought to have been uplifted at least 900 feet as a result of isostatic rebound.

The rebound effect appears to be restricted to large ice caps, and is not usually associated with small alpine glaciers.

As indicated by tidal gage records in Scandinavia, the upwarping there is still going on, the maximum rate of uplift approaching nearly a meter per 100 years near the former center of the ice sheet. This movement is causing lakes in the Baltic region to encroach on their southeastern shores and recede from their northwestern shores (Flint, 1957).

lake fluctuations in nonglacial areas

Lakes in arid and semiarid regions expanded greatly during periods of glaciation, and were reduced in size during interglacial times. Fluctuations of lake levels in Utah, Nevada, and eastern California (Fig. 5-16) are shown by abandoned shorelines, buried soils, interfingering of lake deposits with fluvial sediments, and other lines of evidence. The largest of these lakes was Lake Bonneville, which was over 1,000 feet deep and covered approximately 20,000 square miles in western Utah. Great Salt Lake is a remnant of this lake. Gilbert (1890) found several fluctuations of Lake Bonneville recorded in abandoned shorelines and alluvial-fan deposits. The lake rose to an altitude of about 5,050 feet, nearly 1,000 feet above the floor of Great Salt Lake, and developed a shoreline, but did not overflow. A subsequent rise of lake level of about 85 to 90 feet caused the lake to overflow into the Snake River in Idaho at Red Rock Pass at the northern end of the lake. Rapid downcutting of the outlet caused great floods of water to be discharged into the Snake River drainage. After a bedrock threshold in the outlet was encountered, the lake became stabilized at about 625 feet above Great Salt Lake, where the prominent Provo shoreline was developed. A later stand of lake level at about 330 feet above the present lake resulted in development of the Stansbury shoreline.

That high-water stages of Lake Bonneville were contemporaneous with glacial advances in the nearby mountains was demonstrated by Gilbert. A similar contemporaniety was found between lakes east of the Sierra Nevada in California and alpine advances in the mountains (Putnam, 1950).

Fig. 5-22 Distribution of Pleistocene lakes in the Basin and Range province of Nevada, Utah, and California. *(Modified from Snyder, Hardman, and Zdenek, 1964.)*

cause of pleistocene glaciations

Many theories have been advanced to explain the climatic variations associated with the Pleistocene, but none have been universally accepted. In

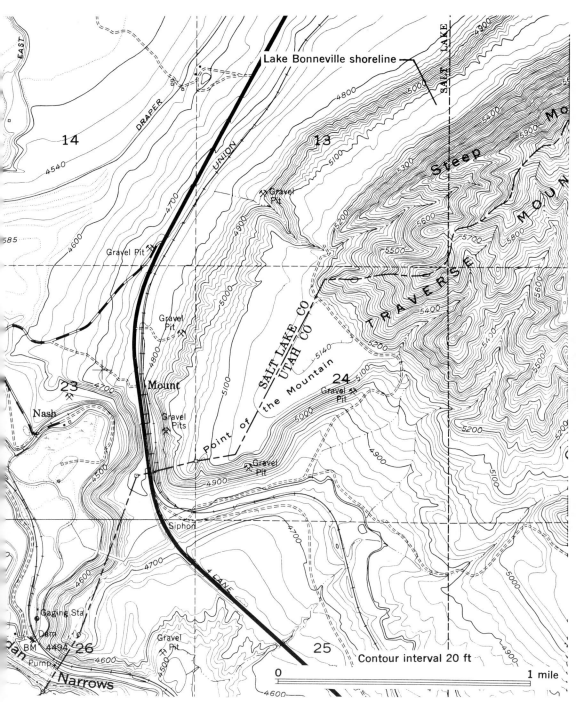

Fig. 5-23 Topographic map of shoreline features associated with glacial Lake Bonneville at Jordan Narrows, Utah. *(U.S. Geological Survey.)*

postulating a cause for Pleistocene glaciations, the following factors must be accounted for:

1. Pre-Pleistocene glaciations are relatively rare in the geologic record, although some have been recognized.

2. Pre-Pleistocene climates were nonglacial for many millions of years prior to the Pleistocene glaciations.

3. The Pleistocene was characterized by multiple glaciations, not just a single glaciation.

4. Climates during interglacial episodes were apparently somewhat similar to present-day climates.

5. Pleistocene glaciations were synchronous on several continents in the Northern Hemisphere and probably were synchronous in the Southern Hemisphere.

6. As indicated by Pleistocene ocean-water temperatures, climates during glaciations were characterized by lower temperatures, in addition to increased snowfall.

Theories of the cause of glaciation include both solar and terrestrial factors. Variation in the earth's climate could obviously be caused by change in the amount of solar radiation, but might also be caused by changes in the earth's atmosphere, which influence the amount of radiation received at the earth's surface. Space does not allow a full discussion here of the many complex problems related to the various theories, and only a brief summary is given.

The amount and type of radiation received by the earth from the sun are known to vary somewhat, and some of the variations are cyclic. One of the best-known variations is associated with the 11-year sunspot cycle, during which the number of dark spots visible on the sun's surface varies from a maximum to a minimum. Times of high average intensity of various particle emissions and ultraviolet radiation are associated with high numbers of sunspots, and these variations may be correlated with minor atmospheric fluctuations (Flint, 1957). Fluctuations of solar radiation over longer periods of time or fluctuations of different magnitude thus could cause marked climatic changes on the earth.

Other factors which could affect distribution of solar radiation received by the earth include the periodic changes that occur in the inclination of the earth's axis with respect to the sun and in the relationship between the inclination of the axis and position in the earth's orbit. They would not change the total amount of solar radiation received, but would change the distribution of solar energy on the earth's surface. The position of the earth in its orbit when it is closest to the sun (perihelion) presently occurs about January 3 each year. A shift in perihelion, having a period of about 21,000 years, affects the distribution of energy received from the sun, resulting in a tendency for relatively long cold winters and short hot summers for a time, then shorter, less cold winters and longer, somewhat cooler summers. In addition, two other variations, the eccentricity of the earth's orbit, which changes with a period of about 91,800 years, and the angle of the earth's axis relative to the plane of its orbit, which changes with a period of about 40,000 years, were used by Milankovitch (1941) and others as a basis for calculating temperature changes. The temperature changes which may have resulted from these changes in the earth's orbit explain many of the features associated with Pleistocene glaciation, but the magnitude of the temperature changes may be too small to have caused glaciations, and the timing is not quite right to fit observed evidence. Since pre-Pleistocene glaciations are so rare, these changes by themselves do not seem adequate to cause glaciations. Such periodic changes should have resulted in periodic glaciations in earlier times if they were the cause of Pleistocene glaciations.

Variation in the character of the earth's atmosphere has been called upon by some geologists to explain glaciation. Changes in the amount of volcanic dust or carbon dioxide in the atmosphere affect the amount of solar radiation received and retained and thus affect temperature. Although there was a good deal of volcanism during the Pleistocene, there is no reason to believe it was periodically distributed in such a manner as to cause the temperature fluctuations associated with glaciations. Similarly, there is no reason to believe the carbon dioxide content of the atmosphere was substantially different during the Pleistocene from today's.

Ewing and Donn (1956) have suggested that Pleistocene glaciations could be explained by (1) migration of the north pole from a point in the Pacific Ocean to the Arctic Ocean in the Pleistocene and (2) periodic freezing and thawing of the Arctic Ocean in response to changes in circulation of water between the Atlantic and Arctic Oceans. A shallow threshold separates the Arctic Ocean from the Atlantic. A small lowering of sea level would greatly reduce exchange of water between the Atlantic and Arctic Oceans, resulting in freezing over of the Arctic Ocean and consequent loss of the source of moisture necessary for continental glaciers. Rise of sea level as a result of melting of the continental glaciers would once again allow exchange of water between the Atlantic and Arctic Oceans across the shallow threshold, the Arctic sea ice would melt, and moisture from the open Arctic Ocean would once again be available as a source of precipitation. Increased snowfall in northerly latitudes then would produce another glaciation. However, polar wandering is required to explain the initial change to glacial climates from the warmer Tertiary climates, and evidence for polar wandering is at present inconclusive. The theory also does not explain Pleistocene alpine glaciations not affected by arctic storms.

Flint (1957) has suggested that unusually high topographic features formed during the Pleistocene plus solar fluctuations which decrease annual temperatures combined to cause high accumulations of snow and ice. According to this idea, similar solar fluctuations in the past may not have caused glaciations because high uplands suitable for snow accumulations may have been absent, and the presence of high upland areas may not have resulted in large glaciers because the solar fluctuations did not coincide. However, there is little evidence that continental glaciers in northerly latitudes were dependent on altitude during the Pleistocene to the extent that they would not otherwise have formed.

At present no completely adequate theory for the cause of Pleistocene glaciations can be demonstrated. Of the theories proposed, most do not fit all the evidence of Pleistocene changes, and some theories which are plausible cannot yet be proved with the data currently available.

references

Carr, D. R., and J. L. Kulp: 1957. Potassium-argon method of geochronometry, *Geol. Soc. America Bull.*, vol. 68, pp. 763-784.

Charlesworth, J. K.: 1957. The Quaternary Era, Edward Arnold (Publishers) Ltd., London, vol. 1.

Clayton, Lee: 1962. Glacial geology of Logan and McIntosh Counties, North Dakota. *North Dakota Geol. Survey Bull.* 37.

——— : 1967. Stagnant glacier features of the Missouri Coteau in North Dakota, in Glacial geology of the Missouri Coteau, *North Dakota Geol. Survey Misc. Ser.* 30, pp. 25-46.

Cox, A., and R. R. Doell: 1960. Review of paleomagnetism, *Geol. Soc. America Bull.*, vol. 71, pp. 645-768.

——— and G. B. Dalrymple: 1965. Quaternary paleomagnetic stratigraphy, in The Quaternary of the United States, Princeton University Press, Princeton, N.J., pp. 817-830.

Curray, J. R.: 1965. Late Quaternary history, continental shelves of the United States, in The Quaternary of the United States, Princeton University Press, Princeton, N.J., pp. 723-735.

Easterbrook, D. J.: 1963. Late Pleistocene glacial events and relative sea level changes in the northern Puget Lowland, Washington, *Geol. Soc. America Bull.,* vol. 74, pp. 1465 - 1484.

———— : 1966. Radiocarbon chronology of late Pleistocene deposits in northwest Washington, *Science,* vol. 152, pp. 764 - 767.

Ewing, W. M., and W. L. Donn: 1956. A theory of ice ages, *Science,* vol. 123, pp. 1061 - 1066.

Faul, Henry (ed.): 1954. Nuclear Geology, John Wiley & Sons, Inc., New York.

Flint, R. F.: 1957. Glacial and Pleistocene Geology, John Wiley & Sons, Inc., New York.

Gilbert, G. K.: 1890. Lake Bonneville, *U. S. Geol. Survey Mon.* 1.

Gravenor, C. P.: 1955. The origin and significance of prairie mounds, *Am. Jour. Sci.,* vol. 253, pp. 475 - 481.

———— and W. O. Kupsch: 1959. Ice-disintegration features in western Canada, *Jour. Geology,* vol. 67, pp. 48 - 64.

———— and W. A. Meneley: 1958. Glacial flutings in central and northern Alberta, *Am. Jour. Sci.,* vol. 256, pp. 715 - 728.

Gutenburg, B.: 1941. Changes in sea level, postglacial uplift and mobility of the earth's interior, *Geol. Soc. America Bull.,* vol. 52, pp. 721 - 772.

Hansen, H. P.: 1947. Postglacial forest succession, climate and chronology in the Pacific Northwest, *Trans. Am. Philos. Soc.,* vol. 37.

Holmes, C. D.: 1941. Till fabric, *Geol. Soc. America Bull.,* vol. 52, pp. 1299 - 1354.

Lemke, R. W.: 1958. Narrow linear drumlins near Velva, North Dakota, *Am. Jour. Sci.,* vol. 256, pp. 270 - 283.

Lewis, W. V.: 1949. An esker in the process of formation: Boverbreen, Jotunheimen, 1947, *Jour. Glaciology,* vol. 1, pp. 314 - 319.

Libby, W. F.: 1955. Radiocarbon Dating, The University of Chicago Press, Chicago.

Meier, M. F.: 1951. Recent eskers in the Wind River Mts. of Wyoming, *Iowa Acad. Sci.,* vol. 58, pp. 291 - 294.

Milankovitch, M.: 1941. Kanon dar Erdestrahlung und seine Anwendung auf des Eiszeitproblem, *Acad. Royal Serbe,* spec. ed., vol. 133.

Putnam, W. C.: 1950. Moraine and shoreline relationships at Mono Lake, Calif., *Geol. Soc. America Bull.*, vol. 61, pp. 115-122.

Russell, I. C.: 1895. The influence of debris on the flow glaciers, *Jour. Geology*, vol. 3, pp. 823-832.

Snyder, C. T., G. Hardman, and F. F. Zdenek: 1964. Pleistocene lakes in the Great Basin, *U.S. Geol. Survey Map* I-416.

Wright, H. E., and D. G. Frey (eds.): 1965. Quaternary of the United States, review volume for the International Association for Quaternary Research, Princeton University Press, Princeton, N.J.

3

fluvial morphology

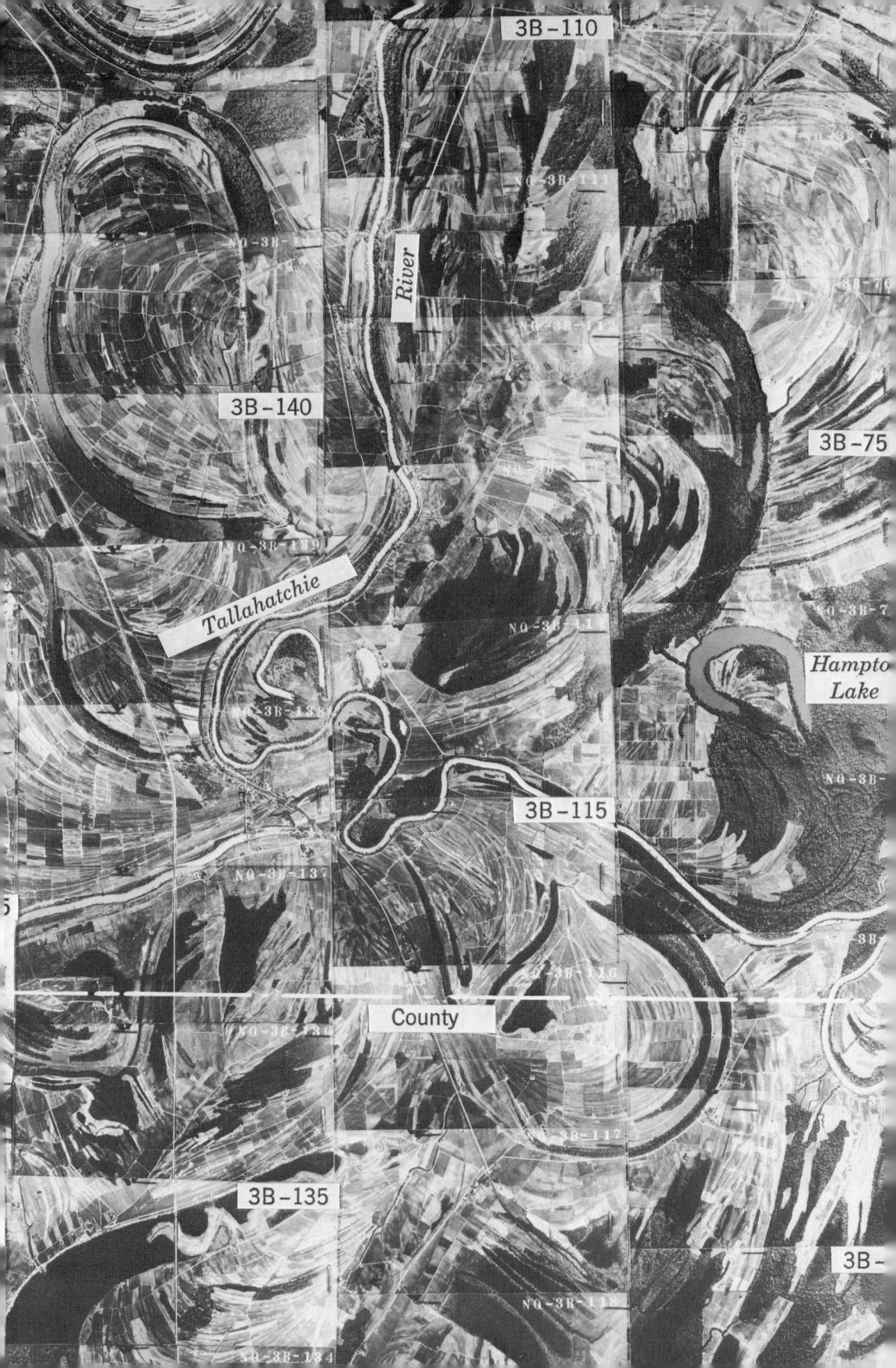

3B-110

River

3B-140

3B-75

Tallahatchie

Hampto
Lake

3B-115

County

3B-135

3B-

stream processes

Of all the processes which sculpture the earth's surface, the single most important agent is running water. Locally, other processes may dominate, but on a worldwide scale, water far outstrips all others.

Approximately 35 to 40 percent of the total precipitation which falls on the earth runs off on the surface. On upper hillslopes sheet runoff becomes concentrated into rills, which in turn feed larger gulleys, and eventually run into larger and larger streams. It has been estimated that in a drainage system such as the Mississippi, roughly 22 trillion cubic feet of water flows every year from the land to the sea. In so doing approximately 517 million tons of rock material is removed from the land, enough to fill a train of freight cars stretching 6 times around the equator. It is thus easy to see why running water is such an important agent in changing the configuration of the earth's surface.

variables of stream action

The ability of a stream to erode its channel and transport sediment depends largely on the velocity of the stream. Water flowing in a stream channel is

Meanders and point bar deposits, Tallahatchie River, Mississippi. *(U.S. Department of Agriculture.)*

impelled forward (downstream) by the force of gravity, or more specifically, by the component of the gravitational force F which is parallel to the stream floor (F sin slope angle). The water is retarded in its movement by friction between the moving water and the floor and sides of the channel and by internal friction between water particles. If there were no friction, water in a stream would accelerate to very high velocities, depending on the angle of slope of the water surface. However there are frictional forces in channels which impede the flow of water, and if there is no acceleration, there must be a balance between the downslope component of gravitational force and internal and external frictional forces.

Friction also affects the distribution of velocities within a stream channel, tending to reduce the velocity near the bed and walls of a stream. The rate of change of velocity within a channel, the velocity gradient, is greatest near the floor and sides of a channel, where frictional forces are largest. Velocity is zero on the channel floor, increasing with distance above the floor. Empirical values for the average velocity indicate that mean velocities usually occur about 0.6 of the distance from the surface of the stream to the bed, i.e., 0.6 of the total depth (Leopold, Wolman, and Miller, 1964). The average velocity may also be approximated by averaging the stream velocities at 0.2 and 0.8 of the depth. The slope of the velocity gradient curve (Fig. 6-4) has much significance with regard to erosion-transportation relationships in a stream because the maximum shearing stresses in the channel coincide with places of maximum difference in velocity. The maximum velocity in a stream usually occurs near the middle of the channel (as seen in plan view) just below the surface, and may be one-half to one-quarter greater than the average velocity of a cross section. In referring to stream velocities care must be taken to specify *which* velocity is meant. Most stream-gaging stations publish average velocities, although many measure other velocities as well. As is pointed out below, average velocities are useful in some considerations of stream processes, but are not always the most significant parameters.

Velocity is controlled by a set of factors which make up a mutually interdependent system. These factors are (1) discharge (volume of water per unit time), (2) gradient of the channel (slope), (3) load, and (4) channel characteristics.

discharge and velocity

The relationship between discharge and velocity is given by $Q = wvd$, where Q is discharge, w is width, d is depth, and v is velocity. The cross-sectional area of a stream channel is $w \times d$, and the wetted perimeter is $w + 2d$ (Fig. 6-1).

Fig. 6-1 Relationship of discharge to width
and depth of a stream.

As discharge increases, both cross-sectional area and the wetted perimeter
increase, but $w \times d$ increases more rapidly than $w + 2d$, so that cross-sectional area increases more rapidly than wetted perimeter, resulting in a relative decrease in frictional retardation by the floor and banks of the channel.
For this reason velocity increases with discharge.

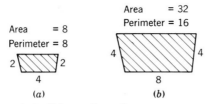

Fig. 6-2 Relationship between cross-sectional area and wetted perimeter in stream
channels. The examples do not represent actual channel shapes.

In Fig. 6-2a the cross-sectional area is $2 \times 4 = 8$, and the wetted perimeter
is $4 + 2(2) = 8$. In Fig. 6-2b the area is $4 \times 8 = 32$, and the wetted perimeter is $8 + 2(4) = 16$. By doubling the width and depth the cross-sectional
area has increased 4 times, whereas the wetted perimeter has increased 2
times, resulting in relatively less frictional retardation of water per unit of
cross-sectional area.

A great deal of empirical data has been obtained on changes in channel
depth, width, and velocity with change in discharge (Leopold and Maddock,
1953; Leopold and Wolman, 1957; Leopold, Wolman, and Miller, 1964). Figure
6-3 illustrates one such accumulation of data for a stream at a single gag-
ing station, and shows that, at that point, as discharge Q increases, channel
width, depth, and velocity increase.

The increase of width w, mean depth d, and mean velocity v at a station
with increase in discharge Q may be written as equations (Leopold and Maddock, 1953).

$$w = aQ^b \qquad d = cQ^f \qquad v = kQ^m$$

The exponents b, f, and m represent the slope of the lines shown in Fig. 6-3,
and a, c, and k are numerical coefficients. Since $Q = wvd$, Q also equals

$aQ^b \times cQ^f \times kQ^m$, and if this true, the sum of the exponents, $b + f + m$, must equal 1.0, and the product of $a \times c \times k$ must equal 1.0. In strong bank-forming material, as in cohesive silts, channels tend to have large values for f and low values for b, whereas in weak bank-forming material, as in loose sand, values are high for b and low for f.

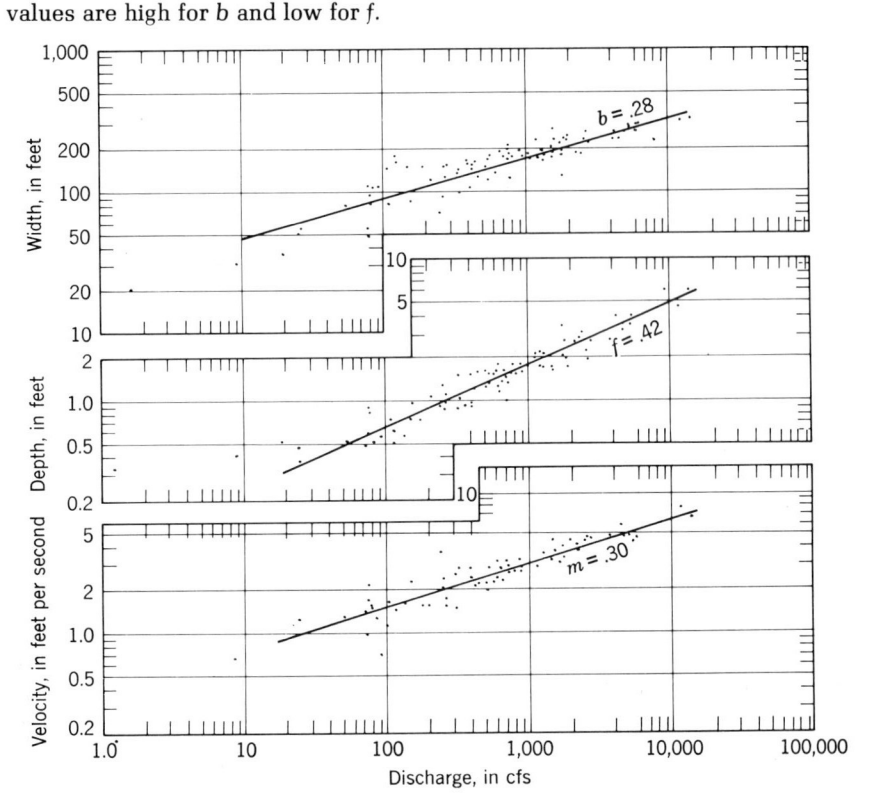

Fig. 6-3   Relationship of discharge to width, depth, and velocity in a stream, Powder River at Locate, Montana. *(Leopold and Maddock, 1953.)*

slope and velocity

Velocity is directly proportional to slope and increases as slope increases. Velocity varies within the stream channel, being greatest near the center of the stream slightly below the water surface and decreasing toward the bottom and sides of the channel, where frictional forces retard the rate of flow. Figure 6-4 shows the typical velocity distribution in a stream. In dealing with stream velocity the average velocity for a given channel cross section is commonly used, but this is not the most meaningful factor. The average velocity of a stream has little meaning with regard to transporting power since

(1) only the velocity near the floor of the channel is usable for transportation of bed load, and (2) much of the bed load is transported during a relatively short period of time during floods.

Both slope and discharge affect velocity, but in slightly different ways. In general, an increase in velocity due to increase in slope is more effective in transportation of bed load than increase in velocity due to increase in discharge (Gilbert, 1914). The discharge of a stream is given by $Q = wdv$. If Q is increased, then depth increases. However, if slope increases at constant discharge, then v increases and, in order that the product wdv remain the same, depth decreases. In both cases average velocity increases, but in the case of increased slope the greater increase in velocity near the bed tends to effect greater transporting power of material along the floor of the stream channel. Another way of saying this is that the maximum velocity that can be developed without causing increased erosion of the channel floor is a function of depth; i.e., with the same velocity a deep channel will not erode its channel as much as a shallow channel since, as depth increases, the water above the bed merely drags ahead more water without affecting particles on the channel floor.

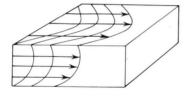

Fig. 6-4 Distribution of velocity within a stream channel.

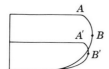

Fig. 6-5 Idealized velocity curves in a vertical section of a stream. *(After Gilbert, 1914.)*

Figure 6-5 shows idealized velocity curves in a vertical section of a stream (Gilbert, 1914). The lower curve is assumed to represent the same average velocity in a stream half as deep. An important difference between the two curves occurs near the bottom of the stream, where the bed velocity of the shallower stream possesses a greater potential for transportation of material along the bed of the channel than does the stream with a deeper channel having the same average velocity.

transportation of load

The load of sand, gravel, silt, and clay carried by a stream may be transported in several ways: (1) solution, (2) suspension, (3) saltation, and (4) traction.

Dissolved Load. Dissolved material liberated from rocks by chemical weathering is carried in solution and is not readily noticeable in streams. It has been estimated, however, that about 136 million tons of dissolved material is carried annually by the Mississippi River and that as much as 2.5 billion tons of dissolved material may be carried to the sea by streams each year. For any individual stream the ratio of dissolved material to water may be only a few hundred parts per million, most of which has been added from inflowing groundwater that has percolated through the ground. In contrast to the other types of transportation in streams, movement of the dissolved load does not depend on the nature of flow since the material in solution is actually part of the fluid.

Suspended Load. A major portion of the total load carried by streams is transported mechanically by *suspension* or along the floor of the channel. Fluids with high viscosities, such as honey or other slow-moving masses, tend to move with smooth streamlined flow, known as *laminar flow.* Water flowing in a stream channel may also move with laminar flow, but because of low viscosity and relatively high velocities, the water tends to swirl and eddy in an irregular fashion, producing *turbulent flow.* If a colored dye is placed in a fluid possessing laminar flow, the colored water moves downstream in a series of parallel paths at relatively uniform speeds. However, as turbulent flow is encountered, mixing occurs in the swirls and eddies.

Turbulent flow is of great importance in the transportation of load in a stream, because, even though the dominant motion is downstream, turbulence creates random movements, some of which have upward components. The upward components of movement in turbulent water allow transportation of particles that otherwise might sink to the floor of the channel.

The suspended load consists mostly of silt and clay carried in suspension by the turbulence of the water. The length of time a particle may be held in suspension depends on the velocity with which a particle falls to the bottom of a stream and the intensity of turbulence which moves a particle upward or prevents it from sinking. The downward rate of movement of a particle in suspension is governed by Stokes' law:

$$\text{Settling velocity} = \frac{\frac{2}{9}(dp - df)\, gr^2}{\text{viscosity}}$$

$$\text{where } dp = \text{density of particle}$$
$$df = \text{density of fluid}$$
$$g = \text{gravity}$$
$$r = \text{radius of particle}$$

Stated in simple terms, Stokes' law says that settling velocity increases with the density and the square of the radius of the particle. Since the density and viscosity of water are fairly constant relative to the other factors, and the density of most rock particles varies only between about 2.6 and 3.3, the size of a particle assumes great importance. Beyond a certain size, the quantity determined by the square of the radius increases rapidly, and turbulence in a stream may be unable to hold a particle in suspension because of the high settling velocity.

As might be expected, the amount of suspended load carried by a stream increases as discharge increases (Fig. 6-6). Leopold and Maddock (1953) found that suspended load increases at a more rapid rate than discharge, i.e., the concentration of suspended sediment increases with discharge, and concluded that the increased concentration was a result of erosion of the drainage basin rather than scouring of the channel.

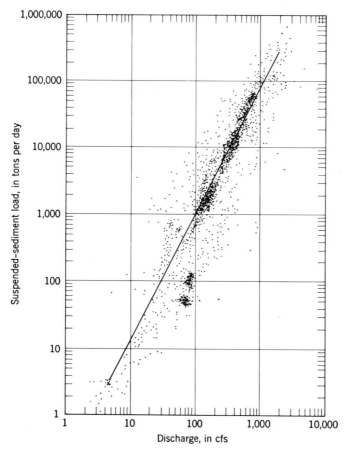

Fig. 6-6 Increase in suspended load with discharge. Powder River at Arvada, Wyoming. (Leopold and Maddock, 1953.)

Saltation and Traction. The portion of the load too large to be carried in suspension may move downvalley as part of the bed load by rolling and sliding along the channel floor. This material usually consists of sand and gravel. Sand grains may be momentarily lifted off the floor of the stream, and in the process are moved downvalley before sinking to the bottom again. Movement in this manner, by a series of jumps, is known as *saltation*. Pebbles and sand grains too large to be lifted off the floor may be rolled or pushed along the channel floor. This type of movement is referred to as *traction*. There is no sharp boundary between movement by suspension, saltation, and traction, since one process may grade imperceptibly into another.

The forces required to move particles on a stream bed have been studied by Rubey (1938), who distinguishes between movement according to the "sixth power law," hydraulic lift, and critical tractive force. The "sixth power law" refers to the relationship between the force of water pushing on the upstream side of a particle and the resistance of the particle to movement. When the force of the water equals the resistance, the particle is on the verge of motion, and the relationship between particle size and velocity is given by $r = kv^2$, where r is the radius of the particle, k is a constant, and v is the velocity of the water. Thus the volume of the largest particle moved is directly proportional to r^3, and thus to v^6. A small increase in velocity will therefore result in a large increase in erosive power.

Since velocity is not distributed equally with depth in a stream, a particle lying on a stream bed in the area of steepest velocity gradient has diminished pressure on the top surface and tends to rise in a manner somewhat analogous to an airplane wing. This lifting force, known as *hydraulic lift*, is probably most effective with small particles.

The column of water above a particle on a sloping stream bed exerts a force which is proportional to the depth of water and the slope.

$$F_t = \tau g d s$$

where F_t = critical tractive force
τ = density of water
g = gravity
d = depth of water
s = gradient of stream

The downslope component of this tractive force tends to move the particle along the stream bed. Rubey (1938) found that the depth-slope product was important in moving smaller particles and the "sixth power law" was important in the movement of large particles.

Both hydraulic lift and movement, according to the "sixth power law," depend on velocity, but it should be emphasized that this refers to the bed velocity near the floor of the channel, not to average velocity. The two velocities may be considerably different, and only bed velocity is important in the transportation of bed load.

The load of clastic particles being transported by a stream is not uniformly distributed throughout the channel, but tends to be concentrated closer to the floor of the channel. As might be expected, the concentration near the bottom of the channel is more pronounced with larger particle sizes. Whereas clay and fine silt may be distributed more or less evenly from top to bottom of the stream, sand and larger particles seldom rise very far from the floor of the channel.

In order to move a particle from a position on the floor of a stream channel, the moving fluid must overcome both gravitational and cohesive forces acting on the particle. For particles of sand size or larger, cohesive forces tending to hold individual particles to each other are generally not as important as gravitational forces. However, for fine material, such as silt and clay, cohesive forces may play an important part in determining whether or not movement takes place. Small clay particles which would ordinarily be transported by a given velocity may not be set in motion on the stream bed because of the cohesive tendency of the particles. Hjulstrom (1935) found that greater velocities were required to initiate movement of some small particles than for larger particles. The upper curve in Fig. 6-7 shows that velocities necessary to move particles of various size decrease with particle diameter to sizes between about 0.1 and 1.0 millimeter, and then rise again for particles less than 0.1 millimeter. The reason for the surprising increase in velocities required to move smaller particles is the greater cohesiveness of the smaller particles and the decrease in surface resistance offered by the smoother surface of the channel floor.

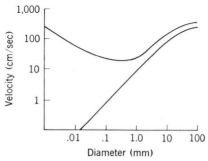

Fig. 6-7 Relationships among erosion, transportation, and deposition for various stream velocities and particle sizes. (Modified from Hjulstrom, 1935.)

The lower curve in Fig. 6-7 shows the relationship between transportation velocities and depositional velocities. For velocities in the region below the curve, deposition will take place; for velocities above the curve, particles already in motion will remain in motion.

channel relationships: meandering and braiding

Channel characteristics include the form of the channel in map view and cross section. In cross section all variations exist between narrow deep channels and wide shallow channels. The shape of the cross section of a stream channel is determined largely by the nature of the material making up the floor and sides of the channel relative to the erosive power of the stream, which is determined by discharge and the quantity and calibre of the bed load. Channels with cohesive silty banks develop narrower and deeper channels than less cohesive sandy banks (Schumm, 1960). Vegetation may also be an important factor in bank erodibility.

In map view all variations exist, from sinuous meandering streams with few islands to relatively straight braided streams having many islands (Fig. 6-8). There appears to be a direct relationship between sinuosity, incidence of channel islands, and the width-depth ratio of the channel. In general, meandering streams having high sinuosity and few channel islands have deeper, narrower channels than braided streams having low sinuosity and many channel islands (Fig. 6-9).

As water flows around a bend of a stream, the force of inertia tending to cause the water to continue to flow in a straight line causes the water to impinge on the outside of the bend with a stronger current than on the inside of the bend. The outside of the channel is thus deeper, and the velocity of the water is greater (Fig. 6-10). There is a tendency for water to "pile up" on the outside of the bend near the surface, and a cross-channel component of flow develops. Water at the surface has a cross-channel component directed toward the outside of the bend, whereas water near the floor of the channel is directed toward the inside of the bend. The flow pattern thus developed has the form of a spiral, or helix, although no single particle of water necessarily crosses completely from one side of the bend to the other. The effect of helical flow is to localize deposition and erosion and promote further development of the bend.

The greater velocity on the outside of a bend causes bank erosion, and the cross-channel component of the helical flow pattern contributes to the building of *point-bar deposits* on the inside of the bend. Matthes (1941) observed that material eroded from the outside of meanders is deposited on point bars on the same side of the channel from which it was eroded. Much of the material eroded on the outside of a bend is not transported all the way across the channel floor before it reaches the next bend downstream, where the helical flow pattern moves it toward the point-bar deposits on the inside of that bend. Continuation of this process leads to further development of a meandering pattern. As curvature of the bends increases, the processes become self-accelerating, until two bends come together to form a *meander cutoff*. The abandoned channel is then left as a horseshoe-shaped remnant of the former bend, often filled with water to form an *oxbow lake* (Fig. 6-11). With time the oxbow lake becomes filled with sediment, and vegetation encroaches on the sides, so that the lake passes eventually to a swamp, then disappears altogether (Fig. 6-12).

Meandering channels have been developed experimentally in flumes from originally straight channels (Fig. 6-13). All that is required to initiate a mean-

Fig. 6-8   Braided pattern of the Platt River near Grand Island, Nebraska. *(John Shelton.)*

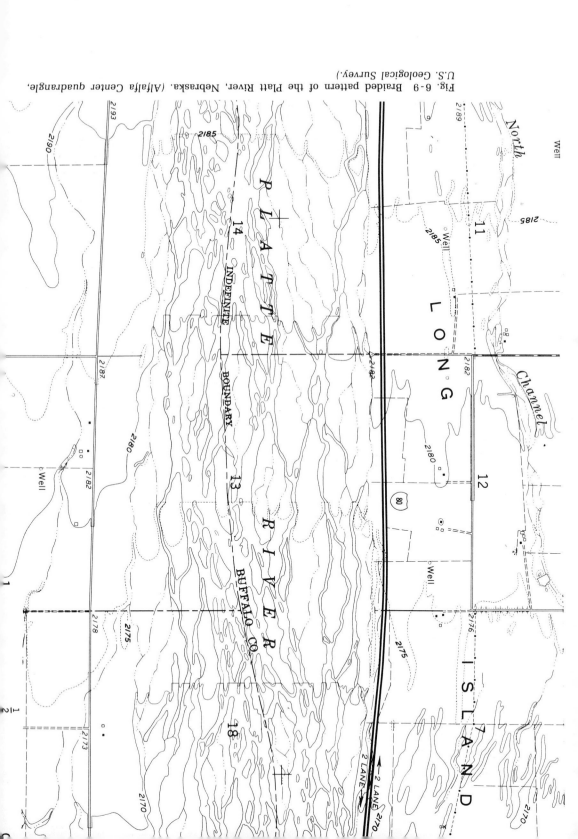

Fig. 6-9 Braided pattern of the Platt River, Nebraska. (Alfalfa Center quadrangle, U.S. Geological Survey.)

dering pattern is for caving of a bank to cause local deposition of a bar in the channel, forcing the flow of water against the opposite bank. Once started, the bend continues to develop.

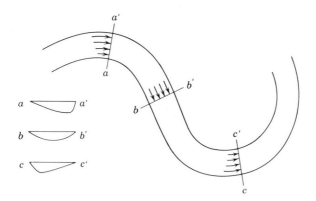

Fig. 6-10 Channel relationships in a meander. Arrows indicate relative velocities. Material eroded on the outside of the bend at a-a' is deposited on the inside of the bend at c-c' as the meander grows.

The size of meanders in a river is related to the discharge of the stream. is obvious from observation of natural streams that large streams have large meanders and small streams have small meanders. The size of meanders produced experimentally by Friedkin (1945) was directly related to discharge: the higher the discharge, the larger the meanders. In the instance when a meandering pattern was developed and the discharge increased, the old meander pattern was destroyed, and a set of larger meanders was produced. Leopold and Wolman (1957) found the length of meanders to be 7 to 12 times the channel width (Fig. 6-13).

Not all rivers develop meandering channels. Some form channels which consist of a number of dividing and reuniting channels anastomosing in such a way as to resemble a braid. Such streams are said to have a *braided* pattern. Large numbers of sandbars and channel islands are typical of braided streams, sometimes giving the impression that the stream is aggrading because it is unable to carry all the sediment load. However, Friedkin (1945), Rubey (1952), Mackin (1956), and others have shown that braided streams are not necessarily aggrading, but rather the braided pattern is related to channel width-depth ratios, which are a function of erodibility of the stream banks.

Why do some streams have a meandering pattern and others a braided pattern? Mackin (1956) suggested that if a stream has a relatively narrow, deep channel, the inertia effect and helical flow localize erosion on the out-

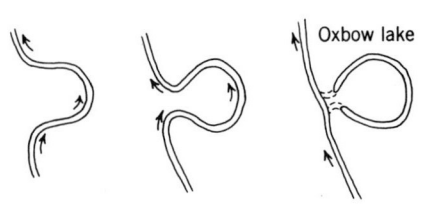

Fig. 6-11   Development of a meander cutoff and oxbow lake.

side of bends and deposition of point bars on the inside of bends, so that meanders develop. If a stream has a relatively wide, shallow channel, helical flow is too weak to maintain order in the whole channel to localize the deposition of point-bar deposits.

Deposition of randomly placed bars and islands occurs, tending to eliminate the inertia factor by breaking up the current into self-baffling channels. Since the width-depth ratio depends on erodibility of banks, whether a stream develops a meandering or braided pattern depends essentially on bank erodibility. Demonstration of this was accomplished in flume studies made by Friedkin at the Vicksburg Experiment Station. Two channels formed by the same flow on the same slope, but in different bank-forming material, produced different results. A narrow, deep channel developed in tough bank material resulted in a meandering pattern, while a broad, shallow channel developed in weaker bank material resulted in a braided pattern.

If two stream segments, one braided, the other meandering, have equal discharges and flow on similar bed material, the braided segment will have a steeper slope. In the braided segment, the sum of the wetted perimeters of two or more channels is greater than the wetted perimeter of a single channel carrying the equivalent discharge. Measurements made on streams which have braided segments in otherwise meandering courses indicate that, in a braided channel, slope increases relative to that in a meandering channel.

concept of the graded stream

Streams are constantly carrying downvalley the material contributed by weathering and mass movement along the valley sides. Whether a stream is capable of eroding its channel and adding new material to its load, is able to transport just the load supplied, or is unable to transport the weathered debris shed from the valley sides depends on a number of factors, most of which bear directly or indirectly on velocity. That which increases velocity increases carrying power; that which decreases velocity decreases carrying power.

The concept of the graded stream has been ably defined by Mackin (1948) as follows:

A graded stream is one in which, over a period of years, slope is delicately adjusted to provide, with available discharge and with prevailing channel characteristics, just the velocity required for transportation of the load supplied from the drainage basin.

As indicated in Mackin's statement of the concept, four variables play important roles: (1) slope, (2) discharge, (3) channel characteristics, and (4) load.

Fig. 6-12 Oxbow lakes developed by meandering of the White River, Arkansas. (J. R. Balsley, U.S. Geological Survey.)

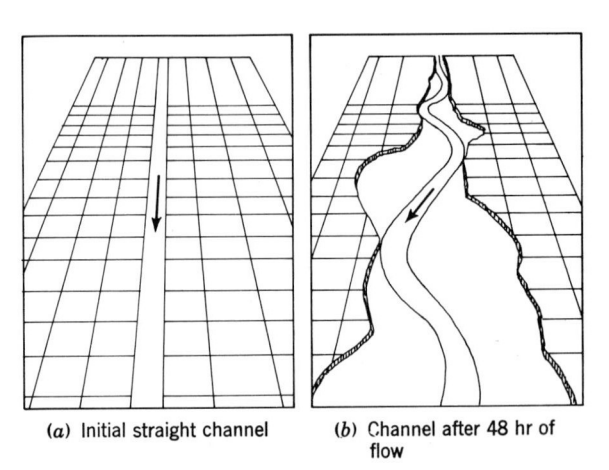

(a) Initial straight channel   (b) Channel after 48 hr of flow

Fig. 6-13 Development of a meandering channel from a straight channel. *(Modified from Friedkin, 1945.)*

slope of the graded profile

The long profile of a graded stream is a concave-upward curve which progressively decreases in slope downstream. The shape of the curve is directly a result of a condition of equilibrium in the stream. It may have minor irregularities below junctions of major tributaries because of changes in discharge or calibre of load, but still be graded throughout. Some streams may have short sections along their courses which are ungraded because of local conditions.

The gradient of a graded stream may be high or low, depending on load-discharge relationships. Some streams, such as the Shoshone River in the Big Horn Basin of Wyoming, have gradients of 30 feet per mile, while others, such as the Illinois River, have gradients as low as 2 inches per mile. Both are graded, the difference in slope being due to differences in discharge and calibre of load transported.

Slope is important as the main factor in the equilibrium which is automatically adjustable by the stream itself to accommodate external changes that call for changes in velocity. The slope of a graded stream is maintained to provide just the velocity required to transport all the load supplied. Load and discharge are important as factors which are in origin independent of the stream itself (Mackin, 1948).

The long profile of a graded stream is essentially a profile of equilibrium:

Its diagnostic characteristic is that any change in any of the controlling factors will cause a displacement of the equilibrium in a direction that will tend to absorb the effect of the change. (Mackin, 1948.)

Fig. 6-14 Meandering pattern of the Yazoo River, Mississippi. Note the point-bar deposits on the inside of the meanders and the size of the meanders of the Yalobusha River as compared with those of the Yazoo. (Greenwood quadrangle, U.S. Geological Survey.)

discharge and the graded profile

An increase in the discharge of a graded stream will result in added velocity, and thus increased carrying power, so that the stream may become able
to move the larger particles which it previously could not. In so doing the
stream channel is eroded and the slope is lessened, which in turn checks
the velocity and limits the ability to erode the channel further. A new equilibrium is thus established. Conversely, if the discharge of a stream is decreased,
velocity is decreased, so that some of the larger particles must be deposited.
This in turn steepens the slope until a velocity is attained which is just sufficient to carry the load.

Short-term changes in discharge, such as the fluctuations occurring sea-
sonally or even daily in a stream, may cause local scouring of the channel
during high-water stages or deposition during low-water stages. However,
it is with regard to long-range changes that the concept of the graded stream
must be concerned. Mackin (1948) states:

> Over a period of years sufficiently long to include all the vagaries of the
> stream, the two independent controls (discharge and supplied load) may
> be essentially constant.

> Scouring and filling with seasonal fluctuations in discharge and velocity
> occur in all streams; it is the peculiar and distinctive characteristic of the
> graded stream that after hundreds or thousands of such short-period fluctua-
> tions, entailing an enormous total footage of scouring and filling, the stream
> shows no change in altitude or declivity. . . . In this long-term sense, there
> is an equivalence of opposed tendencies in the graded stream.

load and the graded profile

The load of a graded stream over a period of years is dependent largely
on the lithology of the type of rock being weathered and eroded from the valley sides, topographic relief, weathering and erosional processes operating
in the drainage basin, and vegetation.

If the load of a stream is increased, as, for instance, when a glacier in the
headwater region introduces large amounts of coarse debris, then a graded
stream will respond by depositing the larger particles which it is unable to
move, thus increasing the slope of the channel. Deposition continues until
the increase in velocity produced by the increase in slope is just sufficient
to move the coarser load, and a new equilibrium is established. If the calibre of the bed load is decreased, the velocity is then in excess of the amount
required to move the smaller particles, and hence the stream begins to cut
down.

downvalley changes in load-discharge relationships

The shape of a graded profile is concave upward, with constantly decreasing slope downstream. The slope of the curve represents an equilibrium system in which two of the governing factors, discharge and load, change progressively downstream.

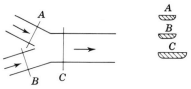

Fig. 6-15 Decrease in wetted perimeter relative to cross-sectional area at the junction of two streams. The cross-sectional area of channel A plus B equals C, but the wetted perimeter of C is less than the total of A plus B.

As each tributary empties its flow of water into the mainstream, discharge increases downvalley. With increased discharge, the cross-sectional area increases relative to the wetted perimeter, and a downvalley segment with a larger discharge can maintain a given velocity on a lower slope than an upvalley segment with a smaller discharge. It is not uncommon to find that downstream from a tributary junction of two graded streams the slope becomes less than that of either confluent as a result of an increase in cross-sectional area relative to wetted perimeter. In Fig. 6-15 the sum of the discharges of the two streams is equal to that of the mainstream, but the wetted perimeter is less than the sum of the wetted perimeters of the two streams above the junction. This reduces the slope required to produce the velocity necessary to transport the load.

A second factor which changes progressively downstream is the downvalley decrease in calibre of load. With reduction in the size of particles transported, slope requirements are less. As an example, the lower Greybull River in Wyoming, which flows from the Absaroka Range through the Big Horn Basin, receives no tributaries in a 50-mile segment below the Wood River, and the discharge remains nearly the same. The calibre of the load, however, decreases downstream, resulting in a decrease in slope from 60 feet per mile at the upper end of the segment to 20 feet per mile at the lower end (Mackin, 1948).

If load increases relative to discharge, as by influx of coarse debris from heavily laden tributaries or from loss of water, local increase in slope occurs. Below its junction with the Platte River, the profile of the Missouri River steepens because of the entry of gravel from the Platte. Both streams are graded, and the profile steepening will remain as long as the same equilibrium conditions prevail.

The decrease in calibre of load downstream is due partly to abrasion of particles during transport and partly to selective transportation. Because smaller particles are able to remain in motion longer and are more frequently moved by saltation or traction, they proceed downstream at a higher rate than larger particles. To distinguish between the effects of abrasion and selective transportation in causing downstream decrease in particle size is a difficult task. In some streams abrasion is the dominant cause, in others, selective transportation, depending on conditions in a particular stream system.

Responses of a graded stream to various changes are listed in the following table.

	response*	slope
increase in load	aggradation	increases
decrease in load	degradation	decreases
increase in discharge	degradation	decreases
decrease in discharge	aggradation	increases

*until a new equilibrium is reached.

floods

An important problem in the study of fluvial processes is the relationship of high discharge to total erosion and transportation performed by a stream. During floods, velocity increases and scour of the bed takes place, putting into motion material at rest on the channel floor since the previous flood (Fig. 6-16). When the flow subsides, sediment carried from upstream comes to rest on the stream bed, replacing the material scoured during the flood. Such scour-and-fill episodes represent temporary, short-term effects, distinctly different from long-term aggradation or erosion.

There is little doubt that scouring and erosion are vigorous and that large amounts of sediment are transported during floods. However, whether or not the amount of work done during short periods of high discharge are more significant than the work accomplished by a stream during longer periods or normal discharge is still only imperfectly known, largely because measurements of the amount of bed load moved during a flood are very difficult to obtain.

The variability in discharge from season to season and year to year poses a problem in the understanding of rivers. For example, if it is true that the most significant amount of erosion and transportation in a stream is accomplished during times of flood, the frequency of flooding becomes a vitally important factor, and the question must be asked, is the stream just "biding time" during periods of normal flow between floods, and are many of the characteristic features observed along stream valleys actually relicts of floods, and not in the process of forming under day-to-day conditions?

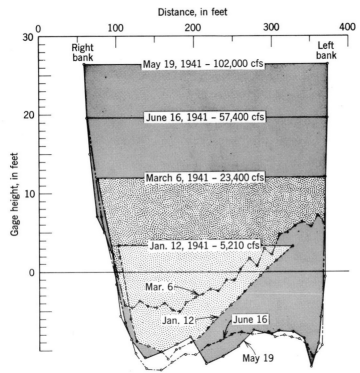

Fig. 6-16 Channel cross sections during progress of flood, January, 1941 to June, 1941, Colorado River at Grand Canyon, Arizona. (Leopold and Maddock, 1953.)

The problem of *misfit streams,* flowing in valleys which exhibit features associated with different discharges from those presently observed, also enters into such questions. During the Pleistocene many streams had much greater discharges than those of the present-day streams, and left features along valleys which are not directly related to present streams in those valleys. The most commonly observed features of valleys containing misfit streams are former meander scars whose radii are much larger than those of present streams in the valleys, obviously produced when discharges were much greater. Dury (1964) has suggested that many meandering valleys were cut by Pleistocene rivers whose discharges were significantly larger than present streams. Although this is undoubtedly true in certain instances, there is some doubt that it is universally the case, at least insofar as the relict nature of the valleys is concerned.

erosional processes

Streams flow in valleys that they themselves have made, implying that streams are capable of cutting their channels both downward and laterally. The processes whereby this is accomplished include abrasion, hydraulic plucking, and solution. The mechanical wear produced by repeated contact between particles being carried by a stream abrades the particles in transit, and also the bedrock floor over which they are carried. Where eddies in the stream current swirl sand and pebbles in circular paths, abrasion between the loose particles and the bedrock floor of the channel may produce potholes in the stream bed. As the loose pebbles which act as the grinding tools are worn down by abrasion or carried away by the current, new ones are added from the load being carried by the stream.

Hydraulic plucking involves the removal of pieces of rock from the stream bed by the force of flowing water. Once detached from the bed, these particles become part of the stream load, and may be further reduced in size by abrasion during transport. Loosening of pieces of rock in the channel usually takes place along joint or bedding planes.

Solution may dissolve rock and mineral material of the stream bed and also of loose particles in transport, but the effects are usually less important than those produced by abrasion and hydraulic plucking in erosion of the stream bed. Most of the dissolved material in the water of the stream comes from solution by groundwater and later addition into the drainage system.

deltas

Deltas are formed where sediment is deposited by streams flowing into bodies of standing water. They tend to have triangular shapes in plan view, somewhat similar to the Greek letter delta (Δ), but many other shapes also occur. The shape of a delta depends largely on the rate of influx of sediment carried by the stream relative to the vigor of wave and current action in the standing body of water. Deltas built into the open ocean are affected by wave and current action, which distributes the influx of sediment laterally along the shoreline.

The density of water and fine sediment relative to the density of water in the standing body of water may differ under different circumstances. Muddy river water discharging into a freshwater lake has a higher density than the lake water, and the muddy water sinks beneath the surface of the lake, flowing along the lake bottom as a turbidity current. Such currents of muddy water may be found in Lake Mead, behind Boulder Dam, where the sediment-laden water of the Colorado River flows into the upper end of the lake and travels along the lake floor for many miles. A similar situation occurs where the Rhone River flows into Lake Geneva. Conditions are often different, however, for freshwater flowing into a saltwater body. The freshwater is less dense and flows on top of the saltwater unless it is heavily laden with silt and clay. Very fine clay particles held in suspension in the freshwater of streams flocculate into larger aggregates in saltwater and become too large to remain in suspension.

Many deltas, such as the Nile Delta, have a triangualr shape in plan view, with a broad arcuate seaward margin. Others, such as the Mississippi Delta, have long extensions of distributory channels and are often described as birdfoot deltas (Fig. 6-17). Deltas built into arms of the sea confined between valley walls are referred to as estuarine, and are similar in shape to those built into freshwater lakes lying in a valley (Fig. 6-18).

Almost all deltas are characterized by splitting of the main channel into numerous branching distributary channels. The internal structure of most deltas is similar in cross section. Sediment carried to the edge of the delta is deposited on the more steeply sloping margin as *foreset beds* (Fig. 6-20). The finer material may be carried farther in suspension, and eventually come to rest as *bottomset beds* on the floor at some distance from the delta margin. As the edge of the delta is extended into the body of water, the level of the earlier-formed portion of the delta is gradually raised by deposition of *topset beds* to give the stream sufficient gradient to flow to the water's edge.

alluvial fans

As tributary streams emerge from their valleys into a larger mainstream valley or into a lowland area, they often encounter reduced slopes, and are forced to deposit material in order to maintain an adequate gradient. The result is an *alluvial fan*, which consists of a low, cone-shaped deposit, with the apex of the cone at the mouth of the valley from which the fan-building stream issues (Fig. 6-21). The fan shape is a result of deposit of material as the stream swings back and forth across the fan from the fixed position at the apex.

Fig. 6-17   The Mississippi delta. (Breton Sound map, U.S. Geological Survey.)

After emerging from the valley at the head of the fan, a stream often separates into numerous distributary channels on the fan. The permeability of material in alluvial fans is generally quite high, so that loss of water occurs by infiltration as the stream crosses the fan. During the dry season, in arid climates, water loss by infiltration may be sufficiently high so that no water flows across the fan. Cloudbursts in the mountains upstream from the fan cause surges in discharge of water, and mudflows may transport boulders and much coarse material onto the fan. Thus a typical fan deposit consists of stratified sand and gravel left by normal stream activity interbedded with poorly sorted mudflows. The upper portions of fans usually contain cobbles and boulders, grading progressively into finer sediments near the edges of the fan.

Fig. 6-18   Delta being built into Peyto Lake, Banff National Park, Alberta, Canada.

Lateral swinging of the mainstream greatly affects conditions on tributary fans. When the mainstream swings to the far side of the valley, the fan is extended downstream and is built up near the apex to supply the necessary gradient (Fig. 6-23). However, when the mainstream swings back to the fan side of the valley, the fan-building stream regrades its channel by downcutting and incises the previously built fan. Therefore incision of fans and development of fan-head trenches do not necessarily mean that uplift of the region or climatic changes have taken place.

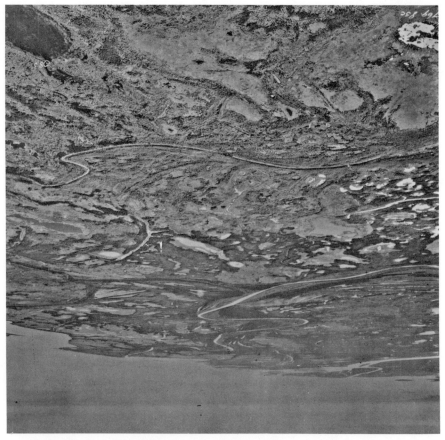

Fig. 6-19 Flood plain and delta of the Slave River, Canada. (*National Air Photo Library, Surveys and Mapping Branch, Department of Energy, Mines, and Resources.*)

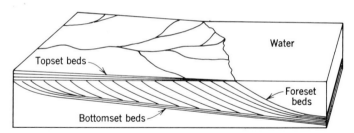

Fig. 6-20   Internal structure of a delta.

Fig. 6-21   Alluvial fans being built by streams emerging from the mountains into Death Valley. *(Aero Service Division, Litton Industries.)*

Fig. 6-22   Trenched fans at foot of Panamint Range, California. *(U.S. Geological Survey.)*

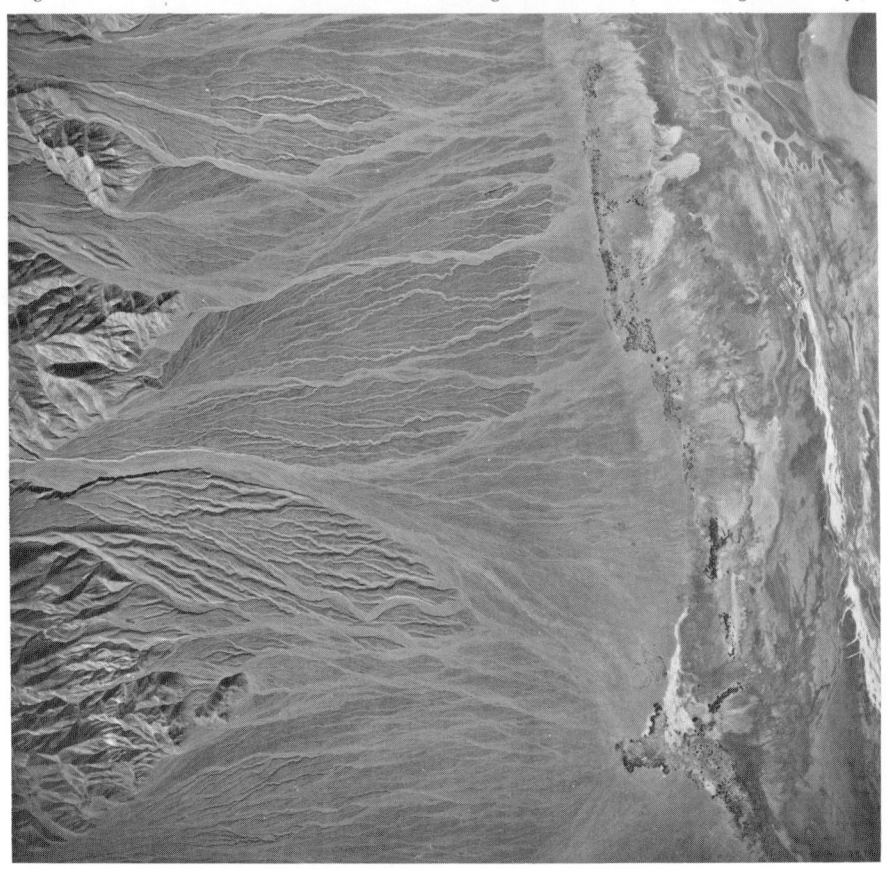

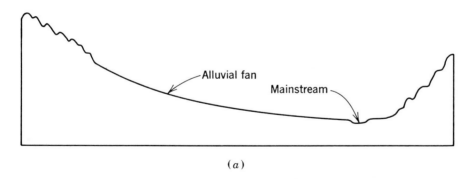

(a)

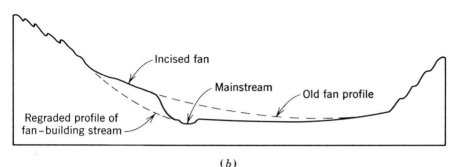

(b)

Fig. 6-23   Incision of alluvial fan by lateral migration of mainstream. (a) Initial conditions; (b) lateral swinging of the mainstream causes regrading of the fan profile.

references

Bagnold, R. A.: 1960. Some aspects of the shape of river meanders, *U.S. Geol. Survey Prof. Paper 282-E.*

Bull, W. B.: 1964. Geomorphology of segmented alluvial fans in western Fresno County, California, *U.S. Geol. Survey Prof. Paper 352-E.*

Davis, W. M.: 1909. Geographical Essays, Ginn and Company, Boston (reprinted 1954 by Dover Publications, Inc., New York).

Denney, C. S.: 1965. Alluvial fans in the Death Valley Region, California and Nevada, *U.S. Geol. Survey Prof. Paper 466.*

Dury, G. H.: 1964. Principles of underfit streams, *U.S. Geol. Survey Prof. Paper 452-A.*

Friedkin, J. F.: 1945. A laboratory study of the meandering of alluvial rivers, U.S. Waterways Experiment Station, Vicksburg, Miss.

Gilbert, G. K.: 1877. Report on the geology of the Henry Mts., Utah, *U.S. Geog. and Geol. Survey, Rocky Mt. Region.*

——: 1914. The transportation of debris by running water, *U.S. Geol. Survey Prof. Paper 86.*

—— : 1917. Hydraulic mining debris in the Sierra Nevada, *U.S. Geol. Survey Prof. Paper 105.*

Hjulstrom, F.: 1935. Studies on the morphological activity of rivers as illustrated by the river Fryis, *Univ. Upsala Geol. Inst. Bull. 25*, pp. 221-527.

Holmes, C. D.: 1952. Stream competence and the graded stream profile, *Am. Jour. Sci.*, vol. 250, pp. 899-906.

Jefferson, M.: 1902. Limiting widths of meander belts, *Nat. Geog. Mag.*, vol. 13, pp. 373-384.

Lane, E. W., and W. M. Borland: 1954. River-bed scour during floods, *Am. Soc. Civil Engineers Trans.*, vol. 119, pp. 1069-1079.

Leliavsky, Serge: 1955. An Introduction to Fluvial Hydraulics, Constable & Co., Ltd., London.

Leopold, L. B., and T. Maddock: 1953. The hydraulic geometry of stream channels and some physiographic implications, *U.S. Geol. Survey Prof. Paper 252.*

—— and M. G. Wolman: 1957. River channel patterns: braided, meandering, and straight, *U.S. Geol. Survey Prof. Paper 282-B.*

—— and —— : 1960. River meanders, *Geol. Soc. America Bull.*, vol. 71, pp. 769-794.

——, ——, and J. P. Miller: 1964. Fluvial Processes in Geomorphology, W. H. Freeman and Company, San Francisco.

Mackin, J. H.: 1937. Erosional history of the Big Horn Basin, Wyo., *Geol. Soc. America Bull.*, vol. 48, pp. 813-894.

——: 1948. Concept of the graded river, *Geol. Soc. America Bull.*, vol. 59, pp. 463-512.

——: 1956. Cause of braiding by a graded stream (abstract), *Geol. Soc. America Bull.*, vol. 67, pp. 1717-1718.

Matthes, G. H.: 1941. Basic aspects of stream meanders, *Am. Geophys. Union Trans.*, vol. 22, pp. 632-636.

Morisawa, M.: 1968. Streams, Their Dynamics and Morphology, McGraw-Hill Book Company, New York.

Quraishy, M. S.: 1944. The origin of curves in rivers, Current Sci. (India)., vol. 13, pp. 36-39.

Rubey, W. W.: 1933. Equilibrium conditions in debris-laden streams, Amer. Geophys. Union Trans., pp. 497-505.

——— : 1938. The force required to move particles on a stream bed, U.S. Geol. Survey Prof. Paper 189-E, pp. 121-141.

———: 1952. Geology and mineral resources of the Hardin and Brussels quadrangles, U.S. Geol. Survey Prof. Paper 218.

Schumm, S. A.: 1960. The shape of alluvial channels in relation to sediment type, U.S. Geol. Survey Prof. Paper 352-B.

Thomson, J.: 1879. On the origin of windings of rivers in alluvial plains, Royal Soc. London Proc., vol. 25, pp. 5-6.

Werner, P. W.: 1951. On the origin of river meanders, Am. Geophys. Union Trans., vol. 32, pp. 898-902.

Wolman, M. G.: 1959. Factors influencing erosion of a cohesive river bank, Am. Jour. Sci., vol. 257, pp. 204-216.

——— and L. M. Brush, Jr.: 1961. Factors controlling the size and shape of stream channels in coarse noncohesive sands, U.S. Geol. Survey Prof. Paper 282-G.

——— and L. B. Leopold: 1957. River flood-plains: some observations on their formation, U.S. Geol. Survey Prof. Paper 282-C, pp. 87-109.

Woodford, A. O.: 1951. Stream gradients and Monterey sea valley, Geol. Soc. America Bull., vol. 62, pp. 799-852.

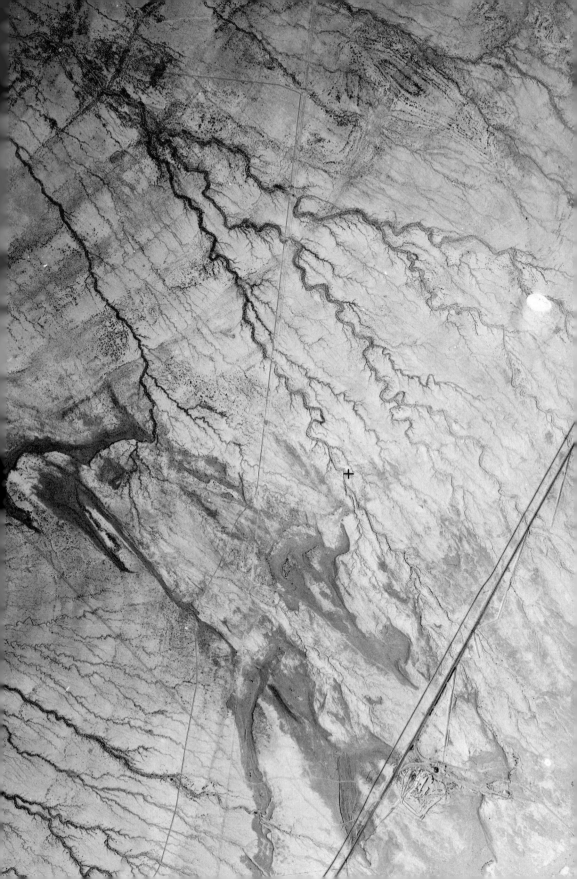

origins of stream valleys and drainage patterns

stream origins

The relation of streams to their valleys was recognized as early as the fifteenth century. Leonardo da Vinci (1452-1519) believed that valleys were cut by streams, and was thus among the first to realize that streams occupy valleys, not because the valleys were there before the streams, but because the streams eroded their channels. Similarly, Demarest (1725-1815) proposed that the valleys of France were the result of the streams in them. This basic idea was perhaps best expressed by Playfair in 1802:

> Every river appears to consist of a main trunk, fed from a variety of branches, each running in a valley proportioned to its size, and all of them together forming a system of valleys, connecting and communicating with one another, and having such a nice adjustment to their declivities that none of them join the principal valley, either on too high or too low a level, a circumstance which would be infinitely improbable if each of these valleys were not the work of the stream which flows in it.

Fundamentally, stream courses are determined by (1) the initial slope of

Consequent drainage with dendritic pattern on recently exposed floor of the Salton Sea. *(U.S. Department of Agriculture.)*

the land, (2) adjustment to differences in rock resistance, and (3) random headward erosion.

A stream whose course developed when water first flowed down a newly created land surface is known as a *consequent stream* (Powell, 1875). The direction of flow is determined by the direction in which such initial land surface slopes. Among common examples of this kind of stream are those which form on (1) newly developed volcanic cones or lava fields, (2) recently deglaciated regions, (3) emergent marine areas, and (4) newly uplifted domes, anticlines, or fault blocks.

In any drainage system tributary streams continually extend their courses by headward erosion. When water falls on the land surface as precipitation, it flows as a thin sheet downslope. Irregularities in the land surface interrupt the sheet flow, causing portions of it to diverge in some places and converge in others. Where convergence of the flow occurs, water is concentrated into distinct channels or rills, resulting in deeper water, greater velocity, and increased transporting power. The convergence of water into the heads of gullies and streams enlarges them and extends them headward. As each rill deepens, it carries a larger discharge, which concentrates the flow of water and erodes the channel downward. In so doing, the lowered channel collects even more water from the nearby slopes, and new rills may form which flow into the deepening rill. Horton (1945), in attempting to quantify relationships between tributary streams, developed the concept of *stream ordering*. The smallest tributaries, from their headward limit to their first junction, are designated as first-order segments; segments below the junction of two first-order tributaries are second-order segments; channels below the junction of second-order segments are third-order segments; and so on. The largest trunk stream in a drainage basin has the highest order. The average slope of streams in a given drainage basin decreases with stream order number, and the average length of a segment increases with stream order number.

In places where rocks vary in resistance to erosion, headward extension of gullies is favored in rocks of least resistance. Streams which have adjusted their courses to weak rock belts by differential erosion are known as *subsequent streams*. Most subsequent streams are developed by selective headward erosion, although other types of adjustment are possible.

Subsequent-stream courses are especially common in areas of folded or tilted sedimentary beds, in areas of faulting or jointing, and in other regions where rocks of differing resistance occur.

Streams whose courses developed by random headward erosion or whose courses are not related to determinable factors are called *insequent streams*. Their courses do not follow regional consequent slopes and are not guided by weak rock belts; so tributary streams develop in random directions. Insequent streams are common on homogeneous rocks or on horizontal stratified sedimentary beds where specific structural control of underlying rocks is not apparent.

A stream which flows in a direction opposite to that of the original consequent drainage is called an *obsequent stream*. A *resequent stream* is one which flows in the same direction as the consequent drainage, but developed at a lower level than the initial slope after other drainage intervened for a time. It has essentially "retaken" a consequent direction.

Obsequent and resequent streams are common in areas of tilted sedimentary beds. If the dip of the beds is the same as the direction of the consequent drainage, resequent streams develop down the dip of ridge-making beds, and obsequent streams develop down the ridge in a direction opposite to that of the dip of the beds. In Fig. 7-1 a consequent stream develops on a newly uplifted sea floor. After tilting occurs, selective headward erosion along weak

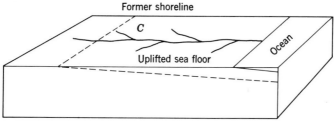

(a) Drainage consequent on newly uplifted sea floor

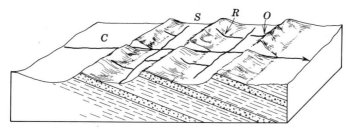

(b) Subsequent drainage after tilting and selective headward erosion along weak rock belts. *C*-consequent stream; *S*-subsequent stream; *R*-resequent stream; *O*-obsequent stream

Fig. 7-1   Evolution of drainage on newly uplifted sea floor.

rock belts results in development of subsequent streams. The stream at *O* which flows down the scarp face of a dipping bed is an obsequent stream because it flows in a direction opposite to the original consequent drainage, *C*. The stream at *R* which flows down the dip of a resistant dipping bed is a resequent stream because it flows in the same direction as the original consequent stream but at a lower level. Since the consequent drainage of a region is not necessarily always in the same direction as the dip of tilted beds, streams flowing down the dip of beds are not always resequent, and streams flowing opposite to the dip are not always obsequent, according to this definition.

drainage patterns

The arrangement of streams in a drainage system constitutes the *drainage pattern*. The pattern may reflect structural or lithologic control of underlying rocks or may be related to other factors. Analysis of stream patterns is frequently helpful in interpretation and identification of geologic structures and rock types in an area.

dendritic patterns

The most common type of drainage pattern is dendritic, characterized by irregular branching of stream valleys in many directions without systematic arrangement. Tributaries join the mainstream at any angle; usually, however, at acute angles (Figs. 7-2 and 7-3).

Dendritic patterns are developed by random headward erosion of insequent streams on rocks of uniform resistance which lack structural controls. The courses of tributary streams are determined by random headward erosion, implying that the underlying material is essentially homogeneous, without systematic differences in rock resistance. Dendritic patterns commonly form upon horizontally bedded sedimentary rocks or massive igneous or metamorphic rocks.

trellis patterns

Trellis patterns are formed by parallel to subparallel streams, with short tributary streams flowing into the mainstreams at nearly right angles (Fig. 7-4). The streams are usually subsequent in origin and tend to follow weak rock belts.

Trellis drainage is especially common in areas of folded sedimentary beds of differing resistance, such as the Valley and Ridge province of the Appalachian Mountains and Rocky Mountain Front Range.

Short tributary streams flowing off ridges of resistant beds into the valleys of the weak rock belts are usually obsequent or resequent in origin.

Under certain circumstances trellis patterns may be formed by consequent

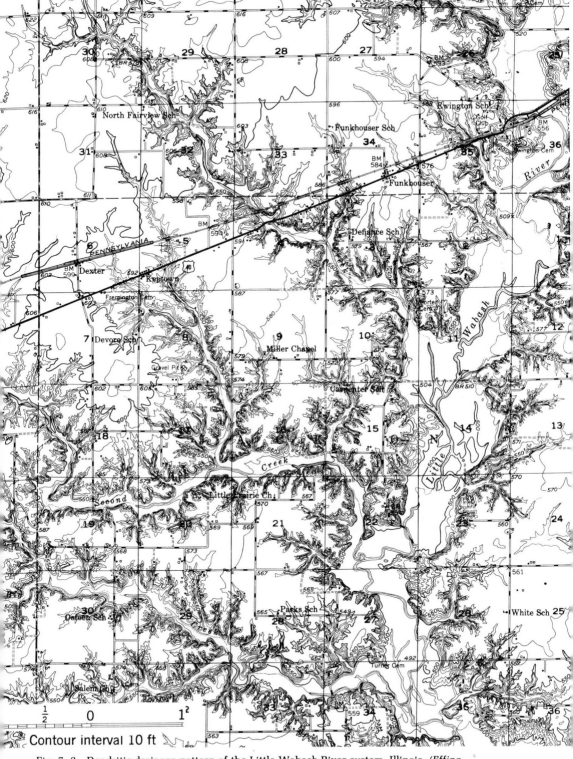

Fig. 7-2 Dendritic drainage pattern of the Little Wabash River system, Illinois. *(Effingham quadrangle, U.S. Geological Survey.)*

streams. Development of elongate ridges or drumlins by continental glaciers may result in a roughly parallel orientation of streams consequent on the deglaciated terrain. Parallel orientation of ridges and streams may also be developed in some areas by wind erosion and deposition.

rectangular drainage patterns

Rectangular drainage patterns (Fig. 7-5) consist of streams which make right-angle bends, usually coincident with intersecting fault or joint systems. The streams are adjusted to the underlying structure, and thus are subsequent in origin.

Fig. 7-3   Dendritic drainage pattern near Green River, Utah. *(J. R. Balsley, U.S. Geological Survey.)*

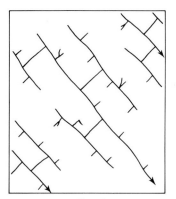

Fig. 7-4 Trellis drainage pattern.

Fig. 7-5 Rectangular drainage pattern.

Rectangular drainage is common in areas underlain by jointed igneous rock, flat-lying sedimentary beds with pronounced joint systems (Fig. 7-6), or faulted regions.

Rectangular drainage patterns differ from trellis patterns in that, although both have right-angle tributary junctions, individual streams in rectangular patterns have right-angle bends in their channels (compare Figs. 7-4 and 7-5). As can be seen in Fig. 7-6, the orientation of streams is to a large extent controlled by selective headward erosion along joints. In some cases the meanders of major streams follow joint or fault planes, resulting in many right-angle bends.

angular drainage patterns

Angular drainage patterns (Fig. 7-7) form where joints and faults intersect at more acute angles. Since not all joint systems intersect at right angles, streams whose courses are related to jointing will often intersect at angles other than 90°. The term *rectangular* may be used to include intersections at high angles but not necessarily exactly 90°.

radial drainage patterns

Radial drainage patterns (Fig. 7-8) consist of streams flowing away from a central area in all directions. Such patterns are usually developed on newly formed topographic highs such as volcanic cones (Fig. 7-9) or structural or intrusive domes. In most cases radial drainage flows down an initial slope, either constructional, as in the case of a volcanic cone, or uplifted, as in the case of domal structure. Streams originating on such slopes are consequent. Drainage on igneous intrusive bodies, such as laccoliths and stocks, is usually radial.

annular drainage patterns

Consequent drainage flowing radially from the crest of a dome may evolve into subsequent drainage adjusted to weak rock belts as the dome is breached by erosion. Since uplifted sedimentary beds in such a structure form a circular outcrop pattern, subsequent streams follow more or less circular courses and develop an *annular drainage pattern* (Fig. 7-10).

Fig. 7-6   Stereopair of air photographs showing rectangular drainage pattern controlled by jointing in sedimentary rock, Emery County, Utah. *(U.S. Department of Agriculture.)*

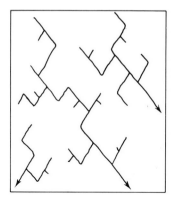

Fig. 7-7   Angular drainage pattern.

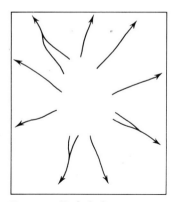

Fig. 7-8   Radial drainage pattern.

centripetal drainage patterns

Centripetal drainage patterns (Fig. 7-11) are formed by streams converging into a central basin or depression. Common examples of this pattern are consequent streams which flow into a structural and topographic basin and local streams which converge into sinkholes formed in limestone terrains.

parallel drainage patterns

Parallel drainage patterns develop where streams flow nearly parallel to one another down a pronounced unidirectional regional slope (Fig. 7-12). Drainage of slopes with parallel to subparallel topographic features, such as various linear glacial landforms, may also result in parallel stream patterns.

drainage density

The number of tributary streams per unit of surface area differs considerably from place to place, depending to a large extent on the nature of material exposed at the surface. Permeability, the ease with which water may pass through material, is of great importance in determining whether most of the precipitation that falls on a region runs off on the surface or sinks into the ground. Runoff is inversely proportional to permeability: the lower the permeability of material, the higher the runoff rate. Generally speaking, permeability is a function of the size, shape, and distribution of particles making up a rock. Fine-grained material, such as clay or shale, tends to have a lower permeability than coarser material. Where permeability is low and the runoff rate is high, many small gullies form, whereas if permeability is high, much of the water infiltrates into the ground, and a larger surface area is required to provide sufficient runoff for maintenance of a channel.

The most intensive gullying occurs in areas underlain by weak clays or shale. Low permeability results in high runoff, and the ease of erosion of material allows development of many channels. Regions of unusually high drainage density are known as badlands (Fig. 7-13). The topography in Badlands National Monument, South Dakota (Fig. 7-14), is a result of a combination of factors which produce extreme gullying: (1) clayey material of low per-

Fig. 7-9   Radial drainage on a volcanic cone in the Aleutians, Alaska. *(U.S. Air Force.)*

meability, (2) little vegetation to impede runoff, (3) rainfall concentrated in widely scattered showers, and (4) downcutting of the drainage system. The closely spaced rills give the topography an appearance which may be described as fine-textured. Low drainage density typical of sandstone or regions of resistant rock results in a coarse texture (Fig. 7-15).

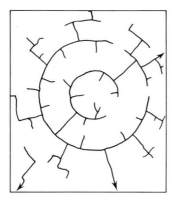

Fig. 7-10 Annular drainage pattern.

A quantitative measure of drainage density may be calculated by determining the total length of stream channels in a drainage basin and dividing this quantity by the surface area of the basin.

$$\text{Drainage density} = \frac{\text{length of channels}}{\text{area of drainage basin}}$$

Drainage densities of less than 10 miles of channel length per square mile of drainage basin are characteristic of coarse-textured drainage developed on sandstone or other resistant or permeable rock, whereas drainage densities of several hundred miles may be found in areas of badland topography.

stream capture

Streams of different drainage systems are separated by ridges or upland areas known as *divides*. The position of a divide remains in the same place only if the rates of erosion on opposite sides of the divide are equal. If, as often happens, rates of erosion in the headwaters of different drainage systems are not the same, the position of the divide will shift in the direction

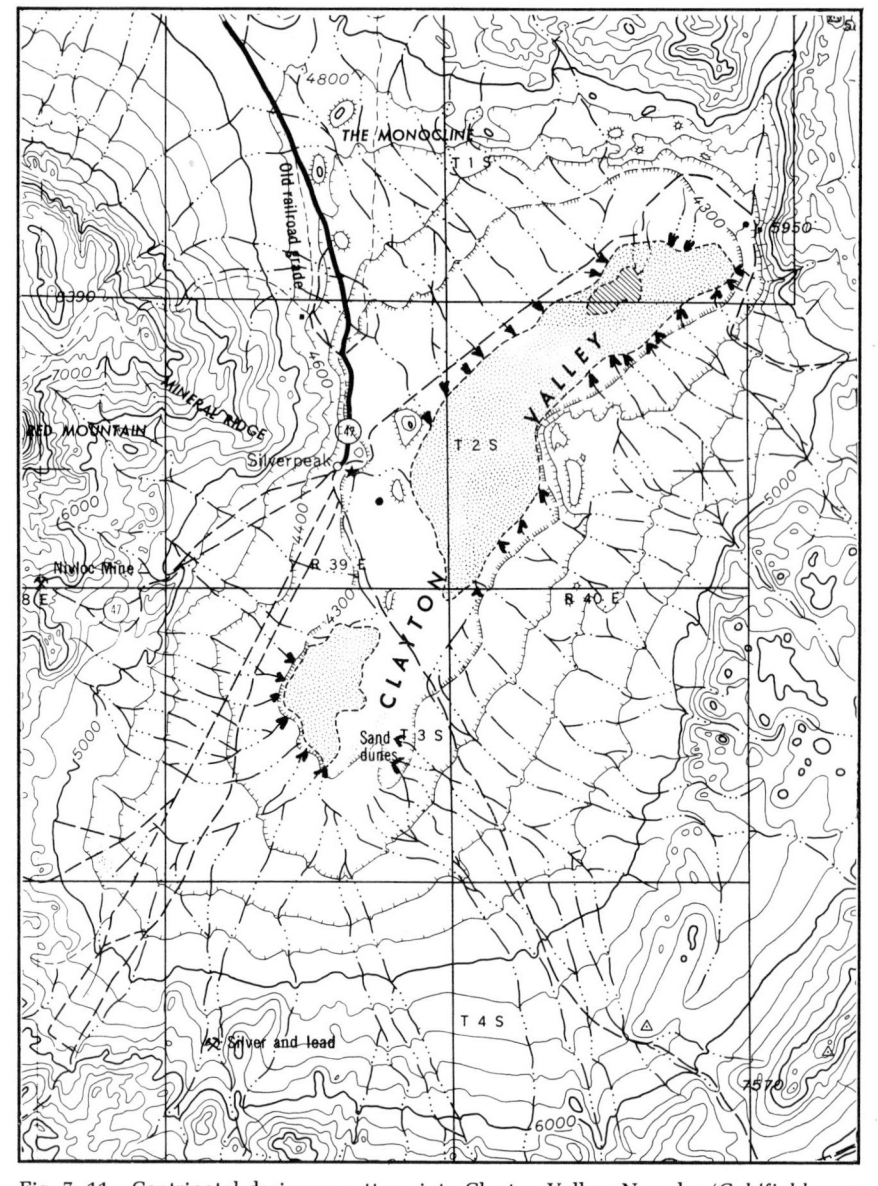

Fig. 7-11   Centripetal drainage pattern into Clayton Valley, Nevada. *(Goldfield map, U.S. Geological Survey.)*

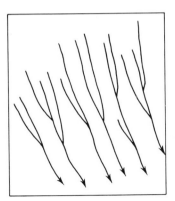

Fig. 7-12 Parallel drainage pattern.

Fig. 7-13 Air photographs of fine drainage texture (high density) in badland topography, South Dakota. *(U.S. Department of Agriculture.)*

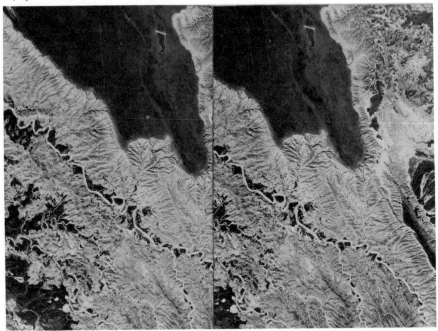

of the less vigorous system. Streams flowing down the steeper side of an asymmetrical ridge erode headward at a faster rate than streams flowing down the gentler slope, and the divide consequently shifts toward the drainage system having the lower slope. In the competition which takes place in the headwaters of drainage systems, streams which exhibit the following qualities have an advantage over others:

1. Steeper gradient

2. Shorter distance to base level

3. Erosion in less resistant rock

Fig. 7-14   Badland topography, Badland National Monument, South Dakota.

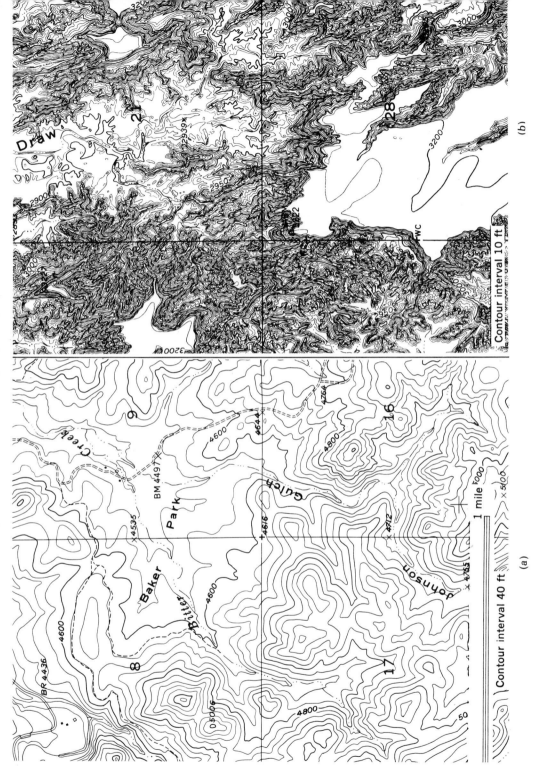

Fig. 7-15 Comparison of (a) coarse-textured, low-density drainage, Mt. Rushmore, South Dakota, and (b) fine-textured, high-density drainage, Cuny Table East, South Dakota. (U.S. Geological Survey.)

Encroachment of a stream into another drainage system results in diversion of drainage by stream capture. The most common type of stream capture takes place by *abstraction*, the headward erosion of one stream until it undercuts and diverts another stream. Such types of capture are common in areas of tilted sedimentary rocks of differing resistance to erosion. In Fig. 7 - 16 a stream eroding its course headward in less resistant rock captures a stream transverse to resistant rock ridges. The former channel through the ridge is left as a *wind gap*, and in the abandoned channel downvalley local drainage can support only a small stream which is *underfit* with regard to the size of the channel. The bend in the channel at the point of capture is known as the *elbow of capture*.

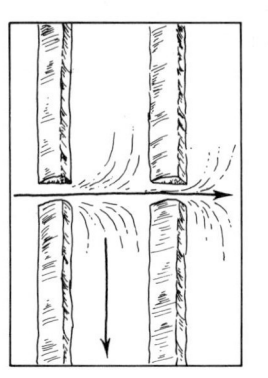 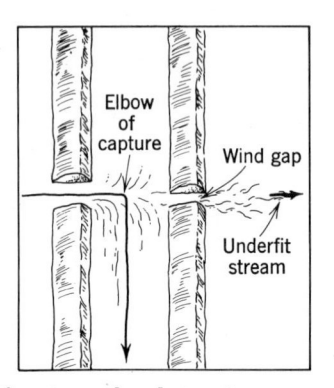

Fig. 7 - 16 Capture of a stream by abstraction.

Under certain conditions, a tributary stream may capture another stream in the same drainage system, or even capture a portion of the mainstream to which it is graded. Mackin (1936) described a set of conditions in the Big Horn Basin of Wyoming where a high-gradient stream was captured by a tributary stream with a lower gradient. The Greybull River (Fig. 7 - 17), a tributary of the Big Horn River, carries a coarse load of detritus from the mountains at its headwaters, and consequently has a rather high gradient of about 40

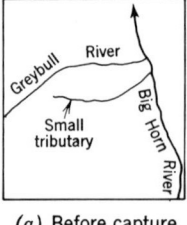 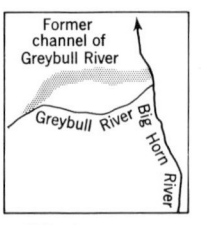

(a) Before capture   *(b)* After capture

Fig. 7 - 17 Capture of the Greybull River by a tributary stream. *(After Mackin, 1936.)*

Fig. 7-18 Capture of the Kentucky River by intercision with Benson Creek. *(Frankfort quadrangle, U.S. Geological Survey.)*

feet per mile. The capturing stream, also a tributary to the Big Horn, had a gradient of only a few feet per mile, because, even though it had a lower discharge than the Greybull, it carried only fine silt and sand derived locally from shales and sandstones. Thus, at the point where the capturing stream intersected the Greybull River, the channel was considerably lower, and the Greybull was diverted. The reversal of the usual condition (i.e., smaller streams having higher gradients) was brought about because of the marked difference in the character of load carried by the competing streams—coarse gravel by the Greybull and fine silt by the captor.

Stream capture is not limited to diversion by headward erosion of streams. Lateral planation by meandering streams may also cause intersection and diversion of streams. In Fig. 7-18 the channel between Frankfort and Thorn Hill was once occupied by the Kentucky River, which presently flows northward past Leestown. Diversion of the Kentucky to its present course could not have been effected by a meander cutoff since the meander lines of the old course do not come together. The most probable solution is that Benson Creek once occupied the channel north of Frankfort, joining the Kentucky River north of Leestown, and at that time the Kentucky flowed from Frankfort past Thorn Hill. Lateral migration of meanders of Benson Creek and the Kentucky River caused them to come together near Frankfort, and the Kentucky spilled into the lower channel of Benson Creek north of Frankfort. Such a stream diversion is known as *intercision*. Elkhorn Creek, in the upper right corner of Fig. 7-18, illustrates the difference between intercision and a meander cutoff. Elkhorn Creek formerly flowed in the abandoned channel around "The Backbone," until two meanders came together at "Elkhorn," and a cutoff occurred, leaving "The Backbone" as a *meander core*. The essential difference between intercision and a meander cutoff is that a meander cutoff involves only a single stream, and the point of cutoff is more obvious.

references

Davis, W. M.: 1909. Geographical Essays, Ginn and Company, Boston (reprinted 1954 by Dover Publications, Inc., New York).

Horton, R. E.: 1945. Erosional development of streams and their drainage basins: hydrophysical approach to quantitative morphology, *Geol. Soc. America Bull.*, vol. 56, pp. 275-370.

Leopold, L. B., M. G. Wolman, and J. P. Miller: 1964. Fluvial Processes in Geomorphology, W. H. Freeman and Company, San Francisco.

Mackin, J. H.: 1936. The capture of the Greybull River, *Am. Jour. Sci.*, vol. 31, pp. 373-385.

Playfair, John: 1802. Illustrations of the Huttonian Theory of the Earth, William Creech, Edinburgh.

Powell, J. W.: 1875. Exploration of the Colorado River of the West and Its Tributaries, Smithsonian Institution, Washington, D.C.

Contour interval 20 ft

evolution of
fluvial landforms

Since the publication of Davis's classic papers near the turn of the century, the idea of a systematic progression of landforms developed by normal surface processes during downwasting and erosion of an elevated land mass has served as one of the basic principles underlying interpretation of surface features of the continents. From the time of publication of the concept, which Davis called the *cycle of erosion,* arguments have been put forth against certain aspects of it, some of which have resulted in clarification and modification of the original idea, thus adding to our knowledge of geomorphic processes. Early objections to the cycle of erosion dealing largely, if not entirely, with the end product of the cycle, the peneplain, were successfully refuted by Davis over sixty years ago, and although much information regarding geomorphic processes has accumulated in recent years, the ideas of Davis and his contemporaries remain valid today. The fundamental idea underlying the concept of the cycle of erosion is that the topography of a stable region evolves through a continuous sequence of landforms having distinctive characteristics at successive stages of development.

Mature landscape sculptured by fluvial processes. *(Campti quadrangle, Louisiana, U.S. Geological Survey.)*

As emphasized by Davis, many of the ideas behind the concepts of pene-planation and the cycle of erosion were not original with him, but were conceived earlier in the works of Powell, Dutton, and Gilbert. From his work in the western United States, Powell developed a classification of stream valleys based on their origin and relationship to geologic structure, but even more significant was his realization that if the processes of erosion were allowed to operate for a long enough period of time, uninterrupted by renewed crustal movement, the land would be eroded to a lowland a little above sea level. To this late phase in the process of erosion Davis later gave the name *peneplain.*

In the Colorado Plateau region an extensive period of erosion which reduced the land to low relief prior to the cutting of the present canyons was recognized by Dutton and referred to as the "great denudation." Dutton referred to Powell's idea of base level and stated, "All regions are tending to base-levels of erosion, and if the time be long enough, each region will, in its turn, approach nearer and nearer, and at last sensibly reach it."

In the nearby Henry Mountains, Gilbert contributed much to geomorphic knowledge by outlining relationships between slope, discharge, and load in streams and recognizing the importance of lateral planation and the processes of stream-divide migration in the development of topography.

In a series of papers published about 1900, Davis incorporated these earlier ideas with his own and developed the idea that during the process of erosion, if an uplifted land mass remained undisturbed by further tectonic movement long enough, a systematic sequence of landforms would occur which could be grouped according to degree of development into stages of youth, maturity, and old age. According to Davis:

> A geographical cycle may be subdivided into parts of unequal duration, each one of which will be characterized by the degree and variety of the relief, and by the rate of change, as well as by the amount of change that has been accomplished since the initiation of the cycle. There will be a brief youth of rapidly increasing relief, a maturity of strongest relief and greatest variety of form, a transition period of most rapidly yet slowly decreasing relief, and an indefinitely long old age of faint relief, in which further changes are exceedingly slow. There are, of course, no breaks between these subdivisions or stages; each one merges into its successor, yet each one is in the main distinctly characterized by features found at no other time.

The terms youth, maturity, and old age were used by Davis in a relative sense, rather than absolute, and thus have no meaning with regard to age in terms of years. Nor did Davis imply that each of the three stages was of equal duration. He considered youth a relatively brief phase and thought that the old-age stage involved a tremendously greater period of time than either of the other two stages.

Once initiated, the cycle of erosion does not necessarily always proceed to completion without interruption. Tectonic deformation of the earth's crust may interrupt the cycle at any stage, and partial cycles are probably more common than completed cycles. Even partial cycles leave imprints upon the land surface by which they may be recognized.

base level

The idea of base level, one of the important concepts basic to the understanding of landscape evolution, was formulated by Powell in 1875.

> We may consider the level of the sea to be a grand base level, below which the dry lands cannot be eroded; but we may also have, for local and temporary purposes, other base levels of erosion, which are the levels of the beds of the principal streams which carry away the products of erosion . . . the base level would, in fact, be an imaginary surface, inclining slightly in all its parts toward the lower end of the principal stream. (p. 203)

Included in Powell's definition of base level are three central ideas: (1) The ultimate limit for subaerial erosion of the continents is the level of the sea, and this limit has an effect on the activity of river systems. (2) Locally resistant rock in the path of a stream, lakes in the stream valley, and certain other features which affect the ability of a stream to lower its channel may produce local base levels. (3) Tributary streams may not erode their channels below that of the main trunk stream, and since the mainstream would always have some slope, however slight, base level is not necessarily a flat surface, but would be a plane inclined toward the mouth of the mainstream.

Sea level may be considered as a general, permanent base level which fluctuates from time to time but which, for the most part, remains within a range of a few hundred feet. Fluctuations in the level of the sea produce changes in the streams graded to a specific level by causing deposition or downcutting of their channels in regrading their profiles to a new base level.

Local base levels such as lakes are temporary. Downcutting of the outlets and filling of lakes with sediment deposited by inlets destroy the lakes in times which are geologically short. After such a local base level is destroyed, streams graded to that level regrade their channels to a new base level.

A change of base level affects principal trunk streams first; then, as tributary streams graded to the mainstream begin to feel the effects induced by the base-level change, regrading of profiles is propagated upvalley throughout the drainage system.

stages of evolution of drainage systems

As fluvial processes continuously alter the topography of a landmass, progressive changes take place which may be envisioned by first considering an idealized case and then dealing with variations. In setting up three stages in the erosional development of landforms, Davis used as a starting point a newly uplifted landmass and traced successive changes which occurred with time. Changes in climatic conditions or crustal warping may complicate the ideal case and, in fact, may be more common than an uninterrupted cycle, but these may be treated as variations of the ideal case.

Fig. 8-1   Incised V-shaped valley in early youth, Owyhee River, Oregon. *(John Shelton.)*

youth

In the youthful stage of erosion of an uplifted landmass, streams have relatively high gradients and are engaged primarily in cutting their channels downward toward a graded profile of equilibrium. Rates of lateral erosion and valley widening are small compared with later stages, and significant portions of the upper reaches of the drainage system may remain ungraded. The cross profile of stream valleys will generally be V-shaped, with little or no floodplain on the valley floor. The width of the stream channel may, essentially, constitute the width of the valley floor (Figs. 8-1 and 8-2). Divides between

Fig. 8-2   Youthful gorge of the Grand Canyon, Arizona. *(J. R. Balsley, U.S. Geological Survey.)*

streams are broad and not well defined. Undrained depressions occupied by lakes and swamps may be present both within the stream valleys and in the interfluves between streams.

Although streams are rapidly downcutting, they may not yet have everywhere been able to erode their channels downward across resistant rock units to attain a graded profile, and waterfalls or rapids may exist at such places. Stream meandering may occur, but the width of the meander belt, as defined by lines tangent to the bends of meanders, is greater than the width of the floodplain.

Topographic relief in the drainage system is moderate to low, depending upon how deeply incised the stream valleys have become. Adjustment of streams to the underlying geologic structure is generally poor or incomplete.

maturity

As streams of the youthful stage become graded, the rate of downcutting relative to lateral erosion decreases, and there is increased meandering. Erosion of valley sides by impingement of meanders widens the valley floor until appreciable floodplains are produced. In the mature stage the width of the valley floors increases to the point where it is approximately equal to the width of the meander belt, and the cross-valley profile is no longer V-shaped. While a meander is eroding into the valley walls on one side of the valley, the opposite side is momentarily under the influence of somewhat different processes. The coarser material associated with the position of the main channel is for a time out of reach, and the channel deposits left when the stream was last against that side of the valley may become covered with finer sediment left as overbank deposits during floods. There is often a marked distinction in texture between channel deposits and overbank deposits, the channel deposits tending to be coarser. As a river meanders from one side of its valley to the other, erosion on the outside of meander bends causes reworking of material on the floodplain, while contemporaneous deposition may be taking place on the inside of the bend. The thickness of deposits left on the inside of meanders for a stream which is not aggrading is limited to the depth of scour of the stream. The thickness of overbank silts relative to channel gravels is determined in part by the frequency with which a stream meanders from one side of its valley to the other. Generally, streams which swing back and forth vigorously have relatively minor overbank silts, whereas streams which meander slowly from side to side allow a greater opportunity for thick overbank deposits to accumulate.

A graded profile of equilibrium is reached by most of the major streams,

and waterfalls and rapids present during the youthful stage have been eliminated. Although the streams are graded, they still continue to erode their channels downward in response to changing load-discharge relationships. An integrated drainage system develops, draining lakes and swamps which may have
been present earlier. Streams become adjusted to variations in rock resistance.

The broad, poorly defined interstream divides of youth become narrow
as tributary streams erode headward. Maximum topographic relief is developed, and the amount of land area consisting of valley sides is greater than
the amount of land area consisting of upland divides and valley floors (Fig.
8-3). Valley sides assume slopes of equilibrium which are just adequate to
transport the rock material on them to the valley floors below.

Fig. 8-3   Mature topography near Beaumont, California. (J. R. Balsley, U.S. Geological Survey.)

old age

Further development of floodplains results in wider valley floors until the width of the floodplain exceeds the width of the meander belt. Many oxbow lakes and swamps are usually present on the floodplain as a result of cutoff meanders. Natural levees which form banks confining stream channels may be built up until the channel is several tens of feet above the general level of the floodplain. Tributary streams have difficulty in breaching levees, and may flow on the same floodplain for many miles before breaking through the levee to join in the mainstream. Such tributary streams are known as *Yazoo streams,* after the Yazoo River in Mississippi, which follows the Mississippi River for many miles before breaking through its natural levee.

Interstream divides have been greatly reduced by mass movement and erosion, and topographic relief is much diminished. Valley sides are much more gently sloping than at any time earlier, and thick soils have been produced by extensive weathering. Slopes of streams and valley sides are graded, but the greatly reduced relief results in production of very fine debris which may be transported on very low slopes. Lowering of relief reduces the calibre of load supplied to streams. Thus slope requirements of graded streams are less, and the gradients of long profiles of streams are gradually reduced.

Fig. 8-4   Monadnock rising above a plain of low relief, Aroostock County, Maine. *(J. R. Balsley, U.S. Geological Survey.)*

Ultimately, a plain of very low relief with low rolling divides is developed, and large areas approach base level. This last phase in the evolution of fluvial landforms was termed a peneplain by Davis, who used the prefix *pene-* (meaning almost) since a region eroded by fluvial processes to base level would still have a small amount of relief because of slope requirements necessary to maintain even the lowest-gradient streams. Occasionally, resistant rock may remain as isolated hills, known as *monadnocks*, standing above the low plain (Fig. 8-4).

Characteristic features of stages in the evolution of fluvial landforms are listed in the accompanying table.

	youth	maturity	old age
steepness of stream gradients	high	moderate	low
relationship of downcutting to lateral erosion	downcutting dominant	shift toward lateral erosion	lateral erosion dominant
cross-valley profile			
width of floodplain relative to meander belt	meander belt greater than floodplain width	meander belt equals floodplain width	meander belt less than floodplain width
graded streams	few streams graded	most streams graded	all streams graded
rapids and waterfalls	present	largely absent	absent
lakes and swamps	may be present both within valleys and on divides	absent	oxbow lakes and swamps present on floodplains
relief	moderate	maximum	minimum
divides	broad, irregular	sharp-crested	broad, rolling

criticisms of the concept

Although Davis's concept of the evolution of landscapes gained widespread acceptance, there have been criticisms concerning certain aspects of the cycle of erosion. Early objections centered around the idea of peneplanation and the manner in which landscapes evolved. Some of the criticisms to which Davis himself replied are discussed in the following paragraphs.

A number of critics maintained that the earth's crust is not stable enough to permit prolonged degradation necessary for the development of a peneplain. To this Davis (1909) replied:

> Stability or instability of the earth's crust can be learned only by comparing the consequences reasonably deduced from one condition or the other with the observed facts. (p. 365)

In other words, the existence of a peneplain is the very evidence by which stability of the earth's crust may be proved.

Some argued that the concept requires initial rapid uplift followed by still-stand—a condition very unlikely to occur. Davis, however, did not imply that the cycle of erosion, once begun, proceeded every time to completion. He clearly recognized that interruptions could, and most likely would, occur.

> Movements of the land mass may . . . occur at any stage in the advance of the cycle. They then interrupt the further progress of sculpturing processes with respect to the former base-level by placing the land mass in a new attitude as referred to the sea. The previous cycle is thus cut short and a new cycle is entered upon. (p. 285)

Nor need the crust be entirely still while a complete cycle is achieved.

> It has never been my intention to imply an absolute stillstand of the earth's crust during an entire cycle of denudation. Any sort of movement that does not cause a distinct dissection of the surface below the peneplain is admissible. (p. 366)

Recognition of surfaces of low relief developed by other processes led to the argument that much low-relief peneplainlike topography may be explained by control of resistant strata, marine planation, altiplanation, and alluviation. Although this is certainly true, it does not invalidate the concept of peneplanation, but merely means that discretion must be used in the interpretation of low-relief features. In fact, it is quite likely that some of these other flat surfaces may merge with true peneplains, and a given surface could be in part peneplain and in part marine plain, structural plain, or alluvial plain.

A few geomorphologists interpreted the appearance of the stage of old age as being dependent on the process of lateral planation. Davis clearly did not imply that peneplains were the result entirely of lateral planation. He states:

There is a vast amount of work performed in the erosion of valleys in which the rivers have no part. It is true that rivers deepen the valleys in the youth and widen the valley floors during the maturity and old age of a cycle, and that they carry to the sea the waste denuded from the land; it is this work of transportation to the sea that is peculiarly the function of a river, but the material to be transported is supplied chiefly by the action of the weather on the steeper consequent slopes and on the valley sides. The transportation of the weathered material from its source to the stream in the valley bottom is the work of various slow-acting processes, such as the surface wash of rain, the action of ground water, changes of temperature, freezing and thawing, chemical disintegration and hydration, growth of plant roots, and the activities of burrowing animals. (p. 266)

Largely as a result of the fact that recognition of a former peneplain in a region presently well above sea level provides a strong argument for crustal uplift, much discussion has centered around the interpretation of flat upland surfaces. The search for accordant upland areas which might have originated by long-continued erosion near base level led to interpretation of "summit peneplains" in many mountain ranges of the world, the idea being that a given region was eroded to a peneplain prior to uplift of a mountain range and that the flat surfaces now seen at high elevations are remnants of such a peneplain. Undoubtedly, many such interpretations were the result of an overly enthusiastic acceptance and application of Davis's ideas, and subsequently skepticism developed among some geomorphologists about the validity of some reportedly uplifted erosion surfaces. Such doubts, however, concern only application of the concept, not the validity of the concept itself.

Quite clearly, all extensive flat surfaces are not peneplains. Among other origins of such surfaces are a variety of constructional plains formed by deposition of alluvium or marine sediments, by eruption of lava, or by accumulation of other surface material. The essential point is that such surfaces are formed by deposition, rather than by erosion, and thus have no necessary relationship to an old-age land surface formed by long-continued erosion.

Extensive flat land surfaces are often developed on flat-lying resistant beds. These surfaces, known as *stripped structural surfaces*, may form anywhere that resistant rocks are overlain by rocks easily removed by weathering and erosion. Among the most common examples are resistant sandstone beds overlain by weak shales. The Kaibab and Coconino Plateaus, covering a broad area in Arizona, Utah, and Colorado, are examples of stripped structural surfaces developed on the resistant Kaibab limestone after several thousand feet of weaker overlying Mesozoic beds had been removed by erosion. Even though erosion has played a role in forming these flat plateaus, they are not considered peneplains because they owe their origin to rock resistance rather than to erosion to a low plain near base level. It is very likely that the Kaibab and Coconino Plateaus were formed not far from their present altitude above sea level, some 6,000 to 8,000 feet.

Somewhat related types of surfaces are *resurrected*, or *exhumed*, surfaces which have been buried and later uncovered by erosion. These surfaces are unconformities in a stratigraphic sense, and if the rocks above the unconformity are more easily eroded than the rocks underneath, a special type of stripped structural surface is formed. The ancient surface may be of any one of a number of origins, i.e., marine erosion, deposition, fluvial erosion, etc., and may be developed on almost any kind of rock.

Extensive marine erosion surfaces may resemble peneplains or late mature erosion surfaces, but usually differ in detail. Marine erosion plains often retain marine deposits such as beach gravels, and there is usually no evidence of prolonged weathering, as might be expected on a peneplain.

In the 1920s Walther Penck, a German geomorphologist, challenged many of Davis's concepts. His principle work, "Die morphologische Analyse," was written in German in a style very difficult to translate and comprehend. Although few geomorphologists accepted Penck's basic ideas, the controversies which developed stimulated thought concerning the evolution of landscapes. Penck was much concerned with rates of uplift relative to erosion, and thought that slopes could be best understood in terms of uplift. He used the term *aufsteigende Entwicklung* for convex upward slopes which he believed were a result of accelerating uplift; *gleichformige Entwicklung* for straight slopes which he thought were due to uniform uplift; and *absteigende Entwicklung* for concave-upward slopes which he believed were caused by declining uplift. These ideas are now considered erroneous because all three kinds of slopes, convex, straight, and concave, may be found in a single slope profile.

More recently some geomorphologists have rejected the idea of landform evolution in concept as well as in interpretation (Hack, 1960). According to such views, landforms achieve a stable state of equilibrium, and may be explained independently of any systematic evolution with time. The fallacy of such a concept lies not in the steady-state, or equilibrium, conditions postulated (these were recognized 70 years ago), but in the inevitable decrease in the rate of downcutting of streams as they lower their channels closer to base level, thus changing the conditions under which equilibrium exists, resulting in reduction of relief. Bretz (1962) and Holmes (1964) have strongly contested the view of Hack (1960) that the end product of prolonged weathering and erosion would be a "mature" rolling landscape rather than a peneplain.

From the foregoing discussion it is obvious that although the concept of evolution of landforms with time is still considered valid, care must be exercised in the interpretation of specific landscapes as belonging to a particular stage of erosion. Not all landscapes presently seen can be fitted neatly into a particular category, especially since much of the topography of northern North America and Europe owes its origin to glacial deposition, and the low rolling topography is a relict feature of glacial processes rather than fluvial erosion.

references

Bretz, J. T.: 1962. Dynamic equilibrium and the Ozark landforms, *Am. Jour. Sci.,* vol. 260, pp. 427-438.

Davis, W. M.: 1909. Geographical Essays, Ginn and Company, Boston (reprinted 1954 by Dover Publications, Inc., New York).

Dutton, C. E.: 1882. Tertiary history of the Grand Canyon district, *U.S. Geol. Survey Mon. 2.*

Gilbert, G. K.: 1877. Report on the geology of the Henry Mts., Utah, *U.S. Geog. and Geol. Survey, Rocky Mt. Region.*

Hack, J. T.: 1960. Interpretation of erosional topography in humid temperate regions, *Am. Jour. Sci.,* Bradley Volume, vol. 258-A, pp. 80-97.

Holmes, C. D.: 1964. Equilibrium in humid-climate physiographic processes, *Am. Jour. Sci.,* vol. 262, pp. 436-445.

Mackin, J. H.: 1948. Concept of the graded river, *Geol. Soc. America Bull.,* vol. 59, pp. 463-512.

Penck, Walther: 1924. Die morphologische Analyse, J. Engelhorn's Nachfolger, Stuttgart.

Powell, J. W.: 1876. Report on the geology of the eastern portion of the Uinta Mts., *U.S. Geog. and Geol. Survey.*

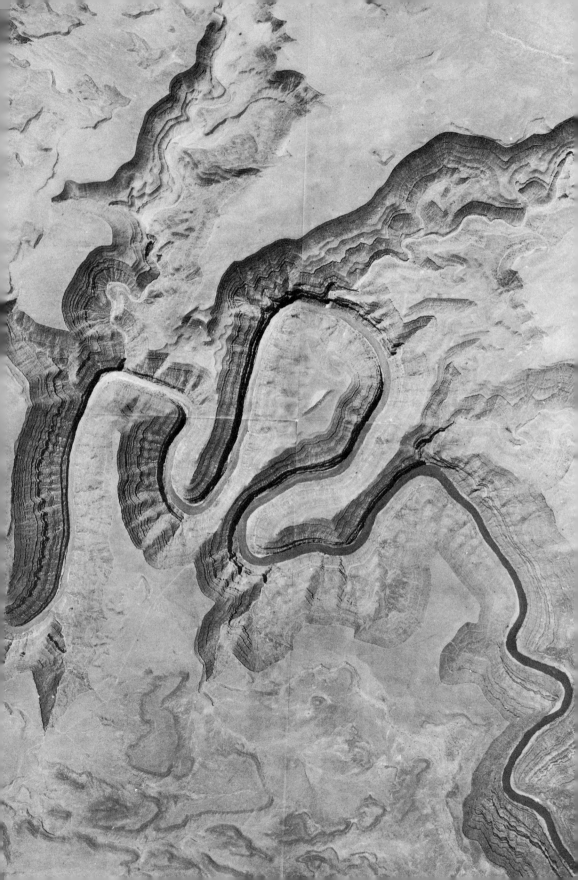

9

rejuvenation

causes of rejuvenation

Changes in conditions during any stage of the cycle of erosion may cause streams to incise their channels with renewed vigor. Such renewed down-cutting is called *rejuvenation*. Changes in any one of three variables is sufficient to cause rejuvenation. These include (1) eustatic sea-level change, (2) tectonic uplift, and (3) climatic change.

Worldwide change of sea level is referred to as a *eustatic change*, implying that sea level rises or drops while the land remains stationary. Such changes are brought about by withdrawal of large amounts of water from the oceans during enlargement of continental ice sheets or may be caused by subsidence of portions of the ocean basins. The effect of eustatic lowering of base level is to cause streams to incise their channels and begin to regrade them to the new base level. A eustatic rise of sea level produces just the opposite effect: streams are forced to aggrade their channels to accommodate the new base level.

Tectonic uplift of the land may produce the same effects as lowering of

Incised meanders, San Juan County, Utah. *(U.S. Department of Agriculture; National Archives.)*

sea level, and tectonic downwarping of the land has the same result as a rise of sea level. The increase in gradient of stream courses in an uplifted area gives rise to increased downward erosion, and channels become incised below their previous levels. Downwarping of the land causes streams to aggrade their channels in order to maintain appropriate gradients.

Climatic changes which affect the load-discharge relationships of a stream also cause rejuvenation. If the annual precipitation in an area increases, the discharge of streams in the drainage basin also increases. With increased discharge, streams are able to maintain a given velocity on a lower slope, and the result is downcutting of the stream. On the other hand, if glaciers enter the headwaters of a drainage basin as a result of climatic changes and contribute large amounts of coarse debris, slope requirements of the stream will increase. The effects of increased discharge and increased load are opposite, one tending to cause downcutting, the other aggradation. The effect which prevails in a given situation depends on whether the added discharge is (1) just able to compensate for the additional load with no increased slope, (2) unable to carry all the added load, resulting in aggradation, or (3) more than able to handle the added load, resulting in downcutting.

Rejuvenation related to any one of the three causes, eustatic, tectonic, or climatic, leaves former valley floors standing above the newly incised stream valleys. If the stage of erosion was well advanced when rejuvenation occurred, large areas of relatively flat uplands may stand above present valleys. If the stage was youthful or mature, terraces may be found along the valley sides, recording the position of former valley floors (Fig. 9-1).

noncyclic surfaces

It is not safe to conclude that all terraces, benches, or flat uplands represent former stages of erosion interrupted by rejuvenation. Perhaps the most common noncyclic flat areas confused with cyclic erosion surfaces are *stripped structural surfaces,* formed when erosion strips less resistant rock from flat-lying resistant rocks, such as sandstone or limestone. The fundamental difference between a flat stripped structural surface and a flat cyclic erosion surface is that the stripped structural surface is directly related to rock resistance.

The benches in Fig. 9-2a correspond to the exposed surfaces of resistant beds, and could be developed with constant downcutting of the stream. The benches in Fig. 9-2b, however, bevel the upturned edges of beds, and must therefore represent pauses in downcutting of the stream sufficiently prolonged to develop wide valley floors, followed by rejuvenation. Figure 9-3 shows an example of a stripped structural surface in the Grand Canyon district of

Arizona. The surface of the Coconino Plateau conforms with the top of the resistant Paleozoic Kaibab limestone, from which less resistant Mesozoic shales and sandstones have been eroded. Although erosion has removed the Mesozoic beds, the reason for the flat Kaibab Plateau is not long-continued erosion to a peneplain, but rather rock resistance. In spite of the extensiveness of the surface, it cannot be considered a peneplain.

Another example of a stripped structural surface occurs also within the Grand Canyon. The Tonto Platform makes a flat bench paralleling the Colorado River. However tempting the idea might be that this is a stream terrace left by a former valley floor, the relationship of the bench to the resistant Tapeats sandstone proves it cannot be. If it were a stream terrace of the Colorado

Fig. 9-1 Terraces along the south bank of the San Juan River between Farmington and Shiprock, New Mexico. *(John Shelton.)*

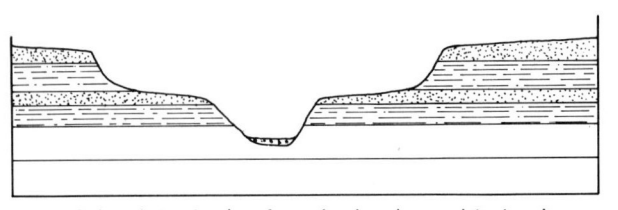

(*a*) Stripped structural surfaces developed on resistant rock

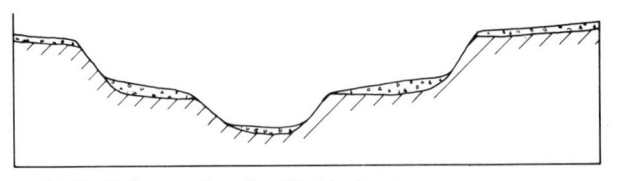

(*b*) Cyclic terraces beveling tilted beds

Fig. 9-2 Comparison of stripped structural surfaces with cyclic erosion surfaces.

River, the bench should have approximately the same slope downvalley (within certain limits) as the present Colorado. Closer inspection shows that the bench in places actually slopes upvalley in conforming to the dip of the Tapeats sandstone.

Another type of noncyclic terrace consists of a set of nonpaired terraces formed by a continuously downcutting stream swinging back and forth across its floodplain. Each time the stream returns to the valley side, where it had left a flat valley floor the previous time there, it is at a lower level than before, and hence will tend to cut away the former floodplain, leaving as terraces those portions which escape complete destruction. In migrating across the valley floor from *a* to *b* in Fig. 9-4, the stream cuts below the level it was when at *a*, and in migrating back across the valley to *c*, it is still lower. The result is a terrace at *a*, *b*, and *c*, made by stream erosion, but the terraces do not match up on opposite sides of the valley as they would if each represented a pause in the downcutting of the stream during which a wide valley floor was developed.

The extent of a former valley fill which escapes destruction during downcutting may be in a large measure fortuitous or may be in part controlled by the uncovering of buried spurs of bedrock which act as "buffers" and protect the terraces from undercutting by the stream (Fig. 9-5). When the latter is responsible for the preservation of the terraces, the terraces may be termed *rock-defended terraces*.

Occasionally, sidestream alluvium will be undercut by lateral swinging of the mainstream, leaving a terracelike remnant.

Fig. 9-3 Topographic map of a portion of the Grand Canyon region. The broad flat area in the lower part of the map is a stripped structural surface on limestone. *(U.S. Geological Survey.)*

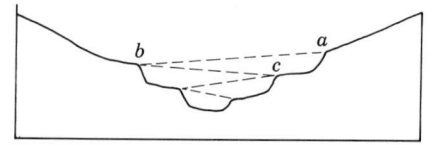

Fig. 9-4 Cross section of nonpaired stream terraces.
The dashed line indicates the path of the stream.

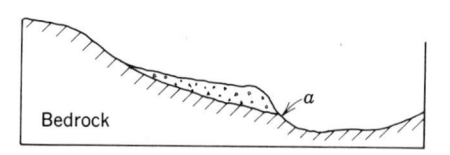

Fig. 9-5 Rock-defended terrace. Bedrock at *a* protects
the terrace above from further destruction by erosion
of the stream at the base.

stream terraces

Stream terraces are flat surfaces along the valley sides of stream courses
marking the level of former valley floors (Fig. 9-6). As opposed to structural
benches, which are controlled by the dip and strike of the bedrock, and up-
faulted segments of flat valley floors, which owe their position to tectonism,
river terraces are independent of bedrock structure, their flatness and degree
of development being functions only of the resistance of the bedrock to abrasion
and the length of time the stream remained graded to one base level under
constant load and discharge.

The downvalley profile of a river terrace is governed by the gradient of
the stream which flowed on it, plus or minus any distortion due to subsequent
tilting or uplifting. Since the terrace marks the position of a former valley
floor, it follows that its downvalley slope must approximate the gradient of
the stream which carved it, if there has been no later modification of it.

At the time of its development, the cross-valley profile is normally relatively
flat, with very minor slopes toward the particular part of the valley floor last
occupied by the stream. However, the cross-valley profile of the terraces
is subject to modification by slope wash and erosion. Thus, depending partly
on the length of time that the terraces have existed since ceasing to be a part
of the valley floor, the cross-valley profile of the terraces need not correspond
exactly to the original valley-floor profile. Slope wash from the valley sides
will tend to obscure the original valley floor, and hence it is necessary to observe the true form of the terrace in transverse tributary stream cuts or by
other means, since profile reconstructions or correlations based on obser-

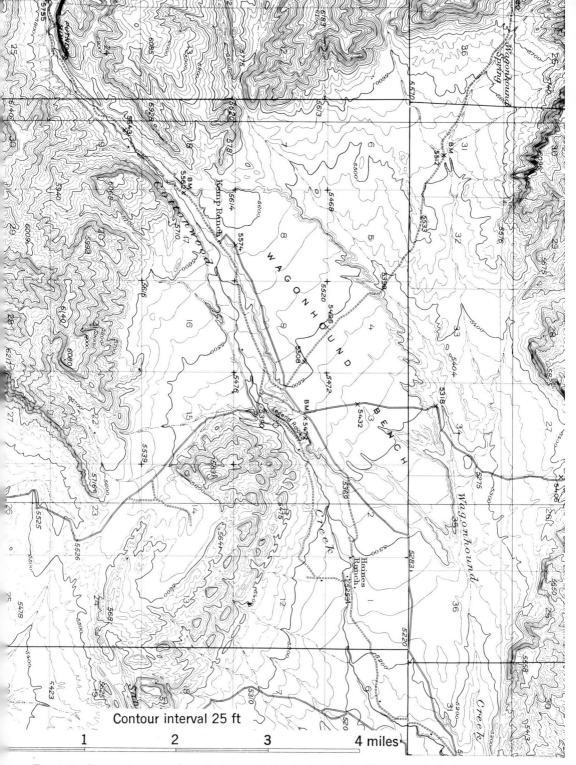

Contour interval 25 ft

1 2 3 4 miles

Fig. 9-6 Stream terraces along Cottonwood Creek, Wyoming. (*Grass Creek Basin* quadrangle, *U.S. Geological Survey.*)

vations of the present surface of the terrace will introduce varying degrees
of error.

Stream terraces may be produced by lateral stream erosion, by valley filling by aggradation, or by combinations of deposition and erosion. *Bedrock
terraces,* also sometimes referred to as *cut terraces,* are formed by erosion
of bedrock (Fig. 9-9a). They are usually mantled with a thin veneer of alluvium
left by the stream that cut the terrace, but the thickness of alluvium does
not exceed the depth of scour of the stream. *Fill terraces* are formed by aggradation of a stream, resulting in the filling of the valley with alluvium, followed
by later rejuvenation (Fig. 9-9b). The surface of a fill terrace is constructional, whereas the surface of a bedrock terrace is erosional. Rejuvenation

Fig. 9-7   Great Terrace of the Columbia River near Chelan Falls, Washington. *(Austin
Post, University of Washington.)*

following a period of valley filling may produce *cut-in-fill terraces* (Fig. 9-9c). In Fig. 9-9c the highest terrace is a fill terrace, but successive lower terraces were formed by erosion of alluvium and are thus cut-in-fill terraces. The upper fill terrace is a constructional feature, whereas the lower terraces are erosional in origin.

Determination of the origin of a terrace is of fundamental importance in the interpretation of the geomorphic history of a region. Figure 9-9a and b illustrates the differences in interpretation between cut and fill terraces. Interpretation of the terrace as a cut terrace (Fig. 9-9a) implies a pause in downcutting at a to develop a wide floodplain, followed by rejuvenation and incision to b. Interpretation of the terrace as a fill terrace (Fig. 9-9b) implies

Fig. 9-8 Stream terrace along the Snake River, Wyoming. *(Wards Natural Science Establishment.)*

downcutting to c, then filling of the valley to a, followed by rejuvenation and incision to b.

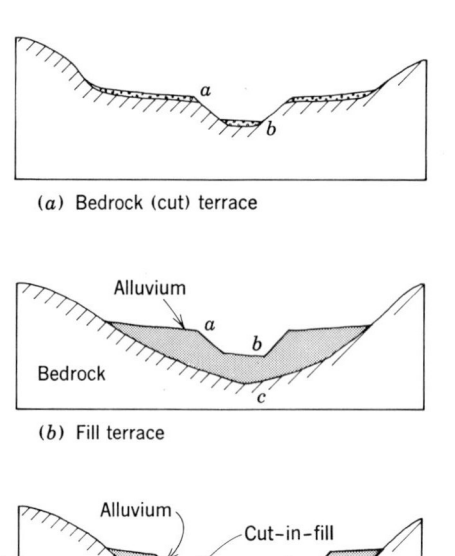

(a) Bedrock (cut) terrace

(b) Fill terrace

(c) Cut-in-fill terrace

Fig. 9-9 Basic types of stream terraces.

correlation of stream terraces

Correlation of terraces is best accomplished by observing the continuity of terraces from place to place and tracing them without interruption, or nearly so, throughout their extent. This may be possible only where terrace discontinuities are of such small magnitude as to preclude the possibility of any doubt as to the identity of terrace remnants. When correlations of terraces in adjacent or nearby stream valleys are desired, or when continuity of a terrace in a valley is not apparent, other methods must be used. The most reliable method is to reconstruct the longitudinal profile of the terrace remnants and extrapolate it to the base-level control which existed at the time of the terrace development. The longitudinal profiles of tributary streams may then be compared with the terrace profiles of the major streams to which they must have been graded. In this respect, terrace altitudes along any particular stream valley will be useful in reconstructing longitudinal profiles, but the altitude of a terrace at any specific point will have little meaning in attempting to correlate terraces from one valley to another, since the altitude of a terrace will change continuously upstream from its controlling base level.

An example of the use of longitudinal profiles for correlation is shown in Figs. 9-10 and 9-11. Emblem Bench and YU Bench, which stand above the Greybull River in Wyoming (Fig. 9-10), are remnants of stream terraces. The terraces slope downvalley at about 40 feet per mile, approximately the same as the present Greybull River, and are cut across the upturned edges of tilted shales and sandstones. Erosion has destroyed much of the original terraces, and correlation of the remaining remnants must be made on the basis of projected longitudinal profiles. Figure 9-11 clearly shows that Table Mountain, a terrace remnant some 8 to 10 miles downvalley, correlates with the terrace at YU Bench and that the lower terrace is the Emblem terrace.

The height of a terrace above the present stream is not dependable as a means of correlation. There need not necessarily be any correspondence between the present stream gradient and the gradient of the stream which was responsible for development of a terrace; thus it is possible for terrace profiles to converge or diverge downstream with the present stream profile, and the height of the terrace above the stream is not constant. The absence of parallelism between terraces and present streams may be caused by differential crustal warping or changes in load-discharge relationships of the streams at different times.

If there are only two or three terrace levels in a valley and if they are prominently developed at rather widely spaced vertical intervals, local correlation of terrace remnants on the basis of height above the present stream may be valid; but with a larger number of terrace levels and shorter vertical intervals between them, correlation becomes more and more tenuous.

Terraces whose development was separated by significantly large time intervals exhibit features which can be used in conjunction with reconstructed profiles for correlation. Degree of weathering is, among other things, a function of time. In a valley which has terrace levels of different age, the older terraces have more mature soil horizons and greater depth of soil development than younger ones. The thickness of weathering rinds on pebbles increases with age and is useful as a correlation tool. For example, basalt pebbles in a terrace gravel having weathered rinds of 2 millimeters would indicate an age greater than basalt pebbles in another terrace gravel having weathered rinds of only 1 millimeter, assuming similar climatic conditions.

Fossils, artifacts, and buried soils in terrace sediments are also useful in establishing age relationships of terraces.

Rejuvenation of a meandering stream in the mature stage of erosion interrupts further development of a floodplain, and the increase in the rate of downcutting relative to lateral cutting results in the formation of *incised meanders*

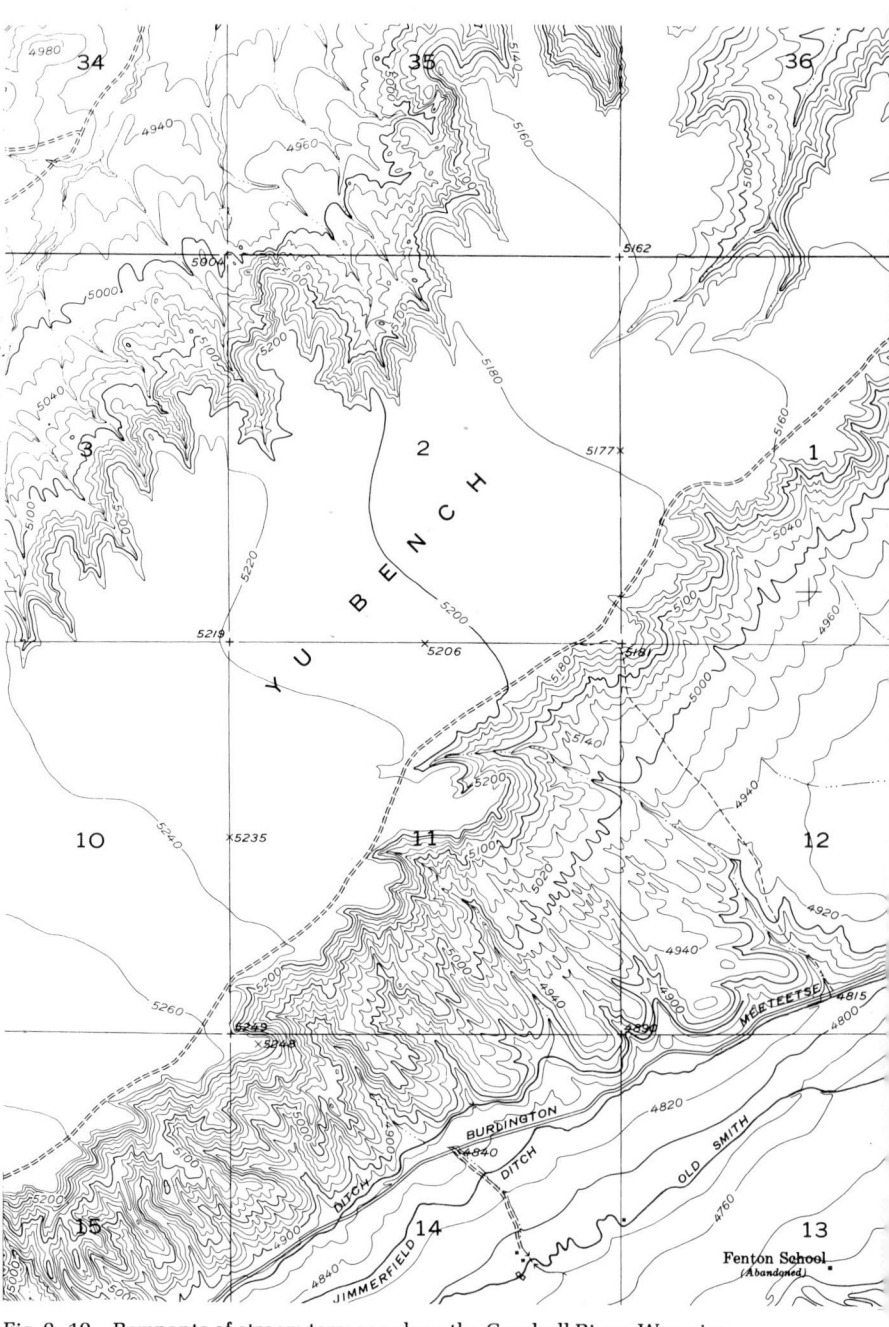

Fig. 9-10 Remnants of stream terraces along the Greybull River, Wyoming.

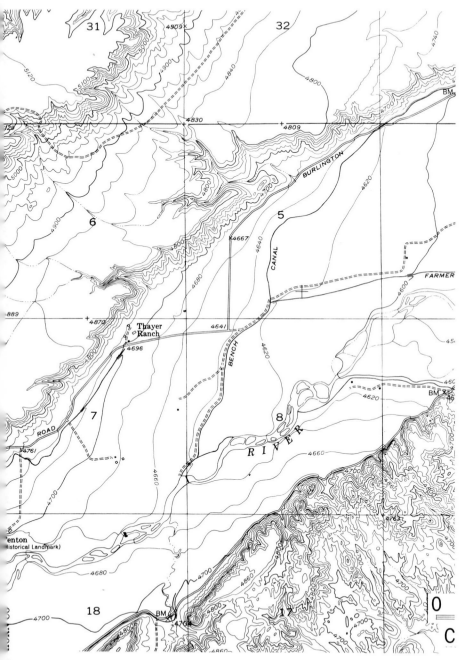

(YU Bench quadrangle, U.S. Geological Survey.)

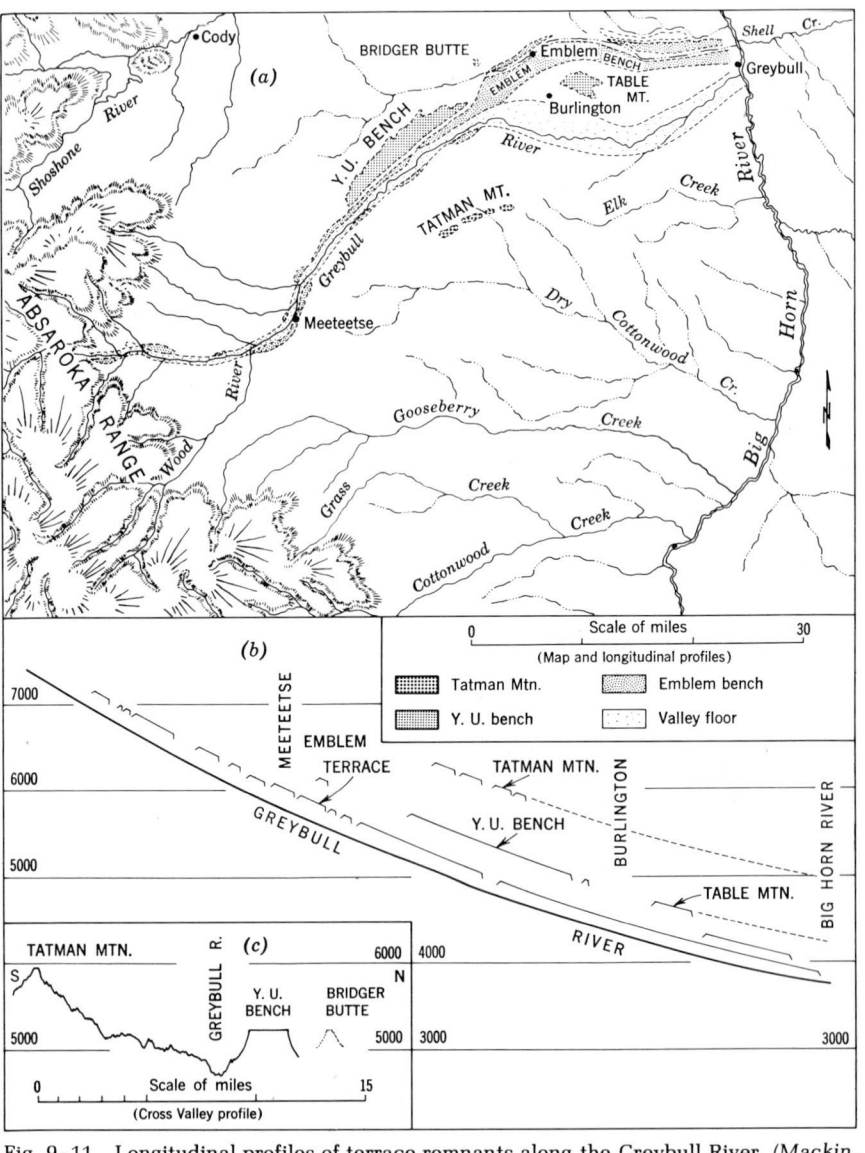

Fig. 9-11 Longitudinal profiles of terrace remnants along the Greybull River. (Mackin, 1937.)

(Fig. 9-14), also known as *entrenched meanders*. Both terms imply that the meandering course was developed in a former cycle of erosion and that rejuvenation has since caused the stream to cut downward while maintaining approximately the same meandering pattern. The cross-valley profile of incised meanders is normally symmetrical.

Not all meanders which occupy narrow V-shaped valleys are inherited from a previous cycle of erosion. Some form by expansion of the meander while downcutting is progressing, and thus are not related to rejuvenation. They typically show asymmetrical cross-valley profiles, with more gentle valley sides on the inside of the meander than on the outside. The term *in-grown meander* has been applied to these meanders. They may or may not have been inherited from a previous cycle.

When there is difficulty in demonstrating whether or not a particular set of meanders in a narrow winding valley is related to rejuvenation, a nongenetic

Fig. 9-12 Cut-in-bedrock stream terrace, Emblem Bench, Wyoming. A higher erosion surface may be seen on the horizon.

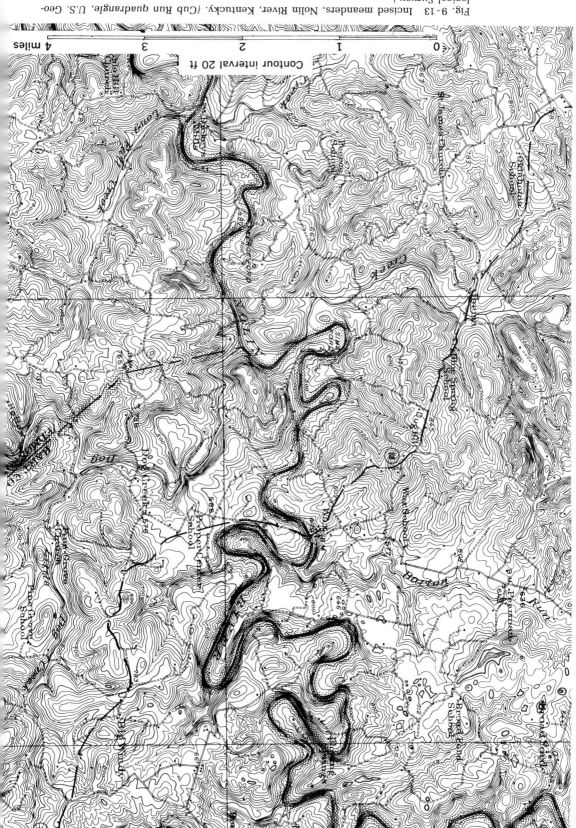

Contour interval 20 ft

term, *inclosed meander,* may be used. An inclosed meander is a meander closely bordered by valley walls and there is no implication as to how or why the inclosure has taken place.

Fig. 9-14 Incised meanders, San Juan River, Utah. *(Wards Natural Science Establishment.)*

references

Baulig, Henri: 1940. Reconstruction of stream profiles, *Jour. Geomorphology,* vol. 3, pp. 3-15.

Cotton, C. A.: 1940. Classification and correlation of river terraces, *Jour. Geomorphology,* vol. 3, pp. 27-37.

———— : 1945. Significance of terraces due to climatic oscillation, *Geol. Mag.* (Great Britain), vol. 82, pp. 10-16.

Davis, W. M.: 1909. River terraces in New England, in Geographical Essays, Ginn and Company, Boston (reprinted 1954 by Dover Publications, Inc., New York).

Johnson, D.: 1938. Stream profiles as evidence of eustatic changes of sea level, *Jour. Geomorphology,* vol. 1, pp. 178-181.

———— : 1944. Problems of terrace correlation, *Geol. Soc. America Bull.,* vol. 55, pp. 793-818.

Mackin, J. H.: 1936. The capture of the Greybull River, *Am. Jour. Sci.*, vol. 231, pp. 373-385.

———— : 1937. Erosional history of the Big Horn Basin, Wyoming, *Geol. Soc. America Bull.*, vol. 48, pp. 813-894.

Rich, J. L.: 1938. Recognition and significance of multiple erosion surfaces, *Geol. Soc. America Bull.*, vol. 49, pp. 1695-1722.

4

valley-side processes

weathering

Weathering is the disintegration and decomposition of rocks and minerals at the earth's surface as a result of physical and chemical action. The rate of weathering is not constant everywhere, but varies according to differences in intensity of processes going on at any given point. The kind of weathering which predominates on the surface also varies from place to place. Since weathering refers only to the breaking up of rock in place, erosion and transportation are not considered weathering processes.

Mechanical weathering refers to the type of weathering which involves disintegration or breaking up of rock by physical processes without changes in chemical or mineral composition. *Chemical weathering* refers to decomposition of rocks by chemical processes which involve chemical or mineral changes.

Exfoliation arch in granitic rock, Yosemite National Park, California. *(F. E. Matthes, U.S. Geological Survey.)*

mechanical weathering

freezing and thawing

When water freezes to ice, its volume is increased by about 9 percent. The expansion of the ice being frozen in a confined space thus causes a great deal of pressure to be exerted against the sides of the material holding the ice. The force exerted on the walls of a container in which ice is being formed may approximate 30,000 pounds per square foot.

In areas of high latitude or high altitude, alternate freezing and thawing of water at or near the earth's surface may occur almost every day under certain conditions. During the daytime water tends to seep into cracks and crevices in the rock, where it may freeze at night. Expansion of the ice then tends to cause wedging apart of the rock and breaking of pieces of rock away from the bedrock. Water seeping into the newly expanded cracks will freeze again and cause renewed wedging apart of the rock. Repeated freezing and thawing, with resulting shattering of rock, is especially common in high mountains above the timberline, where meltwater produced during the day percolates into cracks and refreezes at night. An example is shown in Fig. 10-1, in the Sierra Nevada of California, where intense freezing and thawing has produced an accumulation of broken angular blocks known as *felsenmeer* (rock sea).

gravity

The role of gravity in weathering is an indirect one since movement of material is generally considered erosion or transportation. However, on steep mountain cliffs, blocks loosened by freezing and thawing or other processes may tumble downslope and break off additional blocks upon impact on the way down. The loosening of blocks by the force of impact of falling material contributes to mechanical weathering, but later movement is considered erosion or transportation.

exfoliation

Rocks formed at great depth below the earth's surface exist in an environment of great pressure. If such rocks are uplifted and exposed at the surface by erosion, the reduction of pressure causes expansion of the rocks and development of large concentric fractures (Fig. 10-2). The separation of successive concentric shells from massive rock is *exfoliation*. Expansion of the outer portions of the rock causes thin shells to break away from the inner portions.

Exfoliation domes, such as Half Dome in Yosemite National Park (Fig. 10-3), are thought to have been developed by exfoliation of granite. Matthes (1930) attributes such domes in the Sierra Nevada to concentric exfoliation, resulting from physical expansion due to unloading.

A somewhat different type of exfoliation is caused by *spheroidal weathering,* which results from mechanical effects of chemical weathering. When feldspars are converted to clay by chemical weathering, an increase in volume takes place, and the interlocking texture of the minerals may be broken up. Percolation of water along intersecting joint planes may cause decomposi-

Fig. 10-1 Mechanical weathering of jointed rock by intense freeze and thaw activity. The loose rock fragments have been broken by alternate freezing and thawing of water in fractures. *(Wards Natural Science Establishment.)*

tion of feldspars from the joint planes inward. Corners of joint-bounded blocks are progressively rounded off, and split off in rounded concentric shells when exposed at the surface, resulting in production of spherically weathered forms (Fig. 10-4).

thermal expansion and contraction

Most materials, including rocks, expand when heated and contract when cooled. Rocks heated by the sun in hot climates or by forest fires or by lightning, and then cooled at night or by sudden contact with water, may crack and disintegrate. However, laboratory tests of rocks alternately heated and cooled many times suggest that such changes have a relatively minor effect.

organic activity

Plants play a particularly active role in rock weathering. Given sufficient moisture and suitable growing conditions, plants are to be found almost any-

Fig. 10-2 Exfoliation of granitic rock in Yosemite Valley, California. The exfoliated shell near the hammer has been arched up by expansion to a height of 9 inches. (F. E. Matthes, U.S. Geological Survey.)

where at the earth's surface. As roots grow and extend along fractures or bedding planes, the rocks are pried apart (Fig. 10-5) in a manner somewhat similar to ice expanding in a crack. Overturning and uprooting of trees by windstorms further contributes to the breaking up of the rock in which the trees were rooted.

Animals also play a role in weathering, but with the possible exception of man, their effect is quite minor.

chemical weathering

solution

Minerals composing the common rocks of the earth's crust are soluble to varying degrees in water. The mineral halite (common salt) is readily dissolved in water, and calcite ($CaCO_3$) is soluble in water that contains CO_2. Thus

Fig. 10-3 Exfoliation dome. Half dome, Yosemite National Park, California. *(F. E. Matthes, U.S. Geological Survey.)*

limestone and marble, which are composed almost entirely of calcite, are particularly susceptible to solution (Fig. 10-6), and humid regions underlain by these rocks are affected by solution by groundwater.

The chemical relationships of calcite undergoing solution are given in the following formulas:

$$CaCO_3 + H_2O + CO_2 \rightarrow Ca(HCO_3)_2$$
$$\text{Calcite} + \text{water} + \text{carbon dioxide} \rightarrow \text{bicarbonate}$$

Solution is of course not restricted to calcite-bearing rocks. Even silica, which is relatively resistant to solution, may be dissolved in certain fluids.

Fig. 10-4   Spheroidal weathering in basalt near Dry Falls, Washington.

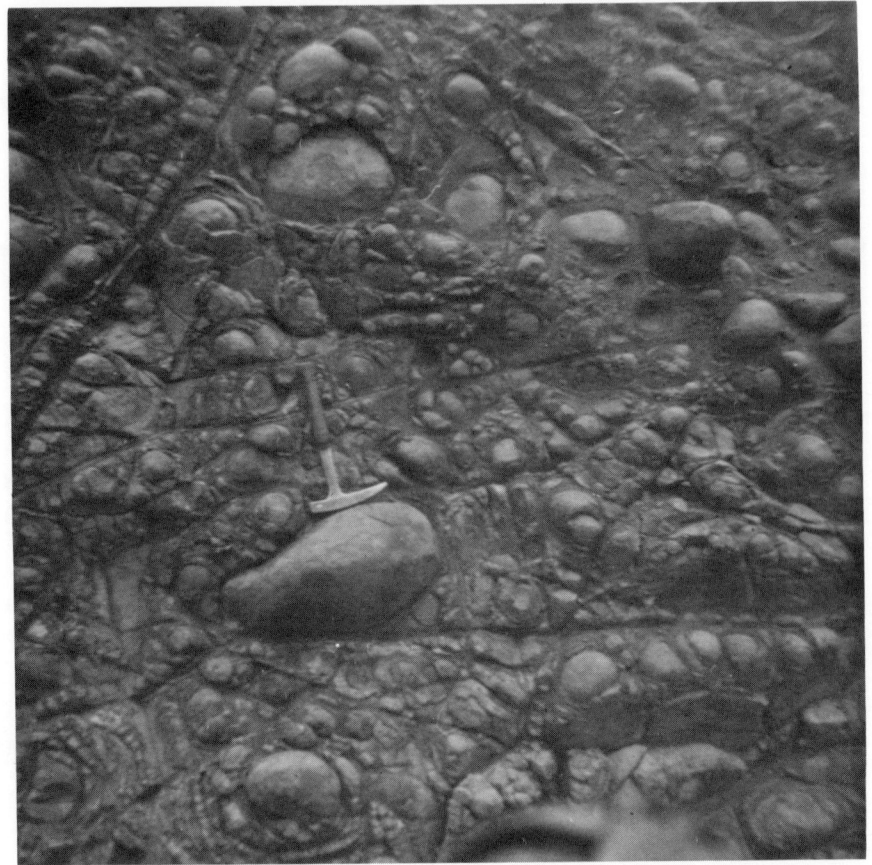

oxidation

The effect of leaving iron objects outside in the wet for any length of time is well known to almost everyone. The reddish-brown coating of rust which appears on the surface of the iron is just one of many types of chemical weathering involving the process of oxidation in which oxygen is added to the material.

$$4Fe + 3O_2 \rightarrow 2Fe_2O_3$$
Iron + oxygen $\rightarrow$ hematite

During chemical weathering ferrous silicates, such as the common rock-

Fig. 10-5 Effect of root growth on weathering of rocks. *(Wards Natural Science Establishment.)*

forming minerals pyroxene, amphibole, olivine, and biotite, form, among other compounds, hematite and hydrous iron oxides (limonite, goethite) by oxidation. As a result, rocks containing these minerals often turn brown, reddish brown, or yellow-brown during weathering. Oxidation takes place more rapidly in the presence of water, and water may enter into the composition of the new minerals, as when hydrous oxides are formed.

$$4Fe + 3H_2O + O_2 \quad \rightarrow \quad 2Fe_2O_3 \cdot 3H_2O$$
$$\text{Iron} + \text{water} + \text{oxygen} \quad \rightarrow \quad \text{hydrous iron oxide}$$

Fig. 10-6 Fretting of calcareous sandstone by solution of calcite, San Francisco Plateau, Arizona, (G. K. Gilbert, U.S. Geological Survey.)

carbonation

The process of chemical weathering in which minerals that contain calcium, sodium, potassium, or magnesium are changed to carbonates by the action of carbonic acid is known as *carbonation*. Nearly all water at or near the surface contains dissolved carbon dioxide which reacts with water to form carbonic acid (H_2CO_3). Water containing carbonic acid is capable of dissolving many compounds more readily than pure water.

Calcium, magnesium, sodium, and potassium carbonates formed by carbonation during chemical weathering are soluble in water, and are usually carried away in solution.

hydration

Hydration occurs in chemical reactions where water is added to a mineral or compound. It often occurs along with carbonation and other chemical processes during decomposition. A very important example of hydration and carbonation operating together is the weathering of feldspar, the most abundant mineral in the earth's crust, to clay. The chemical decomposition proceeds according to the reaction

$$2KAlSi_3O_8 + 2H_2O + CO_2 \rightarrow Al_2Si_2O_5(OH)_4 + K_2CO_3 + 4SiO_2$$

Orthoclase + water + carbon → kaolin + potassium + silica
feldspar   dioxide   (clay)   carbonate

The clay (kaolin) thus formed is a hydrous aluminum silicate produced as a result of hydration. The potassium carbonate is a result of carbonation, and is carried away in solution.

rate and character of weathering

In some areas mechanical weathering may be dominant; in other areas chemical weathering may be dominant. The character of the weathering, whether mechanical or chemical, depends on a number of factors, most of which relate to availability of water necessary for chemical weathering. The rate at which either type of weathering goes on depends also on some of the same factors.

parent material

The nature of the bedrock undergoing weathering may have a pronounced effect on the rate of weathering. Minerals of different composition have differing responses to weathering. Quartz is more resistant to both chemical and mechanical weathering than most other minerals. Hence rocks such as sandstone and quartzite, which consist largely of quartz, weather more slowly than other rock types. Rocks such as limestone and marble, which consist largely of calcite, are susceptible to solution, and hence weather rapidly in wet climates.

Physical features of rocks, such as jointing, bedding, and porosity, also affect rates of weathering, since these factors allow percolation of water through the rock.

A summary of the weathering characteristics of some common rock-forming minerals is given in the following table:

| | | decomposition products formed by: | | |
mineral	composition	oxidation	hydration	carbonation
orthoclase feldspar	$KAlSi_3O_8$		clay	potassium carbonate (soluble)
plagioclase feldspar	$NaAlSi_3O_8$, $CaAl_2Si_2O_8$		clay	calcium carbonate, sodium carbonate (soluble)
biotite, pyroxene, amphibole, olivine	Ca, Mg, Fe, K, Al silicates	Hematite, limonite	clay	calcium carbonate, magnesium carbonate, potassium carbonate (soluble)
quartz	SiO_2	little affected by chemical weathering; some silica in solution		
calcite	$CaCO_3$			calcium carbonate (soluble)

climate

Perhaps of greatest importance in determining the type of weathering which predominates in any area is the climate. The amount and distribution of precipitation annually are especially important since water promotes chemical weathering. In moist tropical regions where there is high rainfall and the precipitation is fairly evenly distributed throughout the year, so that water is in contact with weathering material for long periods of time, chemical weathering produces deep weathering zones. In contrast, arid regions receive little rainfall, and it often falls in a few showers during a year. There is thus limited time during which water remains in contact with rock at or near the surface.

Temperature also plays a role in weathering. Areas which undergo freezing during parts of the year are sites of mechanical weathering by frost action. Since not all parts of the world have freezing temperatures, the effects of freezing and thawing on rock weathering differ from place to place.

An example of the effect of climate on rock weathering is the difference in weathering of limestone in humid and in arid climates. In a humid climate solution causes limestone beds to weather more rapidly than interbedded sandstones or conglomerates, and valleys pitted with solution depressions are formed where the limestone beds outcrop at the surface. However, in an arid climate, where water is not as readily available for solution to take place, limestone beds commonly stand out as resistant ridges.

vegetation

Vegetation is to a large degree dependent on climate, and thus is not a completely independent factor. The ratio of the amount of water which soaks into the ground to the amount which runs off on the surface is affected by the amount and kind of vegetation. Heavily vegetated areas inhibit runoff, and much water soaks into the ground, where it promotes chemical weathering and soil development. In areas of bare slopes with sparse vegetation, there is little to impede runoff, and only a small percentage of precipitation soaks into the ground. In these areas chemical weathering is not nearly so effective.

Vegetation also is an important soil-forming factor. Retention of water by vegetation promotes chemical weathering and root action promotes mechanical breaking up of bedrock. Organic acids chemically attack the rock and organic material from decaying plants is added to the soil.

topography

The effects of topography on weathering primarily concern altitude and slope. In high mountains above the timberline, where there is little vegetation, freezing and thawing is an important weathering agent, and mechanical weathering predominates. At lower altitudes in the same region there is heavier forest cover, less freeze-thaw activity, and a greater amount of chemical weathering.

Where slopes are steep, precipitation runs off, and is not retained for chemical weathering. Vegetation is thus relatively sparse. On more gentle slopes, more moisture is retained in the weathering mantle, weathered products accumulate on the surface, and vegetation is heavier. Soils may develop on gentle slopes, whereas steep slopes or cliffs in the same areas expose only bare rock.

time

The length of time that weathering has been taking place in any region has a direct bearing on the amount of weathering that has occurred. If all the surface of the earth had been exposed to weathering for the same length of time, time would not be a factor. However, stripping away of the weathered mantle by glaciation during the Pleistocene, tectonic uplift of areas previously below sea level, and various other changes of the earth's surface result in differences in the length of time that weathering has been going on in different parts of the world.

Differences in the weathering of bedrock in areas covered by large continental glaciers during the last glacial stage and areas which lay beyond the glaciers are particularly striking. In many parts of the world rock polished and grooved by these glaciers remains relatively untouched by weathering, whereas uninterrupted weathering in regions beyond the ice has produced deep soils.

effects of weathering

topographic expression

Differential weathering of rocks of unequal resistance etches out the weaker rocks (Fig. 10-7) and develops topographic relief. In humid areas in which tilted sedimentary rocks of differing composition underlie the surface, shales and limestones are etched out to form valleys or lowlands, whereas more resistant sandstones and conglomerates remain as ridges.

Differences in topography are typical of different climates. Generally, the topography in arid climates is characterized by angularity and sharp breaks in slope, steep cliffs with accumulations of loose debris at the base, and extensive exposure of bedrock. Topography in humid climates is characterized by smooth, rolling hills, with gently rounded slopes covered with a deep accumulation of weathered material and little exposure of bedrock. These differences are brought about largely because the effects of mechanical weathering are dominant over chemical weathering in arid climates, and effects of chemical weathering are dominant in humid climates.

soils

Perhaps the most important effect of weathering as far as man is concerned is the development of soils. Plants depend on soil for growth, and man depends directly or indirectly on plants for food. The nature of soils changes with the factors which control the rate and character of weathering. In addition, the composition of a soil generally changes with depth in a systematic manner.

Immature soils may reflect the composition of the material from which they were derived. For example, a soil developed on granite may differ from a soil developed on limestone. Climate appears to be a most important factor in development of soils. Russian scientists advanced the idea that similar soils are developed in similar climates more or less independently of the nature of parent

Fig. 10-7 Natural bridge created by differential weathering and erosion, Arches National Monument, Utah.

material, and that soils developed on the same parent material differ if the climate varies from place to place. Soils on a steep slope are usually different from soil on flat ground in the same area. Soil formed under a forest cover is different from soil formed under a cover of grass. Time also affects soil formation. Soils formed very recently show differences from soils developed over a long period of time.

A mature soil is one in which a series of zones having well-defined characteristics has been produced by weathering processes. In such a soil there is a succession of distinctive zones from the surface downward to the parent material. The lowermost zone, known as the C horizon, consists of partially weathered bedrock grading downward into unaltered parent material. In the C horizon the composition of the original material is still identifiable, although changes have occurred. The B horizon lies above the C horizon and consists of clayey material and oxides which accumulated by downward percolation

Fig. 10-8 Balanced rocks produced by differential weathering and erosion of beds of differing resistance. (Wards Natural Science Establishment.)

from the overlying layer. The B horizon is commonly referred to as the *zone of accumulation*. The material in the B horizon has undergone more weathering than in the underlying C horizon, and the nature of the parent material is difficult to recognize. The top layer in a soil profile is the A horizon, which consists of weathered and leached material and organic matter. Percolation of water downward through the soil removes certain elements from the A horizon and carries them to the B horizon. Hence the A horizon is sometimes known as the *zone of leaching*. Organic material is added to the A horizon by decaying leaves and other plant material, giving the A horizon a darker color. A soil profile showing relationships of soil horizons is shown in Fig. 10-9.

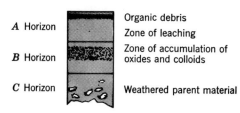

A Horizon	Organic debris
	Zone of leaching
B Horizon	Zone of accumulation of oxides and colloids
C Horizon	Weathered parent material

Fig. 10-9 Typical mature soil profile.

pedalfers

Pedalfers are soils in which iron and aluminum oxides and clay have accumulated in the B horizon. They are usually found in humid, temperate climates under forest vegetation, where downward percolation of precipitation causes strong leaching of the A horizon. Sodium, calcium, magnesium, and potassium are carried from the A horizon downward through the B horizon and carried away in solution by groundwater. Less soluble iron and aluminum oxides and clay accumulate in the B horizon. Among the types of pedalfer soils are *podsols*, characterized by intense leaching of the A horizon to a light-gray color and clayey B horizons rich in iron oxides.

latosols

Latasols are formed in humid tropical climates where large amounts of water percolate through the upper soil zones, removing calcium, sodium, magnesium, and potassium in solution, leaving behind concentrations of iron and aluminum oxides which are not as soluble. The intense leaching of silica, in addition to the bases, may leave only the most insoluble elements behind, notably aluminum oxide, i.e., bauxite (Al_3O_4). Iron oxide often results in brick-red colors for latosols. The process of soil formation under these conditions is referred to as *laterization*.

pedocals

Pedocals are soils which contain accumulations of calcium carbonate. They form in regions of low rainfall and relatively high temperature, often where vegetation consists only of grass or brush. Water in the upper portion of pedocals evaporates, and thus carbonates are not removed downward, as in the pedalfers, but accumulate in the soil. In arid or semiarid climates, where evaporation exceeds precipitation, soil moisture may move upward toward the surface, where evaporation leaves behind calcium carbonate in the form of crusts around pebbles or as layers just below the surface. Deposits of calcium carbonate formed in this manner are known as *caliche.*

references

Blackwelder, Eliot: 1925. Exfoliation as a phase of rock weathering, *Jour. Geology,* vol. 33, pp. 793-806.

Chapman, R. W., and M. A. Greenfield: 1949. Spheriodal weathering of igneous rocks, *Am. Jour. Sci.,* vol. 247, pp. 407-429.

Goldich, S. S.: 1938. A study in rock weathering, *Jour. Geology,* vol. 46, pp. 17-58.

Jenny, Hans: 1941. Factors of Soil Formation, McGraw-Hill Book Company, New York.

Lyon, T. L., H. O. Buckman, and N. C. Grady: 1952. The Nature and Properties of Soils, The Macmillan Company, New York.

Matthes, F. E.: 1930. Geologic history of the Yosemite Valley, *U.S. Geol. Survey Prof. Paper* 160.

Millar, C. E., L. M. Turk, and H. D. Foth: 1958. Fundamentals of Soil Science, John Wiley & Sons, Inc., New York.

Reiche, Parry: 1950. A Survey of Weathering Processes and Products, The University of New Mexico Press, Albuquerque, N. Mex.

Tabor, Stephen: 1930. The mechanics of frost heaving, *Jour. Geol.,* vol. 38, pp. 303-317.

(11)

mass movement

On the evening of Aug. 17, 1959, campgrounds along the Madison River west of Yellowstone National Park were filled with campers. At 1:37 P.M. a severe earthquake jarred the region, setting off a very large landslide and many smaller ones. One family, camped near Rock Creek campground in a small house trailer, was awakened by a loud noise. As they left the trailer a great blast of air struck them. The mother of the family saw her husband "grasp a tree for support, then saw him lifted off his feet by the air blast and 'strung out like a flag' before he let go. Before she lost consciousness she saw one of her children blown past her and a car tumbling over and over" (Witkind, 1964). The cause of the tremendous blast of air was a gigantic landslide from the valley side (Fig. 11 - 1). Approximately 40 million cubic yards of rock and debris slid into the Madison Canyon, burying a mile of the Madison River and the adjacent highway to depths of 100 to 200 feet and causing 26 fatalities. Although the earthquake set off the landslide, the cause of the slide was related to the geologic conditions along the valley side.

Spectacular landslides such as the Madison slide provide ample evidence

Turtle Mountain slide, Frank, Alberta. *(National Air Photo Library, Surveys and Mapping Branch, Department of Energy, Mines, and Resources.)*

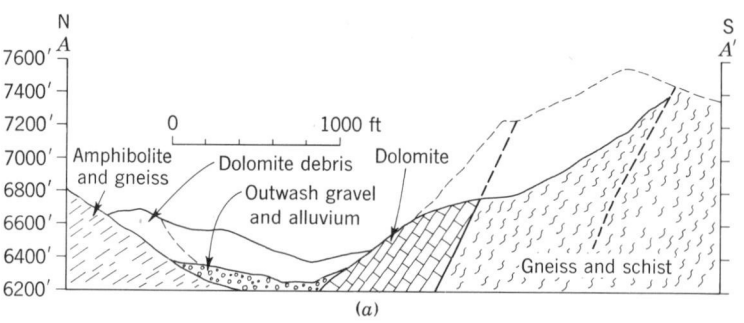

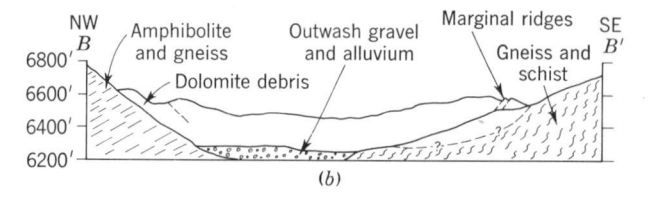

Fig. 11-1 Madison landslide west of Yellowstone National Park, Aug. 17, 1959. *(a)*
Madison landslide viewed from upstream. The lake behind the slide resulted from
damming of the Madison River by the slide. *(J. R. Stacy, U.S. Geological Survey.) (b)*
Geologic relationships of the Madison landslide. *(Witkind, 1964.)*

of the effect of downslope movement of material, but these landslides are not the only types of mass movement. Other, less obvious types of movement are going on all the time.

gradual movements

creep

The slow, imperceptible downslope movement of finely broken up rock material is known as *creep*. Weathered material in a shallow zone near the surface of even gentle slopes migrates slowly and continuously downslope in response to gravity and various surface processes. The effects are readily noticeable in the tilting of telephone poles, fence posts, trees, and other objects which were formerly vertical. Tree trunks often show a bending in their lower portion as a result of a combination of creep, which tends to tilt the tree, and the tendency of the tree to grow vertically (Fig. 11-2). Steeply dipping rocks

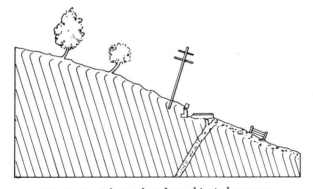

Fig. 11-2   Tilting of surface objects by creep.

may bend sharply downslope in the zone a few feet down from the surface as a result of creep (Fig. 11-3), and distinctive pebbles of weathered rock may often be found some distance downslope from their source. Sometimes weathered pebbles in a zone of creep may be stretched out into lens-shaped particles elongated in a downslope direction. This "stretching" effect is caused by shear developed in the weathering mantle by the component of gravity parallel to the surface.

Movement by creep is the sum of a great many minute displacements of particles, the great majority of which are in a downslope direction because of the effect of gravity. The rate of creep varies considerably from place to place, depending on differences in angle of slope, rate and type of weathering, moisture content in the soil, and a number of other factors.

In regions where cold winters cause repeated freezing and thawing, moisture in the soil becomes frozen and expands. The expansion produces a slight uplift of the land surface, referred to as *frost heaving*, and particles on the surface are moved outward at right angles to the original slope. When thawing takes place individual particles drop vertically under the influence of gravity, rather than returning to their original positions (Fig. 11-4). The result is a component of movement in the downslope direction.

Fig. 11-3 Bending of nearly vertical beds of shale by surface creep, Maryland. *(G. W. Stose, U.S. Geological Survey.)*

The growth of roots, burrowing of animals, uprooting of vegetation, and other similar processes also contribute to the total movement by creep.

A special type of creep occurs in some areas of high latitude or high altitude where the ground freezes to fairly great depths. During warm intervals the upper part of the ground thaws, but the still-frozen ground beneath prevents water from migrating downward, and the thawed zone becomes heavily saturated. Saturation continues until the water-soaked debris near the surface begins to flow slowly downslope. The process of slow flowage under such conditions is known as *solifluction*.

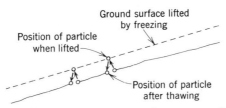

Fig. 11-4   Effect of frost heaving on creep. Particles are moved outward at right angles to the slope during frost heave but are let down with a vertical component during thawing, resulting in a progressive downslope movement.

rock glaciers

Glacierlike tongues of angular rock debris are sometimes found in mountainous areas (Fig. 11-5). Because of the similarity in form to a true glacier, they are called *rock glaciers*, although they are not composed of glacial ice. Typically, they consist of coarse, angular debris, as much as 100 to 200 feet thick, with ice filling the voids between rock fragments.

The origin and mechanism of movement have been studied by many geologists since the early 1900s, but there appears to be no general agreement among those who have worked with rock glaciers. Nearly all agree that (1) they are found in valleys formerly occupied by alpine glaciers; (2) interstitial ice between fragments plays a role in movement of the rock glaciers; and (3) many are moving at the present time. Problems in their interpretation center around whether or not rock glaciers are remnants of former glaciers and related in origin to them.

Capps (1910), who studied rock glaciers in the Wrangell Mountains of Alaska, concluded that there were all gradations from true glaciers to rock glaciers, and that the rock glaciers were remnants of debris-charged glaciers. A study of rock glaciers in the Alaska Range led Wahrhaftig and Cox (1959) to the conclusion "that rock glaciers move as result of the flow of interstitial ice and that they require for their formation steep cliffs, a near-glacial climate cold enough for the ground to be perennially frozen, and bedrock that is broken by frost action into coarse blocky debris with large interconnected voids." They considered that the interstitial ice in rock glaciers may have originated "(1) from compacted snow, as does most glacier ice; (2) by the freezing of water derived from melting snow or rain; or (3) from ground water that rises beneath the talus and freezes on contact with cold air." A tunnel which was

Fig. 11 - 5   Rock glaciers in the Copper River region, Alaska. *(U.S. Geological Survey.)*

driven the length of a rock glacier in the San Juan Mountains of Colorado encountered a few feet of rock debris at the front, then 300 feet of rock debris cemented with ice, 100 feet of clean ice, and finally, solid bedrock at the head of the rock glacier (Brown, 1925).

rapid movements

landslides

A landslide is the downward sliding or falling of a mass of rock, weathered material, or a mixture of the two. The difference between a landslide and creep is that, in a landslide, movement of a well-defined mass takes place in a relatively short period of time along a well-defined surface of failure, whereas creep involves slow, more or less continous movement on slopes over a large area, with no sharp boundary between moving and stationary material (Terzaghi, 1950).

The causes of landslides are many, but most involve lack of support beneath or in front of a slip surface. The lack of support may be brought about by (1) external causes, which produce an increase in shearing stresses along potential surfaces of sliding, or (2) internal causes, such as loss of cohesion in material, which reduces shearing resistance (Terzaghi, 1950). Some examples of external causes of landslides include (1) steepening of slopes by erosion at the base, (2) earthquake vibrations, and (3) overloading of the top of the slope. Examples of internal causes which reduce shearing resistance are (1) increase of water pressure in pore spaces, (2) displacement of air in pore spaces by water, (3) dissolving of cementing material, and (4) liquefaction of water-saturated clay (Terzaghi, 1950).

Inclination of the bedding of sedimentary rocks into a valley produces a potential surface of failure which is responsible for many slides. A classic example of this type of failure is the Gros Ventre slide (Fig. 11-6), south of Yellowstone National Park, Wyoming, in 1925. Figure 11-7 shows the geologic relationships which led to the slide. The resistant Tensleep sandstone was underlain by shale, both dipping about 20° toward the valley floor. The Gros Ventre River had eroded through the sandstone layer, leaving the front of the sandstone unsupported, except for friction at the contact between the sandstone and underlying shale. After a period of very heavy rains which wetted the shale and reduced its shearing resistance, a large mass of sandstone slid parallel to the contact, carrying some 50 million cubic yards of rock debris into the valley and producing a dam 225 feet high which impounded a lake some 4 to 5 miles long. The momentum of the slide carried debris 400 feet up on the opposite side of the valley. The slide was witnessed by several ranchers in the valley, one of whom "whipped his horse and dashed up-river to his ranch, dismounting to open his gate. At that moment the slide shot past him at the rate of 50 miles an hour, by his estimation, and went through his fence not 30 feet from where he was standing. His ranch was completely buried, but his new . . . ranch house still stood. It was to float away on the newly-formed lake a few days later." (Hayden, 1956, p. 20)

Fig. 11-6 Gros Ventre landslide, Wyoming. *(Austin Post, University of Washington.)*

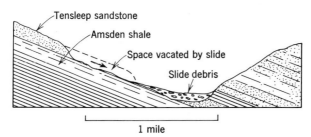

Fig. 11-7 Geologic cross section of the Gros Ventre
slide, Wyoming. (After Alden, 1928; Keefer and Love,
1956.)

Joint planes dipping into a valley may develop slide conditions similar to
those produced by bedding planes. An example of this type of slide is the
Turtle Mountain slide, which destroyed much of the town of Frank, Alberta,
in 1903. Figure 11-8 shows the geologic relationships of the area. Joints cut
across the limestone beds and act as sliding surfaces. Mining of coal near
the base of the mountain probably also contributed to the removal of support.

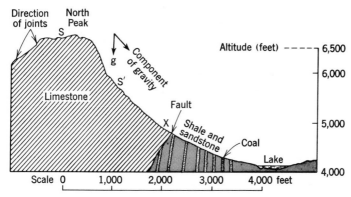

Fig. 11-8 Geologic cross section of the Turtle Mountain slide,
Alberta. (Geological Survey Canada Mem. 27, 1912.)

When shearing stresses along the surface of failure exceeded the shearing resistance, approximately 40 million cubic yards of rock slid down the mountain.

Landslides may also be produced in relatively homogeneous material as
a result of increased shearing stress produced by undercutting of slopes by
streams or waves or by reduction in shearing resistance brought about by
loss of cohesion, often because of water saturation.

rockfall

Rockfalls consist of individual blocks of rock which have broken loose from bedrock, falling through the air to the slope below. Continued rockfall produces *talus,* an accumulation of loose fragments at the base of steep slopes (Fig. 11-9). Talus is normally at or near the angle of repose, the maximum slope angle assumed by loose material, which ranges from about 25 to 35°.

Talus slopes are found in great abundance in mountainous regions where steep cliffs are common, vegetation is sparse, and active freezing and thaw-

Fig. 11-9   Talus at the base of cliffs, Glacier National Park, Montana. *(Wards Natural Science Establishment.)*

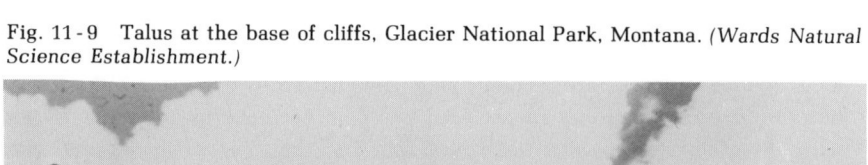

ing supply much broken rock material. Near the base of a talus slope, weathering of the loose blocks may have proceeded to the point where soils appear on the talus and further downslope movement is by creep (Fig. 11-10).

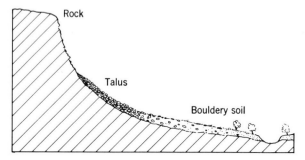

Fig. 11-10 Cross section of a talus slope.

slump

Slump is the downward and outward sliding of a mass of rock or unconsolidated material along a curved shearing plane, usually with some backward rotation of the land surface (Fig. 11-11). Near the base of the slump flowage may take place, and a bulge in the ground surface may appear at the toe of

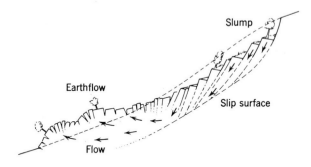

Fig. 11-11 Cross section of a slope undergoing slump.

the slump (Fig. 11-12). This particular type of slope failure is common in oversteepened slopes developed in more or less homogeneous material, especially clays, which have become saturated with water. Areas of potential slumping may sometimes be recognized by the development of cracks in the ground and backward tilting of the ground surface near the head of the incipient slump block.

During movement of a slump block, secondary slumps may develop and produce a stair-step-like series of parallel slump blocks.

mudflow

Flowage of a heterogeneous mixture of rock debris saturated with water produces a *mudflow* which often begins as a landslide or slump along the

Fig. 11-12 Slump and flowage on a hillslope in Berkeley Hills, California. Material excavated by slumping appears as a bulge at the toe of the slump. *(G. K. Gilbert, U.S. Geological Survey.)*

upper parts of hillslopes. When the material becomes thoroughly mixed with water, it flows as a viscous mass. The rate of the movement depends in part on the angle of slope and in part on the amount of water mixed with the mud. There are all gradations, from fairly stiff mud to what is essentially muddy water. Because of their relatively high density, mudflows are capable of transporting boulders several feet in diameter.

The size of mudflows varies from a few inches to several miles. A flow of unusual size occurred on the flanks of Mt. Rainier in Washington about 4,800 years ago. The flow originated above the Emmons Glacier and flowed for about

Fig. 11-13 Reed Terrace above Grand Coulee Dam before slide on Apr. 10, 1952. *(F. O. Jones, U.S. Geological Survey.)*

55 miles, terminating in the Puget Lowland (Crandell and Waldron, 1956). This mudflow varied in thickness from 0 to 75 feet and covered an area of about 65 square miles in the lowland, involving an estimated 1.3 billion cubic yards of material (Fig. 11-17). It is thought that the mudflow originated as a very large landslide, partly induced by stream eruptions on the east flank of Mt. Rainier, and as the material came down the White River Valley, it became mixed with sufficient water to produce a huge mudflow.

Mudflows are particularly common in arid regions where vegetation is scarce and heavy rain showers occur between periods of dryness. Where

Fig. 11-14 Reed Terrace after slide on Apr. 10, 1952. Compare with Fig. 11-13. (F. O. Jones, U.S. Geological Survey.)

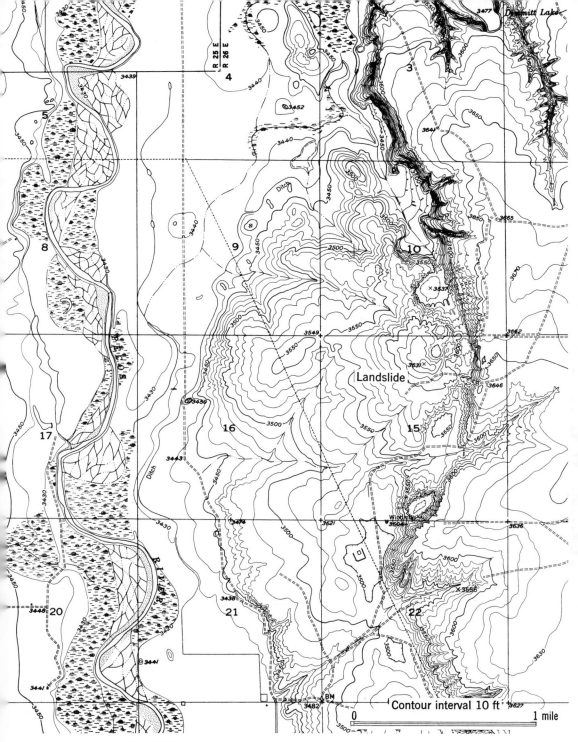

Fig. 11-15 Map of landslide topography. *(Bottomless Lakes quadrangle, New Mexico,
U.S. Geological Survey.)*

weathered material becomes saturated by heavy rains, it moves down even moderate slopes, picking up still more material. Where it leaves the confines of stream valleys, it spreads out in lobes.

A special type of mudflow occurs occasionally on slopes covered with volcanic ash and other types of pyroclastic material, when the material becomes saturated with water after heavy rains or melting of snow. A flow of volcanic debris known as a *lahar* is then produced.

Fig. 11-16 Landslide area of Point Fermin, California, 1954. *(Wards Natural Science Establishment.)*

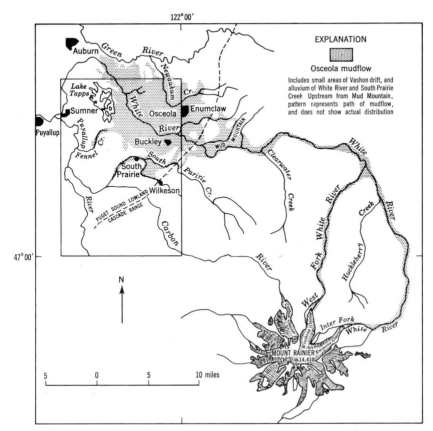

Fig. 11-17 General distribution map of the Osceola Mudflow, Washington.
(After Crandell and Waldron, 1956.)

earthflow

A slow flow of unconsolidated materials saturated with water but not possessing enough water to make a mudflow results in an *earthflow*. Because of its
lower mobility, its movement is generally slower and more erratic than that of
a mudflow. During periods of saturation, flowage may occur on fairly low
gradients, but upon loss of water, the flow may cease until the material again
becomes saturated.

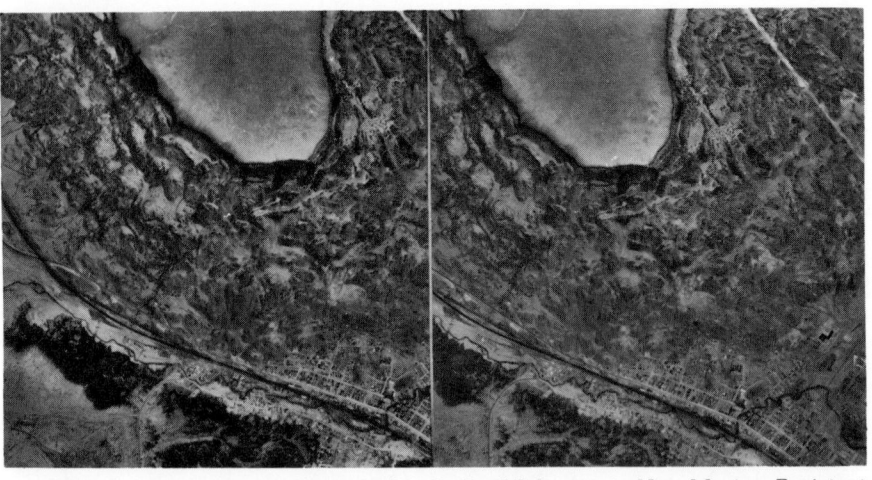

Fig. 11-18 Stereoscopic air photographs of a landslide area in New Mexico. Resistant basalt caps the mesa, and is underlain by relatively weak shale and sandstone in which the sliding occurs. Note the hummocky topography typical of landslide areas. *(U.S. Geological Survey.)*

references

Alden, W. C.: 1928. Landslide and flood at Gros Ventre, Wyoming, *Amer. Inst. Mining and Metall. Engineers Trans.*, vol. 76, pp. 347-361.

Anderson, J. G.: 1906. Solifluction, a component of subaerial denudation, *Jour. Geology*, vol. 14, pp. 91-112.

Blackwelder, Eliot: 1928. Mudflow as a geologic agent in semi-arid mountains, *Geol. Soc. America Bull.*, vol. 39, pp. 465-480.

Brown, W. H.: 1925. A probable fossil glacier, *Jour. Geology*, vol. 33, pp. 464-466.

Capps, S. R.: 1910. Rock glaciers in Alaska, *Jour. Geology*, vol. 18, 359-375.

Crandell, D. R., and H. H. Waldron: 1956. Recent mudflow of exceptional dimensions from Mt. Rainier, Washington, *Am. Jour. Sci.*, vol. 254, pp. 349-362.

Flint, R. F., and C. S. Denny: 1958. Quarternary geology of Boulder Mountain, Aquarius Plateau, Utah, *U.S. Geol. Survey Bull.* 1061-D, pp. 103-164.

Hadley, J. B.: 1964. Landslides and related phenomena accompanying the Hebgen Lake earthquake of Aug. 17, 1959, *U.S. Geol. Survey Prof. Paper* 435-K, pp. 107-138.

Hayden, E. W.: 1956. The Gros Ventre Slide (1925) and the Kelly Flood (1927), Wyoming Geological Association 11th Annual Field Conference Guidebook, pp. 20-22.

Holmes, C. D.: 1955. Geomorphic development in humid and arid regions—a synthesis, *Am. Jour. Sci.*, vol 253, pp. 377-390.

Howe, Ernest: 1909. Landslides in the San Juan Mountains, Colorado, *U.S. Geol. Survey Prof. Paper* 67.

Ives, R. L.: 1940. Rock glaciers in the Colorado Front Range, *Geol. Soc. America Bull.*, vol. 51, pp. 1271-1294.

Jones, F. O., D. R. Embody, and W. L. Peterson: 1961. Landslides along the Columbia River Valley, Northeastern Washington, *U.S. Geol. Survey Prof. Paper* 367.

Keefer, W. R., and J. D. Love: 1956. Landslides along the Gros Ventre River, Teton County, Wyoming, Wyoming Geological Association 11th Annual Field Conference Guidebook, pp. 24-28.

Kesseli, J. E.: 1941. Rock streams in the Sierra Nevada, *Geog. Rev.*, vol. 31, pp. 203-227.

Miller, D. J.: 1960. Giant waves in Lituya Bay, Alaska, *U.S. Geol. Survey Prof. Paper* 354-C.

Schumm, S. A.: 1956. Evolution of drainage basins and slopes in badlands at Perth Amboy, New Jersey, *Geol. Soc. America Bull.*, vol. 67, pp. 597-646.

———: 1956. The role of creep and rainwash on the retreat of badland slopes, *Am. Jour. Sci.*, vol. 254, pp. 693-706.

———: 1962. Erosion on miniature pediments in Badlands National Monument, South Dakota, *Geol. Soc. America Bull.*, vol. 73, pp. 719-724.

Sharpe, C. F. S.: 1938. Landslides and Related Phenomena, Columbia University Press, New York.

Terzaghi, Karl: 1950. Mechanism of landslides, in Application of Geology to Engineering Practice, Geological Society of America, Berkeley volume, pp. 83-121.

Varnes, D. J.: 1958. Landslide types and processes, chap. 3 in Landslides and Engineering Practice, Highway Research Board Special Report, Washington, D.C.

Wahrhaftig, Clyde, and Allan Cox: 1959. Rock glaciers in the Alaska Range, *Geol. Soc. America Bull.*, vol. 70, pp. 383 - 436.

Witkind, I. J.: 1964. The Hebgen Lake, Montana, earthquake of Aug. 17, 1959, Events on the night of Aug. 17, 1959: the human story, *U.S. Geol. Survey Prof. Paper* 435 - A, pp. 1 - 4.

Contour interval 20 ft

0   1   2

Miles

evolution of slopes

The form of the profile for any slope depends on the rate at which weathering produces rock debris and the rate at which the weathered material is moved. The most important type of mass movement of weathered material on the great majority of hillslopes is creep, aided by transportation by surface runoff. According to Gilbert (1909):

A layer of unconsolidated material resting on a gentle slope holds its position (1) because the particles are arranged so as to support one another, and (2) because one particle cannot slide on another without developing friction. Whatever diminishes friction promotes flow. Whatever disturbs the arrangement of particles, permitting any motion among them, also promotes flow, because gravity is a factor in the rearrangement and its tendency is down the slope.

Smooth hillslopes and valley-side slopes are graded so that the position of any point in the profile depends at any moment on that of all others, and any changes which occur have repercussions both up- and downslope, much in the same manner as in a graded stream (Cotton, 1952).

Divide areas composed of unconsolidated material are typically convex

Hillslopes near Irontown, Missouri. *(U.S. Geological Survey.)*

upward in profile, in contrast to the concave-upward profiles of streams. Several factors control the upper convex profile of a slope: (1) Weathering of bedrock produces debris which becomes reduced to finer and finer sizes downslope. Chemical weathering becomes more effective where the layer of weathered debris covering the bedrock is thick, partly because retention of water in this material promotes chemical weathering processes. The increasing fineness of material downslope decreases permeability so that surface wash becomes an important factor in producing the lower concave part of a slope profile. (2) Gravity manifests itself through creep, the rate of movement being at a maximum near the surface and decreasing to zero at the base of the weathered mantle.

The form of the profile which is developed on a slope then depends on the rate at which debris is produced and the rate at which it is moved by creep. In Fig. 12-1 the material creeping into any given zone, as from A into B, is equal to the amount of material moving by creep at A plus the amount added between A and B by weathering and robbing of bedrock. Thus creep must accelerate downslope in order to transport an increasing amount of material from A to B to C, etc., or else the weathering mantle between these points must thicken.

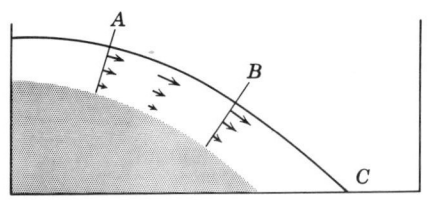

Fig. 12-1 Zone of creep on a convex hillslope.

At the lower part of a hillslope the cumulative amount of runoff from the upper slopes produces a concave-upward profile. Whereas creep dominates the upper convex part of the profile, running water dominates the lower part. Baulig (1957) states:

> In terrains that are approximately homogeneous (at least as regards the ensemble of rocks) a hillslope that is fully mature exhibits a compound curve, convex on the upper and concave on the lower part of the slope. The convex segment is shaped by creep and by rain wash that is as yet so feeble that it is likely to be stopped by the smallest obstacle and is thus described as "unconcentrated"; it remains discontinuous and diffused. Steepening of the slope downhill is a result of the increasing effect of gravity on an increasing mass of material with small mobility. Concavity of slope, on the other hand, results mainly from the activity of a more abundant runoff

now gathered into rills which pick up and carry off finely divided particles from materials that are becoming progressively less permeable and more easily removed. . . . There is, therefore, a tendency to develop the concave longitudinal profile characteristic of running water, and the incipient rill channels or gutters impose their own profile, howbeit with reduced curvature, on the lower slope as a whole, as they gather in from it the runoff and carry away the debris. The relative extent of convexity and concavity varies with humidity of climate and permeability of the rock, or, more correctly, of the debris furnished by the rock (e.g., granite sand), and this proves that the processes dominant on the upper slopes decrease in their relative value downhill, and vice versa. (p. 917)

Schumm (1956) found two distinctive types of topography developed in the badlands of South Dakota. On areas of permeable material which tended to impede runoff, broadly rounded, convex slopes were formed by the action of creep, whereas on less permeable material where runoff was higher, concave slopes were formed by running water.

One of the most controversial topics with regard to the evolution of slopes is what happens to the slope profile with time. Since the publication of Davis's classic papers, the concept of a systematic progression of landforms developed by erosion of landmasses has served as the basis for interpretation of surface features. Basically, Davis's ideas involve evolution of the topography of a landmass elevated above sea level, beginning with the initial stages of stream incision, and progressing toward the stage when the landmass has been eroded to a nearly flat plain.

In the 1920s Davis's ideas were challenged by Penck, a German geomorphologist who assumed that the slopes of valley sides were determined by relative rates of uplift of a landmass. However, few geologists accepted this idea. Johnson (1940), commenting on it, wrote:

Penck's conception that slope profiles are convex, plane, or concave, according to the circumstances of the uplifting action, is in my judgment one of the most fantastic errors ever introduced into geomorphology. One may in the same region, which has experienced the same history throughout (as for example in the Bad Lands of South Dakota or in certain portions of the Black Hills), photograph both convex and concave slopes within a short distance of each other.

However, a few of Penck's concepts still persist. Much discussion in geological literature has been devoted to the manner in which slopes retreat during erosion of an elevated landmass, particularly around the point of whether slope angles gradually decrease with time as erosion takes place or tend to retreat parallel to themselves. Two schools of thought developed, largely as a result of the contrasting views of Davis (1909) and Penck (1924). Whereas Davis maintained that slopes "lie down" as a result of long-continued erosion, Penck and others concluded that as slopes retreat, a constant angle is maintained. Figure 12-2 shows diagrammatically successive stages of slope retreat according to each concept.

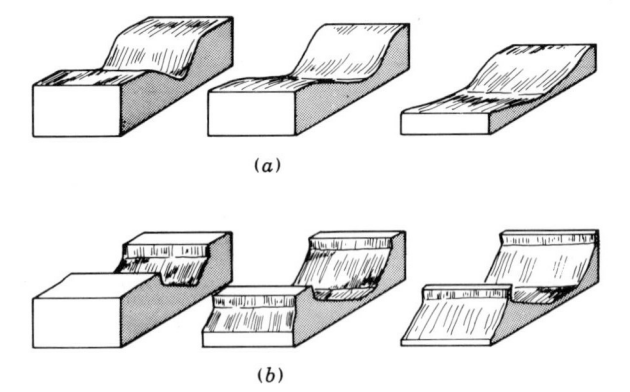

(a)

(b)

Fig. 12-2 Successive stages in the evolution of slopes (a) according to Davis and others; (b) according to Penck and others.

Penck considered that slopes developed an upper steep slope and a lower, more gentle slope which intersected each other at a distinct angle. Meyerhoff (1940) referred to the upper slopes as gravity slopes, immature, ungraded slopes controlled primarily by the effects of gravity, and referred to the lower slopes as wash slopes, mature, graded slopes controlled by the effects of running water. As erosion acts on a hillside, the slopes were thought to maintain a constant angle by parallel retreat of the existing slopes. Examples of this kind of slope retreat are difficult to find in humid climates, but are more common in arid regions under certain conditions. If this kind of slope retreat tends to operate universally on all slopes under conditions which are controlled by geomorphic processes, the same kind of slope retreat should occur in rocks of homogeneous composition. However, it appears that this kind of parallel retreat of slopes occurs only under certain conditions. These include (1) presence of a resistant, flat-lying bed overlying a less resistant layer so that sapping at the base of the resistant layer maintains a steep cliff or (2) undercutting of slopes by streams or wave action (Johnson, 1940). It thus appears that development of gravity slopes and wash slopes with parallel slope retreat at constant slope angles occurs only under special and temporary conditions.

Figure 12-3 suggests the evolution of slopes controlled by a resistant cap rock overlying less resistant material. Sapping at the base of the resistant layer maintains a steep gravity slope, and a constant wash slope is maintained to transport the material shed from above. However, once the resistant cap rock is consumed by erosion, the source of material which keeps the wash slope graded at the same angle is cut off. With coarse debris no longer being shed from above, regrading of the wash slope occurs, and a lower slope is developed on the less resistant material. Thus, although parallel retreat occurs for a time, eventually the slope "lies down."

Fig. 12-3 Evolution of slopes controlled by resistant cap rock.

Another kind of parallel retreat not implied in the Penck concepts may occur under certain conditions. Starting from an initial condition of downcutting on homogeneous rock, a slope of equilibrium will be attained, the slope of which depends on the rates of weathering, wash, and creep. Graded slopes are essentially equilibrium systems such that, once attained, a slope of equilibrium will maintain the same angle as long as no changes occur in conditions under which the slope became graded. The angle of slope once developed under a particular set of variables will remain the same so long as no change in the variables occurs. Two important kinds of changes may occur: (1) random changes related to changes in climate, base level, etc., which are not systematic, and (2) systematic changes such as those related to progressive changes in rates of downcutting as base level is approached. Since it is not possible for the variables of equilibrium under which any particular slope is developed to remain indefinitely the same, it must be concluded that parallel slope retreat is maintained only under certain conditions, for relatively short periods of geologic time. As a stream valley is lowered closer and closer to base level, the rate of downcutting must decrease and eventually approach zero, so that the initial equilibrium conditions progressively change and the angle of slope of the valley sides progressively decreases with time. The end result, downwasting of the slopes to a surface of low relief, follows as an inescapable conclusion.

In the last twenty-five years an increasing interest in quantitative approaches has taken place (Horton, 1945; Strahler, 1950, 1956; Melton, 1960; Rapp, 1960; Schumm, 1966). Many of these studies deal with statistical analyses of relationships involving slope angles, stream ordering, drainage density, and other measurable parameters. Of considerable interest are studies by Schumm of erosion at Badlands National Monument, South Dakota, where measurements of slope changes have been made over a number of years. From these studies Schumm (1966) concludes:

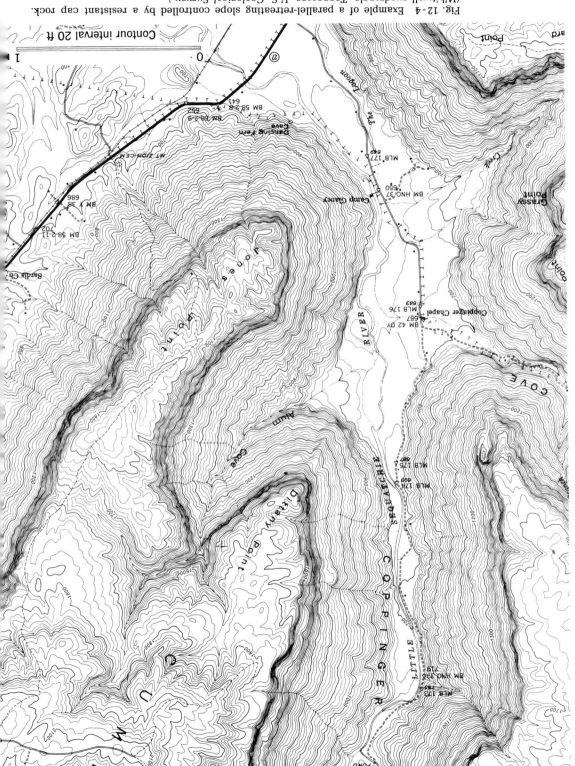

Fig. 12-4 Example of a parallel-retreating slope controlled by a resistant cap rock.
(Whitwell quadrangle, Tennessee, U.S. Geological Survey.)

Contour interval 20 ft

0   1

... Steep slopes composed of siltstone of low permeability also retreat in parallel planes under the action of rainwash, whereas in the same area hillslopes that are composed of a shale that weathers to yield a permeable slope retreat by downwearing under the action of creep.

These studies demonstrate that some hillslopes retreat in parallel fashion under the action of rainwash, but that other hillslopes will decline in angle with retreat under the action of creep. However, the removal of the hillslope sediment from the base of the slope appears to be a prerequisite for its parallel retreat. If colluvium accumulates at the slope base, this material, in effect, anchors the base of the slope, and a decline of the hillslope angle will occur. (p. 101)

Although there is presently a general qualitative understanding of slope processes, much quantitative work on relationships between relief, lithology, climate, and other factors remains to be accomplished, and detailed studies of slope changes with time need to be made, before a thorough understanding of this complex subject can be reached.

references

Baulig, Henri: 1957. Peneplains and pediplains (transl. by C. A. Cotton), *Geol. Soc. America Bull.,* vol. 68, pp. 913-930.

Bryan, Kirk: 1940. The retreat of slopes, *Assoc. Am. Geog. Annals.,* vol. 30, pp. 254-268.

Cotton, C. A.: 1952. Erosional grading of convex and concave slopes, *Geog. Jour.,* vol. 118.

Davis, W. M.: 1892. The convex profiles of bad-land divides, *Science,* vol. 20, p. 245.

————— : 1909. Geographical Essays, Ginn and Company, Boston (reprinted 1954 by Dover Publications, Inc., New York).

Gilbert, G. K.: 1909. The convexity of hill tops, *Jour. Geology,* vol. 17, pp. 344-350.

Holmes, C. D.: 1955. Geomorphic development in humid and arid regions, *Am. Jour. Sci.,* vol. 253, pp. 377-390.

Horton, R. E.: 1945. Erosional development of streams and their drainage basins: hydrophysical approach to quantitative morphology, *Geol. Soc. America Bull.,* vol. 56, pp. 275-370.

Johnson, Douglas: 1940. Comments on the geomorphic ideas of Davis and Penck, *Assoc. Am. Geog. Annals,* vol. 30, pp. 228-232.

Melton, M. A.: 1960. Intravalley variation in slope angles related to micro-climate and erosional environment, *Geol. Soc. America Bull.*, vol. 71, pp. 133-144.

Meyerhoff, H. A.: 1940. Migration of erosional surfaces, *Assoc. Am. Geog. Annals*, vol. 30, pp. 247-254.

Penck, W.: 1924. Die morphologische Analyse, J. Englehorn's Nachfolger, Stuttgart.

Rapp, Anders: 1960. Recent development of mountain slopes in Karkevagge and surroundings, north Scandinavia, *Geografiska Annaler*, vol. 42, pp. 71-200.

Schumm, S. A.: 1956. The role of creep and rainwash on the retreat of badland slopes, *Am. Jour. Sci.*, vol. 254, pp. 693-706.

_____ : 1966. The development and evolution of hill slopes, *Jour. Geol. Education*, vol. 14, pp. 98-104.

Strahler, A. N.: 1950. Equilibrium theory of erosional slopes approached by frequency distribution analysis, *Am. Jour. Sci.*, vol. 248, pp. 673-696, 800-814.

_____ : 1956. Quantitative slope analysis, *Geol. Soc. America Bull.*, vol. 67, pp. 571-596.

Von Engeln, O. D.: 1942. The Walther Penck geomorphic system, in Geomorphology, The Macmillan Company, chap. 13, pp. 256-268.

$\left(5 \right)$

groundwater

subsurface water

origin of groundwater

Water that occurs beneath the land surface is known as groundwater, the source of which is primarily from the atmosphere in the form of rain or snow. When precipitation falls on the land surface, approximately one-third runs off into streams and rivers. The remainder evaporates back into the atmosphere, is taken in by plants or animals, or soaks into the ground. In addition, a certain amount of *connate water,* water trapped during deposition of marine sediments, may be added to the groundwater. A minor amount of *primary,* or *juvenile, water* may be incorporated into the groundwater from hydrothermal solutions and steam emanating from igneous intrusions or volcanic eruptions.

Precipitation soaking into the ground passes through a *zone of aeration* (Fig. 13-1) in which most of the pore spaces between solid particles are dry and filled only with air. Water percolating downward in this zone is called *vadose water.* Eventually, it passes into the *zone of saturation,* in which voids are filled with water. The upper surface of the zone of saturation is the *water table.*

Karst topography, Mammoth Cave, Kentucky. *(U.S. Geological Survey.)*

The distribution and movement of groundwater depend primarily on (1) the porosity and permeability of material through which the water passes, (2) the amount and annual distribution of precipitation, and (3) the surface slope.

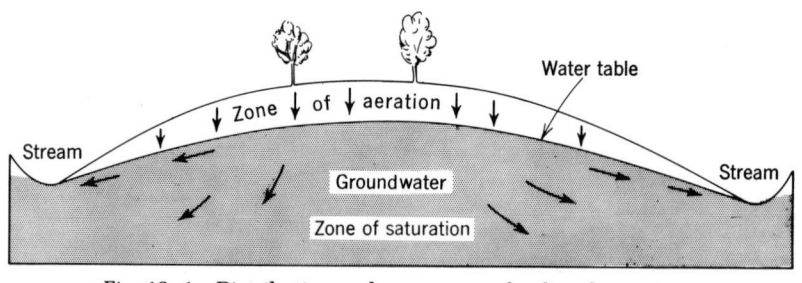

Fig. 13-1   Distribution and movement of subsurface water.

The volume of void space relative to the total volume of a material is its porosity, which may be expressed quantitatively by the equation

$$\text{Porosity} = \frac{\text{volume of pore space} \times 100}{\text{total volume}}$$

Material which has a porosity of 50 percent consists of one-half void space and one-half solid particles. If all the voids in a material are filled with liquid, the material is said to be saturated.

Rocks and unconsolidated sediments near the earth's surface vary greatly in porosity (Fig. 13-2). The porosity of a rock or sediment is governed by (1) degree of sorting of particles, (2) the shape and arrangement of particles, (3) the amount of compaction and cementation, and (4) the degree of fracturing.

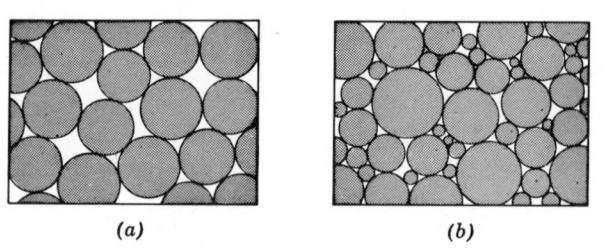

(a)   (b)

Fig. 13-2   Comparison of porosities of (a) well-sorted sediments and (b) poorly sorted sediments.

If all particles are nearly spherical, porosities are higher than if a wide variety of shapes are present. Well-sorted sand composed of particles having about the same diameter has a higher porosity than when the particle sizes vary considerably (Fig. 13-2), because the finer particles can occupy spaces between larger grains which might otherwise remain as voids. Arrangement of particles is also important. Spherical grains packed together as in Fig. 13-3a produce a porosity of about 47 percent. However, if the grains are shifted by one-half of a diameter (Fig. 13-3b), the porosity drops to about 26 percent.

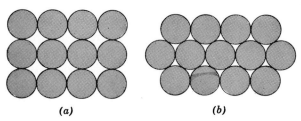

(a) (b)

Fig. 13-3   Effect of packing of spherical grains on porosity.

Compaction of sediments by the weight of overlying material or the filling of voids with cementing material may greatly reduce porosity. For example, the porosity of unconsolidated sediments overridden by glacial ice has been found to be less than similar sediment not compacted by the weight of over-riding ice (Easterbrook, 1964).

Even the densest of rock may become porous by fracturing. Samples of basalt, typically, have a low porosity, but in the field basalt flows may be fairly porous because of the presence of columnar joints. The same is true for granite, which may be rendered porous by sets of intersecting joint planes.

permeability

The permeability of material is a measure of its capacity to allow passage of a fluid through it and is dependent largely on the amount of connected voids in material rather than the total amount of voids. Materials with high porosity do not necessarily have high permeability. A good example of the difference between porosity and permeability is shown by comparing sand and clay. Recently deposited clay may have porosities as high as 70 to 90 percent, but water poured on a clay bank rapidly runs off the surface rather than penetrating the clay, indicating a low permeability. On the other hand, water poured on sand having a porosity of only 30 percent will readily permeate it. The reason for the difference in permeability of sand and clay is related largely to the size of the pore spaces. In a fine-grained material such as clay, voids are very small and often not connected, and water movement is inhibited by molecular attraction between particles and thin films of water. In sand the voids are larger and interconnected.

movement below the water table

The water table is not level, even in homogeneous material, but tends to stand higher beneath hills, and approximately mimic surface topography. Movement of groundwater takes place laterally in the direction of slope of the water table. The slope of the water table h/l (Fig. 13-4), known as the *hydraulic gradient*, depends on the permeability of material and the quantity of water being supplied. Water tables usually slope toward valleys where water is discharged into streams, but in very dry regions the slope is sometimes away from streams, and water is lost from the streams to groundwater. The discharge of groundwater into streams is the principal reason why streams are able to maintain flowage during dry periods.

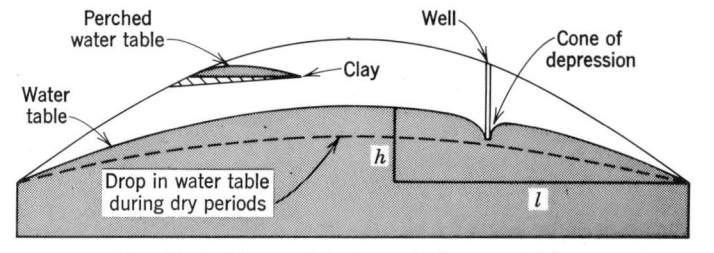

Fig. 13-4   Regional and perched water tables.

The relationship between permeability of material, slope of the water table, and amount of water moving through an area may be expressed by a formula based on Darcy's law for the flowage of water through porous material.

$$Q = PIA$$

where Q = volume of water moving per unit time
   P = a coefficient of permeability
   I = slope of water table
   A = cross-sectional area through which water is moving

Water tables fluctuate between dry and rainy seasons. During periods of low rainfall the water table drops. As shown by the equation above, a decrease in slope of the water table I results in a decrease in discharge Q of groundwater into nearby streams, and the discharge of water in streams also drops. If the water table drops below the level of water wells in an area, the wells go dry, but may produce water again when the water table rises during the rainy season.

The pumping of water from a well which penetrates the water table draws down the level of the water table in the vicinity of the well and produces a *cone of depression* (Fig. 13-4). If a well is pumped excessively, the slope of the cone flattens, thus decreasing the hydraulic gradient and decreasing the discharge of the well. If pumping is stopped, the water table is gradually restored to its original position.

springs

The permeability of material below the earth's surface varies considerably. A layer of permeable material through which water may readily pass is referred to as an *aquifer*. A layer of material which will not transmit water fast enough to furnish an appreciable amount of water is known as an *aquiclude*.

In a region where permeable and impermeable material are interbedded, secondary water tables are created. Figure 13-4 shows an aquifer underlain by an aquiclude. The aquiclude inhibits percolation of water downward, creating a *perched water table* above the normal water table. Intersection of the aquiclude with the side of a valley results in the formation of springs. However, not all springs are formed in this manner. Massive rocks of low permeability, such as granite, may contain numerous intersecting fractures along which groundwater flows. Intersection of fractures with the valley side produces springs.

Thick sequences of lava flows sometimes contain aquifers between flow contacts. Often the top of a lava flow consists of vesicular lava and rough, irregular blocks made by the breaking up of the crust of the flow as it advanced. Subsequent flows, if not too fluid, may cover the earlier flow without destroying the permeable contact zone, and as water percolates downward through fractures or columnar joints in the overlying lava, the zone becomes an aquifer. Intersection of such an aquifer with valley sides can also cause springs (Fig. 13-5).

Occasionally, aquifers are capped by aquicludes in such a way as to prevent normal migration of groundwater, and a *confined aquifer* is formed. In Fig. 13-6 water in the aquifer is prevented from rising to the surface by the capping impervious shale bed. If the impervious layer is cut by fractures or if a well is drilled through it, hydraulic pressure forces the water to the surface, where an *artesian well*, or spring, is developed. Artesian water will rise to the level of the *piezometric surface* (Fig. 13-6). Water poured into one end of a U-shaped tube will rise in the other part of the tube to the same level. The level to which it rises is the piezometric surface. In artesian wells this surface is not quite horizontal because of frictional loss of energy as water flows through the aquifer.

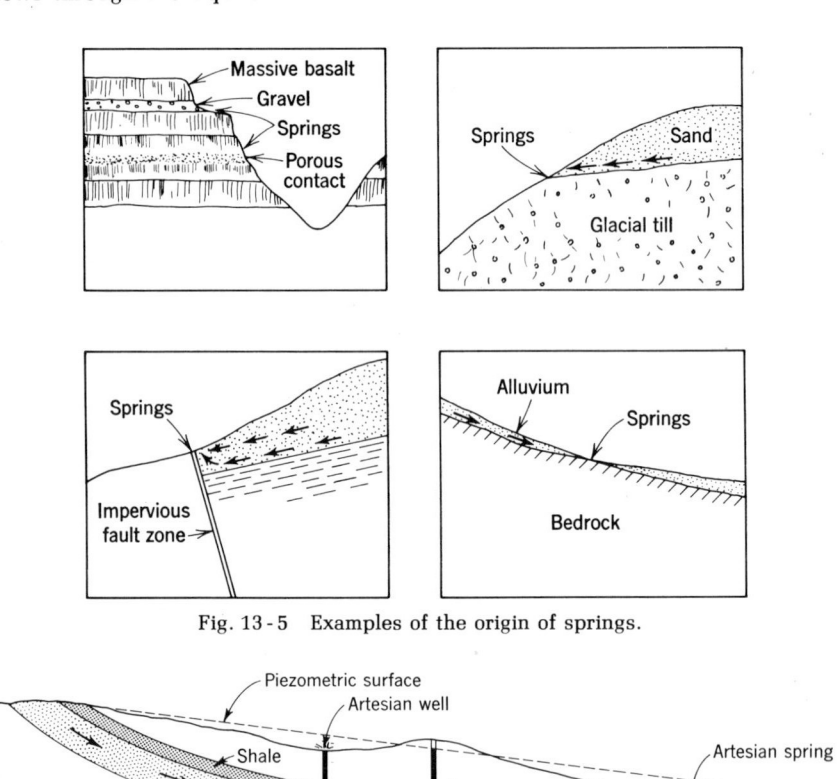

Fig. 13-5   Examples of the origin of springs.

Fig. 13-6   Artesian water resulting from a confined aquifer.

hot springs and geysers

Hydrothermal activity, such as that displayed in Yellowstone National Park and various other parts of the world, provides spectacular effects for tourists and geologists alike. Most of these effects are a result of groundwater coming in contact with a source of heat at depth, and are most commonly observed in regions of recent volcanism. In the case of Yellowstone, a period of extensive volcanic eruptions built a plateau of volcanic rocks, and the present hot springs and geysers that have made the park famous throughout the world probably originate from groundwater heated from a source below.

Although Yellowstone has hundreds of geysers, the most famous of all is Old Faithful (Fig. 13-7), which erupts at approximately one-hour intervals. The beginning of each eruption is preceded by the bubbling out of water from the geyser opening, followed by the main eruption of steam and water. After the eruption the water runs off on the surface and drains back into the ground. The amazing periodicity of Old Faithful apparently stems from the superheating of water at the base of a column of water. Water boils under atmospheric pressure at 212°F, but at greater pressures the boiling point is higher. At

Fig. 13-7   Old Faithful geyser, Yellowstone National Park, Wyoming.

2 atmospheres of pressure, water does not boil until it is heated to 246°F, and at 3 atmospheres, boils at 270°F. Near the base of a column of water the weight of the water above increases the pressure, so that the water may be heated above 212°F without boiling. If, however, the pressure is decreased, as occurs when water bubbles out at the surface, pressure is reduced, and when the hot water from depth reaches the pressure at which boiling can take place, the water flashes rapidly into steam, and the geyser erupts. Other geysers which do not erupt with the regularity of Old Faithful do not have the particular set of conditions necessary for periodic recycling and heating of water below the surface.

In addition to the geysers common in regions of hydrothermal activity, hot springs are also numerous. As the water emerges on the surface, travertine ($CaCO_3$) and silica (SiO_2) may be deposited during cooling and evaporation, some as a result of biochemical activity. Mammoth Hot Springs (Fig. 13-8) is one example among many in which the hot water quietly comes to the surface, collects in pools, and precipitates travertine and silica. Calcareous algae, which are able to withstand the high temperature of the water, aid in the precipitation of $CaCO_3$. Much of the silica deposited around the hot springs in Yellowstone probably is derived from decomposition of rhyolites as the hot water rises through them on the way to the surface.

karst topography

An important effect of groundwater on topography is produced by the dissolving of soluble rocks, transportation of the dissolved material in solution, and precipitation of it elsewhere. Rainwater and groundwater are slightly acidic and react chemically with the rocks in which they are in contact. Especially vulnerable to solution are carbonate rocks such as limestone or marble, which are composed of calcite ($CaCO_3$). Dolomite, another carbonate rock, $(CaMg)CO_3$, is also susceptible to solution, but is not as readily affected by slightly acidic water as are limestone and marble. The term *karst* is given to topography developed by solution of carbonate rocks by groundwater, the name being derived from the Karst district of Yugoslavia along the Adriatic Coast, where an assemblage of landforms related to solution of limestone is unusually well exemplified. Other regions where karst features are well developed include parts of southern France, Greece, Cuba, Puerto Rico, Jamaica, Yucatan, Indiana, Kentucky, Virginia, Tennessee, and Florida.

Not all areas underlain by limestone develop karst topography. A number of factors contribute to the degree of development of solution features (Thornbury, 1954). Among them are (1) a humid to subhumid climate to supply sufficient rainfall in order to allow solution to proceed at a relatively rapid rate, (2) limestone or other soluble rock at the surface, (3) numerous joints and thin bedding to allow percolation of water along well-defined avenues, and (4) entrenchment of major streams so that groundwater descends through the soluble rock to emerge into streams. Full development of karst features is greatly restricted in arid climates because low rainfall and sporadic distribution of rain in showers limit the water in contact with soluble rock. With few exceptions the best-developed karst topography is found in humid regions. The presence of joints and bedding in dense limestone is cited by Thornbury (1954) as of greater importance than high porosity in limestone, largely because in a dense, jointed limestone, water is concentrated along restricted planes rather than disseminated throughout the rock. Karst features are

Fig. 13-8 Mammoth Hot Springs, Yellowstone National Park, Wyoming.

often only poorly developed in areas that are underlain by porous chalk.

As surface water percolates downward through limestone or other carbonate rock, solution takes place along joints and bedding planes, gradually enlarging the openings, until appreciable subsurface caverns are formed. Collapse of the roof of a cavern or downward solution of limestone from the surface produces depressions at the surface, commonly referred to as *sinkholes*. Sinkholes occur by the thousands in regions of karst topography, and are responsible for giving such regions much of their topographic expression (Fig. 13-9). The size of sinkholes varies considerably, ranging in diameter from a few feet to over half a mile, and in depth from a few feet to over 100 feet. Typically, they are a few hundred feet in diameter and 20 to 50 feet in depth (Fig. 13-10). The bottom of sinkholes, which occasionally becomes

Fig. 13-9 Numerous sinkholes developed in the karst topography east of Bowling Green, Kentucky. *(John Shelton.)*

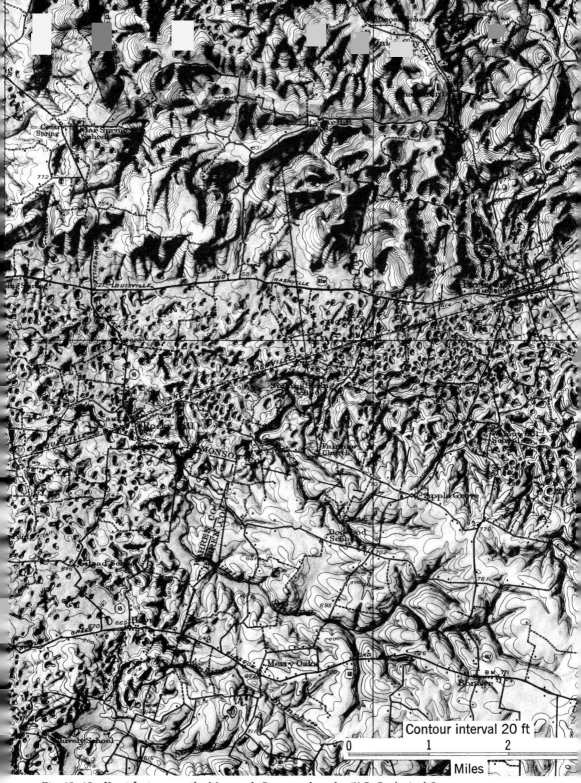

Fig. 13-10　Karst features on the Mammoth Cave quadrangle. *(U.S. Geological Survey*

clogged with clay and silt washed in from surface drainage or left behind as a residue during solution, may fill with water to form *sinkhole ponds,* or *karst lakes* (Fig. 13-11). The surfaces of sinkhole ponds or lakes may sometimes correspond to the top of the water table in an area.

Coalescence of several individual sinkholes results in the formation of *compound sinkholes.* Coalescence of a series of sinkholes in a line is common in some parts of karst regions (Fig. 13-10, e.g., one-half mile west of Rocky Hill). The alignment is in some cases caused by the collapse of portions of the roof of a subsurface cavern, but in other instances may be due to sinkhole formation along the former course of a stream which has been diverted underground.

Fig. 13-11   Pond developed in a sinkhole near Park City, Kentucky. *(Wards Natural Science Establishment.)*

Surface streams in a karst region are usually restricted by the diversion of much of the surface water underground. Only rarely does a surface stream cross an area of extensive sinkhole development. Typically, streams which flow on noncarbonate rocks terminate their surface courses in a sinkhole when they reach a karst area, and the drainage becomes diverted underground. Examples of such streams, known as *sinking creeks,* include Gardner Creek, Little Sinking Creek, and Sinking Creek, shown in Fig. 13-10. Caving in of parts of the roof of underground streams results in the development of *karst windows,* in which a part of an underground stream may be seen from the surface through the deroofed portion. Collapse of a significant portion of the roof of a cave or tunnel eventually leads to development of *natural bridges,* which represent remnants of the former roof. Sinking creeks which terminate in a sinkhole for a prolonged period of time may lower the upper part of their courses by erosion below the original level of the sinkhole, so that the valley ends abruptly against steep walls at the sinkhole, forming a *blind valley.* The lower portion of Sinking Creek (Fig. 13-10) is an example of such a valley.

Prolonged solution of limestone from the surface downward may result in the accumulation of a reddish clay soil which mantles the topography and may extend downward into solution openings and depressions. In most cases the clay appears to have originated in the limestone, and has been left behind as a residue after the limestone has been dissolved. Because it characteristically has a reddish color, the name *terra rossa* has been applied to it, the color coming from the oxidation of iron.

The topography in the upper part of Fig. 13-10 consists of ridges capped by sandstone beds which are underlain by limestone. Incision of streams through the sandstone unit into the underlying limestone initially produced normal V-shaped valleys, but upon intersection of the limestone, solution developed a series of sinkholes in these valleys, and surface drainage eventually was diverted underground. Examples of such valleys, known as *solution valleys,* include Woollsey Hollow and Cedar Spring Valley (Fig. 13-10). In the course of solution of limestone by groundwater, extensive cave systems often develop. Among the most famous ones are Mammoth Cave, Kentucky, and Carlsbad Caverns, New Mexico. The caverns at Mammoth Cave, Kentucky, consist of a vast system of interconnected caves and galleries which contain lakes, streams, and waterfalls. Although not yet fully mapped and explored in detail, this vast cave system appears to total many miles of connected caverns. An example of an unusually large gallery is to be found in Carlsbad Caverns, New Mexico, where the Big Room is about 4,000 feet long, about 625 feet wide, and about 300 feet high.

Although the origin of limestone caves has been long debated, the question as yet remains without a conclusive answer. The principal point of controversy revolves around whether or not the solution of large amounts of limestone to develop subterranean caverns and various other features associated with karst topography occurs above or below the water table. Some of the various theories of cavern formation have been summarized by Thornbury (1954). Prior to 1930 it was commonly assumed that subsurface caverns in limestone were developed by solution primarily above the water table and that downcutting of nearby surface streams allowed the progressive development of caverns at several different levels. Accordingly, it was assumed that little or no solution of limestone occured below the water table. However, in 1930, W. M. Davis proposed the idea that the most significant portion of cavern development took place below the water table and that formation

Fig. 13-12 Air photographs of a karst region in Warren County, Kentucky. (U.S. Department of Agriculture.)

Fig. 13-13 Topographic map of karst topography. (*Smith Grove quadrangle, Kentucky, U.S. Geological Survey.*)

of various cave deposits took place only after downcutting of nearby surface streams had resulted in the lowering of the water table, leaving the previously dissolved caves standing above the water table. Bretz (1953) came to a similar conclusion, following studies of limestone caverns in the Ozark region.

The common occurrence of deposits of clay, silt, and gravel in caves led some geologists to the conclusion that at least some caves were formed by mechanical erosion of subterranean streams, rather than by the solution of limestone by groundwater. However, caverns seldom display the normal graded profiles that surface streams do, and often are composed of three-dimensional galleries and rooms which would be very difficult to form by the mechani-

Fig. 13-14 Stalagmite in Carlsbad Caverns, New Mexico. *(Wards Natural Science Establishment.)*

cal abrasion of running water. The presence of some caverns whose walls are lined with calcite crystals has been used as evidence that they were developed below the water table.

In caverns which are above the water table today it appears that the process of deposition of calcite dominates over solution. As water drips from the ceiling of caverns, evaporation of water and escape of CO_2 in solution increases the saturation of $CaCO_3$ until it precipitates. Gradually, an iciclelike form, known as a *stalactite*, is constructed, hanging down from the ceiling. When water drips from the tip of the stalactite to the floor of the cavern below, calcium carbonate may be deposited there, and gradually an inverted iciclelike form, known as a *stalagmite* (Fig. 13-14), is built upward toward the stalactite. Eventually, if precipitation of calcite continues long enough, the stalactites and stalagmites meet and fuse to become a column, or pillar.

references

Bretz, J. H.: 1949. Carlsbad Caverns and other caves of the Guadalupe Block, New Mexico, *Jour. Geology*, vol. 57, pp. 447-463.

————: 1953. Genetic relations of caves to peneplains and Big Springs in the Ozarks, *Am. Jour. Sci.*, vol. 251, pp. 1-24.

Davis, W. M.: 1930. Origins of limestone caverns, *Geol. Soc. America Bull.*, vol. 41, pp. 475-628.

Dicken, S. N.: 1935. Kentucky karst landscapes, *Jour. Geology*, vol. 43, pp. 708-728.

Easterbrook, D. J.: 1964. Void ratios and bulk densities as means of identifying Pleistocene tills, *Geol. Soc. America Bull.*, vol. 75, pp. 745-750.

Horberg, Leland: 1949. Geomorphic history of the Carlsbad Caverns area, New Mexico, *Jour. Geology*, vol. 57, pp. 464-476.

Hubbert, M. K.: 1940. The theory of ground water motion, *Jour. Geology*, vol. 48, pp. 785-944.

Meinzer, O. E.: 1942. Hydrology, McGraw-Hill Book Company (reprinted by Dover Publications, Inc., New York).

Swinnerton, A. C.: 1932. Origin of limestone caverns, *Geol. Soc. America Bull.*, vol. 43, pp. 663-694.

—— : 1942. Hydrology of limestone terraces, in O. E. Meinzer, Hydrology, McGraw-Hill Book Company (reprinted by Dover Publications, Inc., New York), pp. 656-677.

Thornbury, W. D.: 1954. Principles of Geomorphology, John Wiley & Sons, Inc., New York, chap. 13.

6

desert and
eolian landforms

desert landforms

The same processes that are operative in humid climates are also responsible for the development of landforms in arid climates, but the intensities of these processes are often different, and topographic forms in arid regions are thus not identical with those of humid areas.

Although deserts are characterized by high temperatures, perhaps their most diagnostic feature is low precipitation and sporadic annual distribution of precipitation. Because of high daily temperatures and low rainfall in arid regions, the rate of evaporation is such that the amount of water lost by evaporation is greater than the amount of precipitation. Partly because evaporation exceeds precipitation, vegetation is sparse in arid regions, and weathering and soil development are much diminished. Streams are largely intermittent and usually flow only during rainy periods.

origin of deserts

Most of the world's deserts are found in mid-latitude regions, in the lee of mountain ranges, and along cold-water coasts.

Pediments, Sacaton Mountains, Arizona. *(Soil Conservation Service.)*

The global circulation of air results from various combinations of solar radiation and rotation of the earth. Near the equator, air warmed by the sun rises and is carried poleward at high altitudes. Near the "horse latitudes" (about 30°) the air descends to the surface from high altitudes, and in so doing is compressed by the higher pressure at lower altitudes. When air is compressed it is heated adiabatically because of the increase in molecular collisions. Since the ability of air to hold moisture increases with temperature, warm descending air has a drying effect, and the area beneath is likely to have a low rainfall and a high rate of evaporation. The Sahara Desert, perhaps one of the world's best-known deserts, lies in a belt of descending air about 30° north of the equator, which produces an arid climate across northern Africa from the Atlantic to the Indian Ocean.

Deserts are also found in the lee of high mountain ranges. Warm, moist air moving across a mountain range is forced to rise, and as it does so, it expands and is cooled adiabatically, causing moisture in the air to condense as rain or snow. Moving down the lee side of a mountain range, however, the descending air, having lost much of its moisture on the windward side, is heated adiabatically, further drying the air. As air rises, it cools 5.5° for each 1,000 feet until condensation occurs, and thereafter cools 3.2° per thousand feet. For each 1,000 feet that air descends, the temperature rises 5.5°. Thus air descending from 10,000 to 4,000 feet in the lee of the crest of a mountain range would warm 6×5.5, or 33°, and its ability to hold moisture would be much increased.

Coastal regions affected by cold ocean currents are often characterized by desert climates. Examples of this type of desert may be found along the west coast of South America, where the Humboldt Current brings cold water northward toward the equator. As air moving across the cold current is cooled, condensation occurs, and heating of the air as it moves onto the land increases its ability to hold moisture.

erosional landforms

Surprisingly, in spite of the fact that the most characteristic feature of deserts is paucity of water, the single most important agent of erosion in deserts is running water. The lack of vegetation, concentration of rainfall in short periods of time, and intermittent stream flow contribute to intensification of erosion during heavy rains, interspersed with periods of little active erosion.

Chemical weathering in arid climates usually is much diminished compared with humid climates because moisture necessary for chemical decomposition is absent from the zone of weathering for substantial periods of time.

This in turn leads to poor soil development and sparse vegetation, which results in high rates of surface runoff during rainy periods. The general deficiency of water does not, however, mean that no weathering takes place. Mechanical weathering produces a supply of broken pieces of rock to be transported by mass movement and running water.

pediments

Pediments (Fig. 14-2) are gently inclined plains at the foot of mountain fronts formed by degradation and retreat of the mountain front, typically developed in arid or semiarid regions across rocks of varying lithology. They are

Fig. 14-1 Desert of southwest Utah viewed from Granite Mountain, Toole County, Utah. *(W. D. Johnson, U.S. Geological Survey.)*

erosion surfaces thinly veneered with alluvium, representing graded slopes of transportation with concave-upward longitudinal profiles similar to those of graded streams. The principal agencies responsible for the development of pediments are (1) lateral planation of streams, (2) sheet and rill wash, and (3) back weathering of the mountain front. Nearly all geologists agree on the definition of pediments as outlined above, but there is no general agreement concerning the processes which are dominant in producing them.

For purposes of discussion, the problems associated with pediments may perhaps best be separated into two categories: (1) the problem of recognizing pediments and distinguishing them from other landforms and (2) the problem of determining the origin of pediments and the relative importance of each of the three main processes (lateral planation, rill wash, and back weathering).

Pediments may be confused with surfaces of deposition or surfaces which owe their form to structure. In areas of pediment formation, basin filling is common, and in order to correctly term a surface a pediment (and hence a surface of degradation), it is necessary to observe the field relationships of the surface. For example, alluvial fans may be built out from a fault block and coalesce to form a surface which might be similar in form to a pediment,

Fig. 14-2   Pediment east of the Mojave River near Oro Grande, California. *(John Shelton.)*

although in fact being entirely depositional. Johnson (1932) has shown that "rock fans" may be developed by degradation near mountain fronts and have profiles identical with those of alluvial fans. The depth of alluvial material covering bedrock as seen in stream cuts and surface outcrops or as determined from well data seems to be the best criterion to use in identifying pediments. For erosion surfaces such as pediments, the thickness of alluvium is not greater than depth of scour.

The problem of the origin of pediments dates back to the late nineteenth century. Gilbert (1877), in his classic work on the Henry Mountains, recognized certain beveled surfaces developed on tilted strata. He referred to these as "hills of planation," and attributed them to former stream planation. McGee (1897) first used the term pediment for erosion surfaces in Arizona, but he attributed them to sheet-flood erosion rather than lateral planation. Bryan (1922) adopted the term pediment for a number of erosion sufaces in Arizona and regarded them as being formed by a combination of lateral stream planation, rill cutting at the base of the mountain slopes, and weathering of outliers with transportation of debris by rills. He noted that pediment slopes were steeper opposite smaller canyons because the gradients of smaller streams are steeper than those of larger streams.

Blackwelder (1931) defined a pediment as a "gently inclined plain cut by stream erosion indifferently across rocks of varying lithology and structure," and ascribed their origin to active sidewise cutting of desert torrents.

Johnson (1932), a strong proponent of lateral corrasion as the dominant factor, delineated three zones in arid regions: (1) an inner zone, the zone of degradation, corresponding closely to the mountainous highland, in which vertical downcutting of streams reaches its maximum relative importance, (2) an intermediate zone surrounding the mountain base, the pediment, in which lateral cutting of streams attains its maximum relative importance, and (3) an outer zone, the zone of aggradation, where upbuilding by deposition of alluvium has its maximum relative importance. The pediment encroaches upon the zone of degradation as the mountain front recedes and the zone of aggradation encroaches upon the pediment, covering it with alluvium. The supply of debris from the mountain decreases, and the pediment is slowly lowered. As evidence supporting lateral planation as the dominant process in pediment formation, Johnson showed that "rock fans," having the same shape and slopes as alluvial fans, except that they were cut in bedrock, are formed by lateral corrasion of streams issuing from canyon mouths, and suggested that such rock fans evolve into pediments.

Gilluly (1937) favored a composite theory which included "cooperation of lateral corrasion, rill wash, and weathering with subsequent removal of detritus by rills. One process may be dominant in one place and another else-

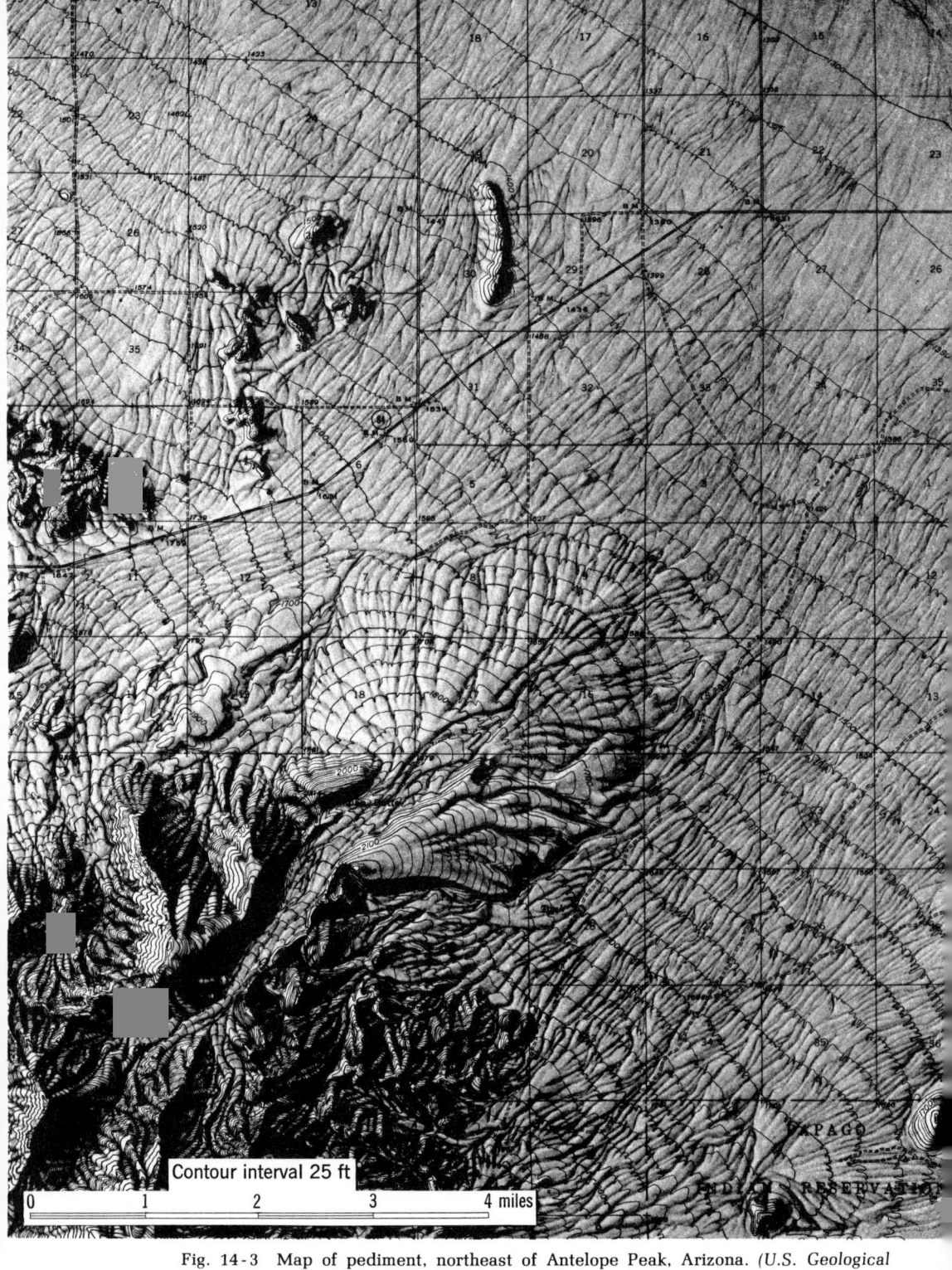

Contour interval 25 ft

0   1   2   3   4 miles

Fig. 14-3 Map of pediment, northeast of Antelope Peak, Arizona. *(U.S. Geological Survey.)*

where, but all are operative." He observed that the slope of pediments was related to the calibre of debris, coarse debris yielding steeply sloping pediments and fine debris yielding gently sloping pediments.

Mackin (1937) has shown that many of the pediments of the Big Horn Basin in Wyoming were formed by lateral planation of streams. Erosion surfaces truncate structures of differing resistance, and are covered with river gravels that could not have been derived locally, making it possible to identify the streams responsible for the cutting of the surfaces. However, conditions there differ from those in the Southwest, for local base levels are controlled by through-flowing streams, resulting in the absence of the thick alluvial wedges found in the Southwest. The Big Horn Basin surfaces were cut by stable, slowly

Fig. 14-4 Air photograph of pediment northeast of Antelope Peak, Arizona. *(U.S. Geological Survey.)*

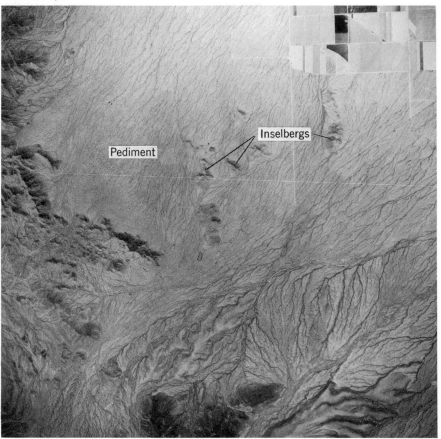

degrading streams graded to the master through-flowing streams and were mantled with gravel to scour depth.

Mackin considered pediments to be formed both by lateral planation and by weathering, wash, and creep, and that competition between the rate of lateral planation and the rate of weathering determined which played the dominant role for any given pediment. The rate of weathering is largely determined by factors of climate and terrain, whereas the rate of lateral planation is controlled by the erosive power of streams relative to rock resistance. Streams on a pediment which issue from a high mountain range with a high discharge and coarse load of detritus swing vigorously back and forth across the pediment, and lateral planation is likely to play a dominant role. However, if streams on a pediment issue from a low mountain range with minor discharge, they are not likely to swing back and forth vigorously enough to effectively erode laterally, and thus weathering, rill wash, and sheet wash may play dominant roles in development of the pediment. In either case, load-discharge relationships determine the general slope of pediment.

Other problems associated with the formation of pediments center around the sharp break in slope between the pediment and the mountain front, which is typical of nearly all pediments. Since the expected condition of erosion would be for a relatively continuous slope from mountain front to pediment, a rather abrupt change in the processes of erosion at the junction of the two is required to explain the sudden break in slope. Bryan (1922) believed that the break in slope was due to the difference in size of weathered materials on the two different slopes and to the concentration of rill cutting at the mountain base. Others have attributed it to trimming by lateral corrasion of streams as they impinge against the mountain front when swinging back and forth.

depositional landforms

The thin veneers of gravel on pediments often grade imperceptibly downslope into thick alluvial fills, commonly known as *bahadas*, which consist of a series of confluent alluvial fans composed of gravel, sand, and mudflows. As seen from the ground, on air photos, or on topographic maps, there is no distinct boundary between a pediment, which is essentially a surface of erosion developed on bedrock, and a bahada, which is a constructional surface of deposition.

Bahadas do not necessarily merge with pediments. They also may be found directly adjacent to mountain fronts, where streams issuing from the mountains deposit cone-shaped alluvial fans (Fig. 14-5). The slope of a fan is determined by the load-discharge relationships of the stream responsible for sediment deposition. Low-discharge streams with loads of coarse detritus build higher-gradient fans than do high-discharge streams carrying smaller-size material.

When pediments and bahadas terminate in closed basins of internal drainage, nearly level plains, known as *playas,* are formed (Fig. 14-6). After a heavy rain, shallow lakes may be formed in the playas by the centripetal drainage. But even though such lakes may be several miles in diameter, they are very shallow, and often evaporate in a short period of time, leaving behind deposits of mud and a variety of salts which precipitate from the water. Playas are among the flattest surfaces found in nature (Fig. 14-7).

Fig. 14-5   Alluvial fans along a fault scarp at the foot of a mountain range in the Death Valley region. *(Aero Service Division, Litton Industries.)*

evolution of desert landforms

Equilibrium slopes developed on pediments and alluvial fans are adjusted to conditions prevailing at a particular time, but these conditions do not forever remain the same. As relief in the headwaters becomes progressively smaller, the calibre of debris shed from the mountains diminishes, causing pediments and alluvial fans to regrade their profiles. If drainage is internal into closed basins, as is the case in much of the southwestern United States, sediment brought from the adjacent mountains cannot escape, and an ever-thickening deposit is built up which progressively raises the local base level.

Successive stages in the evolution of desert landforms are shown in Figs.

Fig. 14-6 Playas developed in intermountain basins near Steens Mountain, Oregon. *(John Shelton.)*

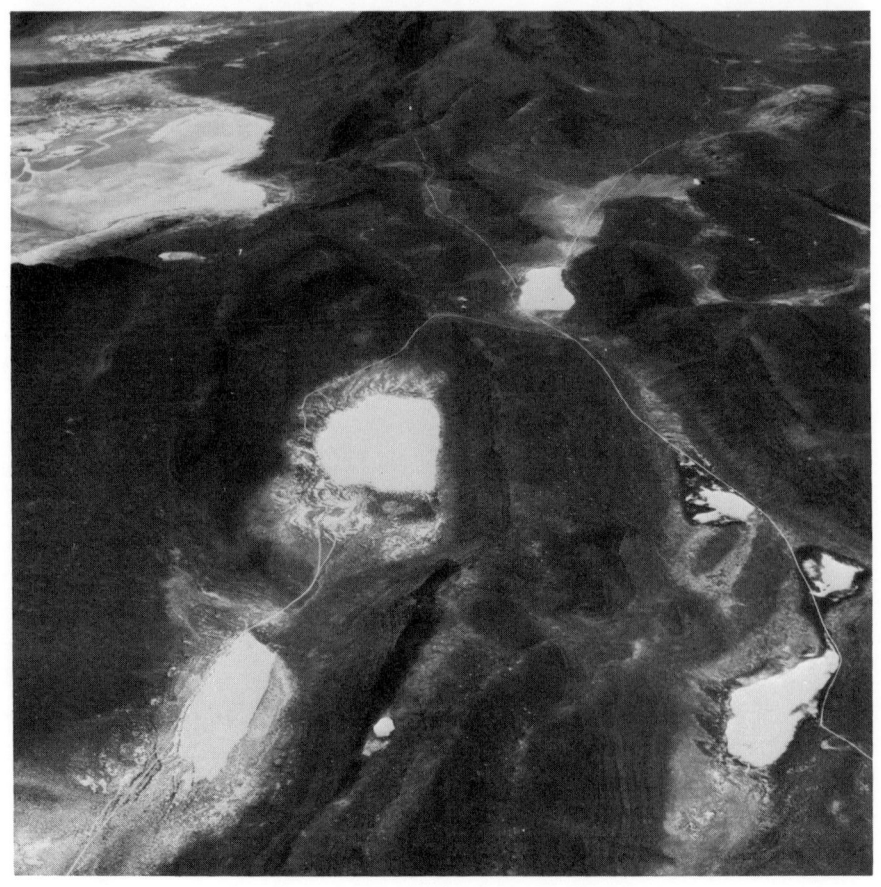

14-8 to 14-10. In the early stages of development of streams, downcutting and headward erosion of streams in the mountains combine with deposition of fans in adjacent basins as the dominant processes. As erosion proceeds, pediments begin to develop next to the mountain front, merging with bahadas in their lower portions. The concave-upward profile of a pediment with a progressively decreasing gradient downslope is essentially a graded slope of transportation, whose gradient is controlled by the coarseness of rock debris, which must be transported across the pediment under prevailing discharge conditions. As each pediment increases in area at the expense of the retreating mountain front (Fig. 14-10), the sharp break in slope between the pediment and mountain front remains.

Fig. 14-7   Surface of playa in eastern California. *(Wards Natural Science Establishment.)*

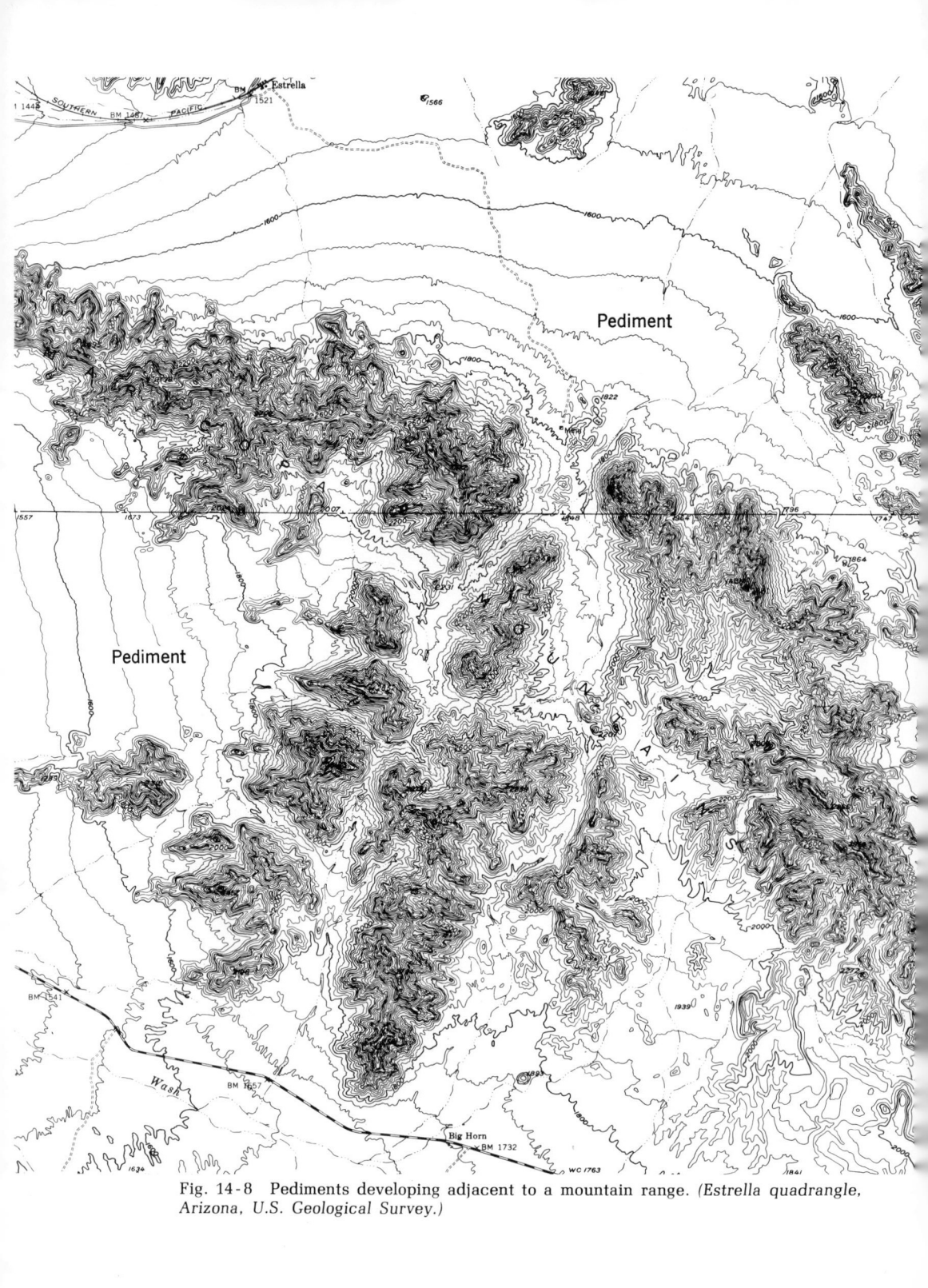

Fig. 14-8 Pediments developing adjacent to a mountain range. *(Estrella quadrangle, Arizona, U.S. Geological Survey.)*

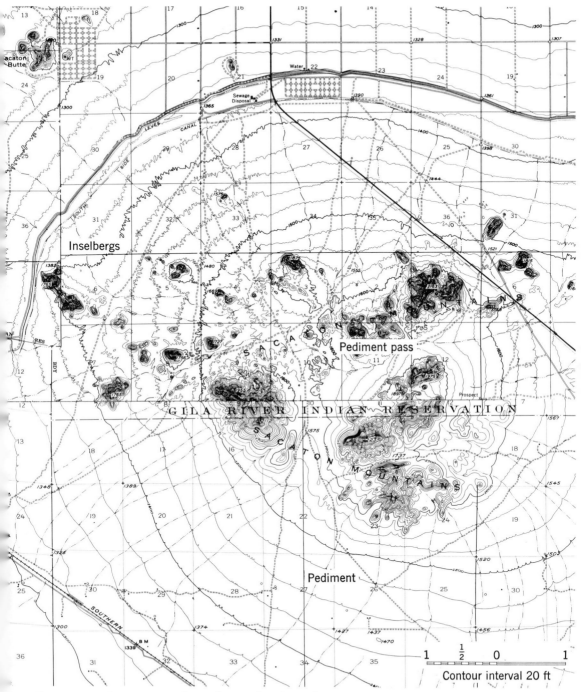

Fig. 14-9 A later stage in the evolution of a desert landscape. Note the pediment passes, inselbergs, and more advanced destruction of the mountain range. Compare with the air photograph at the beginning of the chapter. *(Gila Butte and Casa Grande quadrangles, Arizona, U.S. Geological Survey.)*

As pediments on opposite sides of a mountain range extend themselves headward, narrow tongues of the main pediments may encroach through the mountains and intersect one another, forming *pediment passes* (Fig. 14-9). Enlargement of pediment passes leads to further dissection of the mountains and coalescence of pediments. Residual rock masses left standing above the pediments as isolated remnants are known as *inselbergs,* from the German word meaning "island mountains," referring to the islandlike form, surrounded by the flat pediment surface. Some inselbergs develop miniature pediments around their base.

Retreat of the mountain front by encroachment of a pediment leads to change in slope requirements on the upper part of the pediment, and regrading may result in incision of part of the pediment. A similar effect also occurs on alluvial fans, producing *fan-head trenching.*

The steepest part of the profile of a pediment occurs near the mountain front. As the pediment encroaches on the mountains, a given point on a pediment becomes farther and farther away from the mountain front, and slope

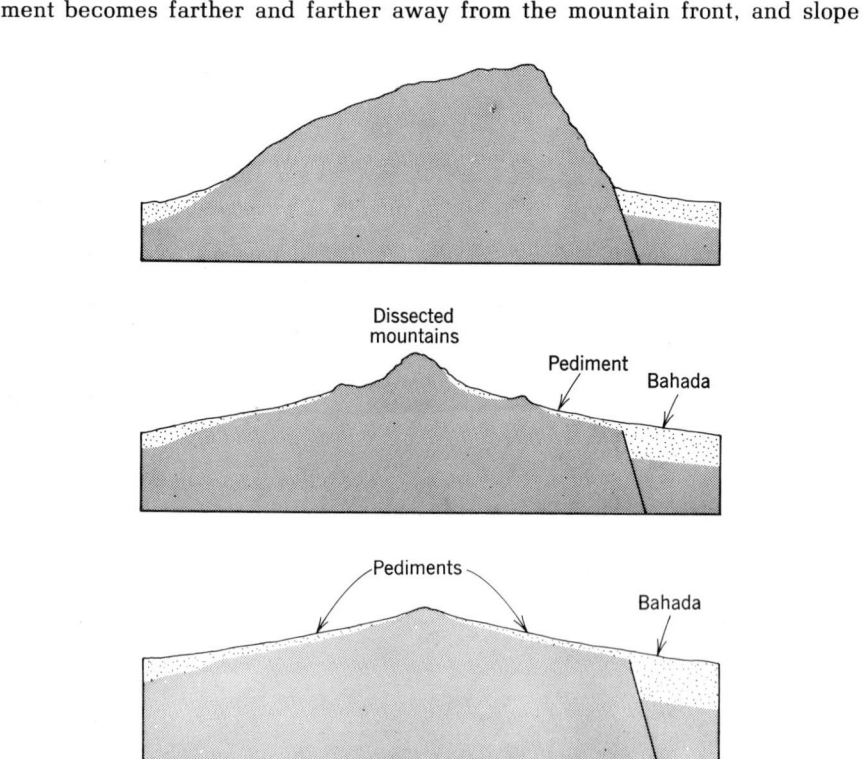

Fig. 14-10 Diagrammatic cross section of successive stages of evolution of desert landforms.

requirements are diminished as a result of the smaller calibre of debris carried across that particular part of the pediment.

Coalescence of pediments on opposite sides of a mountain range and destruction of all but a few isolated remnants of the range results in development of a broad plain, somewhat analogous to the peneplain of humid regions. The term *pediplain* has been suggested for such advanced stages of evolution of desert landscapes.

Pediments, as graded slopes of transportation, are governed largely by conditions at their headward area and at their lower portion. The influence of the size of material introduced in their headward area affects the gradient of a pediment, and the local base level at the lower margin of a pediment affects

Fig. 14-11   Dissected pediment north of Wellington, Utah. *(U.S. Geological Survey.)*

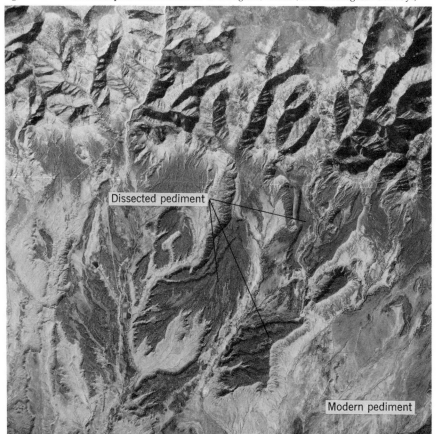

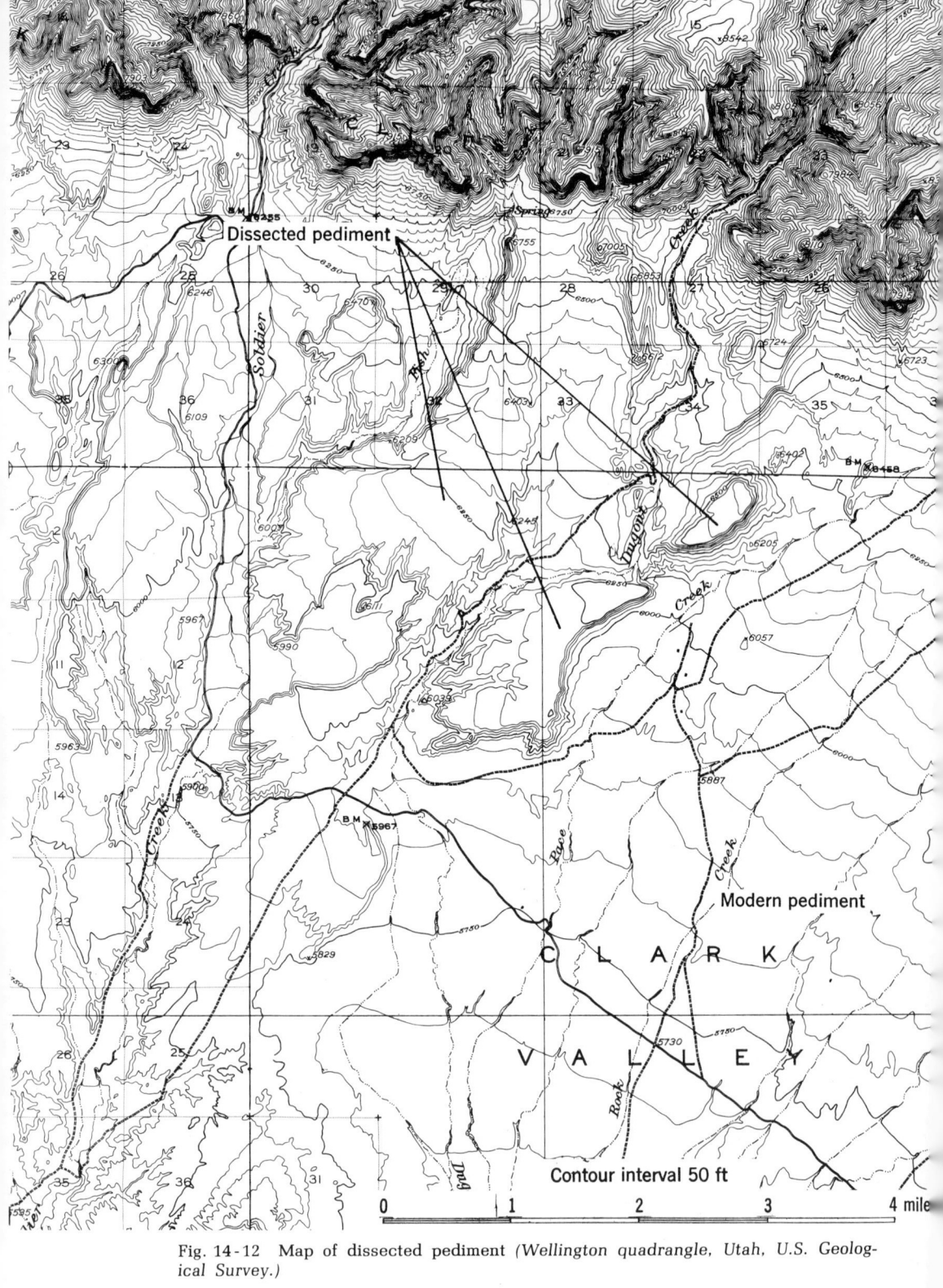

Fig. 14-12 Map of dissected pediment (Wellington quadrangle, Utah, U.S. Geological Survey.)

equilibrium conditions on the pediment. Pediments may be graded to (1) a closed basin, which acts as a steadily rising base level as the basin is filled with detritus, or (2) a through-flowing stream, which may act as a stable, but slowly degrading, base level. In the case of pediments graded to a closed basin, aggradation in the bahada may eventually fill the basin to a level where drainage is able to spill out across a divide, thus affecting a change in base-level conditions. Once this happens, the outlet stream may cut its channel downward and reverse the previous base-level trend. If the local base level consists of a through-flowing stream, pediments will be affected by changes in the base level to which the stream is graded or by other changes which affect the stream regimen.

references

Blackwelder, E.: 1931. Desert Plains, *Jour. Geology*, vol. 39, pp. 133-140.

Bradley, W. H.: 1940. Pediments and pedestals in miniature, *Jour. Geomorphology*, vol. 3, pp. 244-254.

Bryan, K.: 1922. Erosion and sedimentation in the Papago Country, Arizona, *U.S. Geol. Survey Bull.* 730, pp. 19-90.

_____ and F. McCann: 1936. Successive pediments and terraces of the upper Rio Puerco in New Mexico, *Jour. Geology*, vol. 44, pp. 145-172.

Davis, W. M.: 1930. Rock floors in arid and in humid climates, *Jour. Geology*, vol. 38, pp. 1-27, 136-158.

_____ : 1938. Sheetfloods and streamfloods, *Geol. Soc. America Bull.*, vol. 49, pp. 1337-1416.

Gilbert, G. K.: 1877. Report on the geology of the Henry Mountains, *U.S. Geog. and Geol. Survey, Rocky Mt. Region.*

Gilluly, J.: 1937. Physiography of the Ajo Region, Arizona, *Geol. Soc. America Bull.*, vol. 48, pp. 323-348.

Hadley, R. F.: 1967. Pediments and pediment-forming processes, *Jour. Geol. Ed.*, vol. 15, pp. 83-89.

Holmes, C. D.: 1955. Geomorphic development in humid and arid regions: a synthesis, *Am. Jour. Sci.*, vol. 253, pp. 377-390.

Howard, A. D.: 1942. Pediment passes and the pediment problem, *Jour. Geomorphology*, vol. 5, pp. 1-31, 95-136.

Johnson D.: 1932a. Rock fans of arid regions, *Am. Jour. Sci.*, vol. 23, pp. 389-420.

———— : 1932b. Rock planes of arid regions, *Geog. Rev.*, vol. 22, pp. 656-665.

———— : 1932c. Miniature rock fans and pediments, *Science*, vol. 76, p. 546.

Mackin, J. H.: 1937. Erosional history of the Big Horn Basin, Wyoming, *Geol. Soc. America Bull.*, vol. 48, pp. 813-893.

McGee, W. J.: 1897. Sheetflood erosion, *Geol. Soc. America Bull.*, vol. 8, pp. 87-112.

Miller, V.C.: 1950. Pediments and pediment-forming processes near House Rock, Ariz., *Jour. Geology*, vol. 58, pp. 634-644.

Paige, Sidney: 1912. Rock-cut surfaces in the desert ranges, *Jour. Geology*, vol. 20, pp. 442-450.

Rich, J. F.: 1935. Origin and evolution of rock fans and pediments, *Geol. Soc. America Bull.*, vol. 46, pp. 999-1024.

Schumm, S. A.: 1956. The role of creep and rainwash in the retreat of badland slopes, *Am. Jour. Sci.*, vol. 254, pp. 693-706.

———— : 1962. Erosion on miniature pediments in Badlands National Monument, South Dakota, *Geol. Soc. America Bull.*, vol. 73, pp. 719-724.

Sharp, R. P.: 1940. Geomorphology of the Ruby-East Humboldt Range, Nevada, *Geol. Soc. America Bull.*, vol. 51, pp. 337-372.

Smith, K. G.: 1958. Erosional processes and landforms in Badlands National Monument, South Dakota, *Geol. Soc. America Bull.*, vol. 69, pp. 975, 1008.

Tator, B. A.: 1952. Pediment characteristics and terminology, *Assoc. Am. Geog. Annals*, vol. 42, pp. 295-317.

———— : 1953. Pediment characteristics and terminology, Part II, *Assoc. Am. Geog. Annals*, vol. 43, pp. 47-53.

Tuan, Yi-Fu: 1959. Pediments in southeastern Arizona, University of California Publications in Geography, vol. 13.

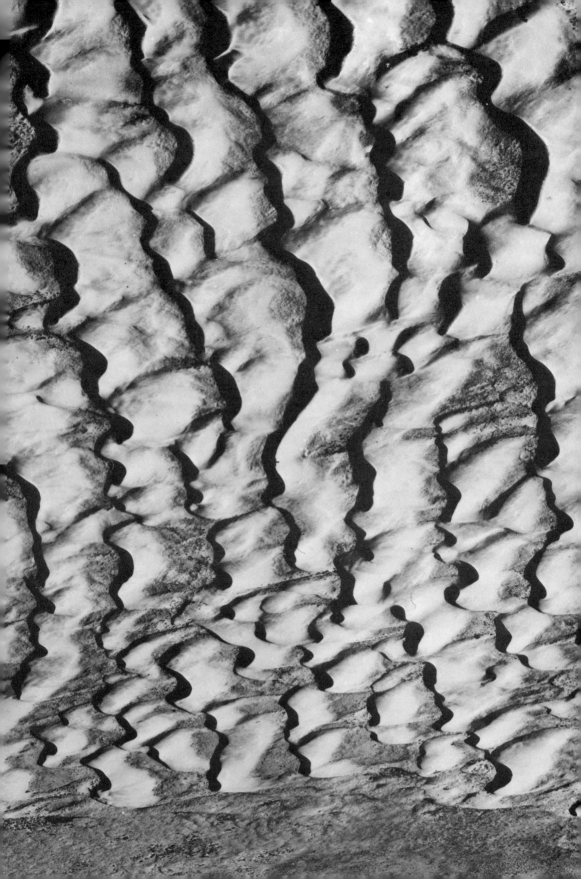

eolian landforms

The wind, although limited in the size of particles that it can transport, may locally play an important role in the development of topography, especially in arid or coastal regions, where there is an abundant sand supply. However, only a very minor portion of most desert regions is covered with dune sand, and in the rest of the desert the effect of running water plays a far more important role in the development of topography. Scarcity of vegetation is an important factor in wind erosion and transportation in arid climates, since plants protect the ground surface and effectively break up wind currents.

movement of particles by the wind

The maximum size of particles which the wind is able to transport depends on relationships between wind velocity and the settling velocity of particles. Settling velocity, the velocity with which a particle will fall to the ground, depends on (1) the net gravitational force, which depends on the difference between the mass of a falling body and that of the fluid which it displaces,

Sand dunes near Delta, Utah. *(Aero Service Division, Litton Industries.)*

(2) the viscosity of the fluid, and (3) the diameter of the particle. Figure 15-1, which shows the settling velocities of a number of different-size particles, illustrates the point that small particles have lower settling velocities and thus are able to remain in the air for a longer period of time for a given wind velocity. The wind is able to keep particles aloft because many of the swirls and eddies, which are invariably associated with turbulence, have an upward component of movement. Bagnold (1941) found that the ratio of the velocity of upward gusts near the ground to mean wind velocity, although quite variable, averaged approximately 1:5. Some of the particles in the air whose settling velocities are lower than one-fifth of the mean wind velocity are thus carried upward by gusts and, while remaining in suspension, are carried downwind. Particles with higher settling velocities remain on or near the ground.

Sand being driven by the wind usually does not rise more than 1 meter above the ground, and most of the grains move in a zone less than about 10 centimeters high. The wind velocity at this height necessary to just set the grains in motion was measured by Bagnold (1941) and found to be about 5 meters per second (11 miles per hour). If the velocity of updrafts averages about one-fifth of this velocity, 1 meter per second, then, from Fig. 15-1, it is apparent that sand grains having a diameter less than about 0.2 millimeter will

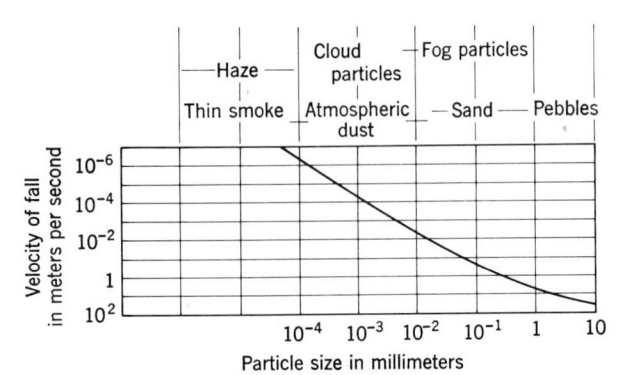

Fig. 15-1   Relationship of settling velocity to particle size. (*After R. A. Bagnold, Methuen & Co., Ltd., London, 1941.*)

be kept in motion and should be winnowed out from the coarser grains. Sieve analyses of dune sands show that, in general, grains having diameters of 0.3 to 0.15 predominate, and even in the finest wind-blown sand, diameters of less than 0.08 are seldom found.

The movement of sand, concentrated in a zone only a few feet above the ground during a typical sandstorm, is described by Bagnold (1941) as follows:

The wind produces for the first hour or so a mist consisting of both dust and sand. Later, although the wind may have shown no signs of slackening, the mist disappears. But the sand still continues to drive across country as a thick, low-flying cloud with a clearly marked upper surface. The air above the sand cloud becomes clear, the sun shines again, and people's heads and shoulders can often be seen projecting above the cloud as from the water of a swimming-bath.

The bulk of the sand movement takes place considerably nearer the ground than the visible top of the cloud. Evidence of this is given by the effects of the sand blast on posts and rocks projecting from the ground; the erosion is greatest at ground level, and is usually inappreciable at a height of 18 inches. (p. 10)

Not all the grains transported by the wind are necessarily lifted into the air by turbulence. Some sand grains which strike the ground at a low angle bounce into the air and travel downwind in a series of hops. In experiments with quartz sand in a wind tunnel Bagnold (1941) found:

When a flying grain, on reaching the ground, strikes the hard surface of a pebble or rock, it may bounce off it with almost perfect resilience, and may reach a height as great as that of the observed top of the sand cloud during sand-driving. When the grain, having risen into the air on rebound, is acted on by the wind, it moves in a curved path whose flatness is quite sufficient to give the impression to the eye of an unsupported horizontal flight. It strikes the ground at a flat angle of between 10° and 16°, depending on the size of the grain, its height of rise, and the speed of the wind. (p. 19)

In other words, the sand grains receive energy from the forward motion of the wind, and when they strike the ground, their horizontal velocity is converted into an upward one by impact, analogous to a number of Ping-Pong balls bouncing downwind. If the ground is covered with pebbles or is bare rock, the grains bounce high into the air when they strike the surface; if the ground is covered with sand, the grains sink into the sand as they strike the surface, ejecting in the process one or more other grains which had been at rest (Fig. 15 - 2). These grains do not rise as high into the air as when a grain strikes a hard surface, but they soon pick up speed as the wind drives them downwind, where they in turn strike other grains. This type of motion is known as *saltation*. Sand grains bouncing along the ground near the surface abrade rocks or other sand grains with which they collide. The results are seen in polishing, pitting, grooving, and faceting of rocks exposed to abrasion and rounding of sand grains in transport. Etched and pitted rocks, known as *ventifacts*, are common in deserts. Pebbles which are occasionally rolled over may develop several faceted sides; those with three faceted sides are called *dreikanters*. If the sand is eventually winnowed away, leaving only pebbles behind, a *desert pavement*, consisting of a residual layer only a pebble or so thick, may be developed.

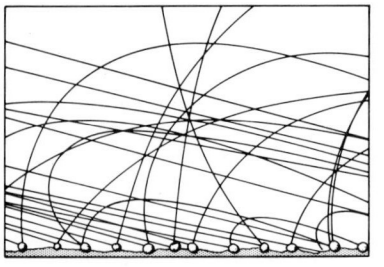

(a) Grain paths over pebble surface

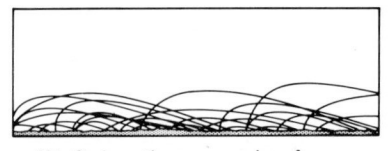

(b) Grain paths over sand surface

Fig. 15-2   Difference in saltation height over pebbly surface and sandy surface. *(After R. A. Bagnold, Methuen & Co., Ltd., London, 1941.)*

Sand grains bouncing along a rocky surface are highly elastic, and are more easily kept in motion than grains striking an accumulation of sand, where momentum is dissipated in the loose surface. Thus, grains bouncing along a rocky surface slow down when striking a dune, and collect in one place rather than spreading out evenly over a relatively barren area.

Grains which are too large to be lifted from the ground surface may be impelled forward by the impact of smaller grains as they strike the surface. A high-speed grain moving by saltation can move a grain 6 times its diameter, or more than 200 times its own weight, by impact. Thus a certain amount of sand may be moved downwind by *surface creep*.

ripples

As shown by Bagnold's wind-tunnel experiments, saltating sand grains of a given size strike a flat surface at approximately the same angle and momentum. However, flat surfaces of sand are not likely to remain stable, because, if for any reason the flatness is disturbed, as, for example, if more grains happen by chance to be moved out of a small area than are moved into it, the angle of incidence changes relative to the surface. In Fig. 15-3 the incidence of saltating sand grains is represented by a series of parallel lines. Once a flat surface is disturbed, fewer grains per unit area strike the upwind side of a hollow *AB* than strike the downwind side *BC*. More grains are driven up the slope *BC* than down *AB*, and the initial hollow is enlarged. Since the bombardment of sand is more intense on slope *BC* than on a level surface, grains excavated from the hollow accumulate at C, forming another lee slope CD, where grain movement is reduced. This in turn causes another hollow, and the process is repeated until the surface is completely rippled.

Fig. 15-3 Development of ripples by differences in intensity of sand bombardment on windward and lee slopes. (After R. A. Bagnold, Methuen & Co., Ltd., London, 1941.)

Once a sand surface becomes rippled, the size and wavelength (distance from crest to crest) of the ripples remain relatively constant for a given average wind velocity. This is because, for a given strength of wind, there is a definite average, or characteristic, path taken by saltating grains. As seen in Fig. 15-4, of all the grains which are ejected from any given small area, a greater number will fall on a second small area situated a distance of one characteristic path downwind than on any other area. Repetition of this situation downwind results in a fairly uniform wavelength. Thus an equilibrium system is established, and the size of the ripple becomes constant when as many sand grains strike the upwind portion of a ripple as are ejected from it. Bagnold experimented with several different wind velocities, and found that ripple wavelengths increased with wind velocity until the speed reached about three times the velocity needed to start grains moving, and then the ripples were destroyed.

Fig. 15-4 Relationship between ripple wavelength and characteristic paths of saltating sand grains. (After R. A. Bagnold, Methuen & Co., Ltd.. London, 1941.)

The size of the ripples is determined largely by the size distribution of the particles. As the crest of a ripple becomes higher, only the larger and heavier grains can remain near the crest because the smaller, lighter grains are removed, leaving the coarser grains to protect the crest and allow it to rise into a region of stronger wind than would otherwise be possible.

sand dunes

When sand accumulates in sufficient quantity, the flow of air over and around it becomes disturbed. Once the pile of sand is high enough to interfere with air flowing over the surface, a wind shadow is produced which has an important effect on the further accumulation of sand (Fig. 15-5). Sand transported up the windward side of the dune encounters the wind shadow at the crest of the dune, and is deposited in the lee of the crest, until the slope reaches the angle of repose for loose sand, at which point the sand slides down the lee side and a *slip face* is produced, which further accentuates the efficiency of trapping sand. As sand moves up the windward slope and is deposited on the slip face, the dune progressively advances downwind.

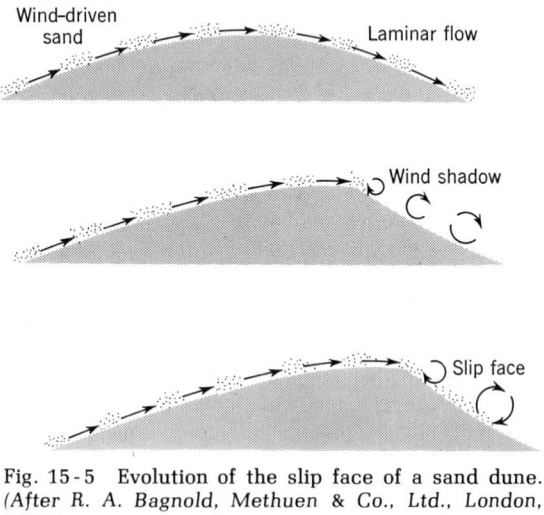

Fig. 15-5 Evolution of the slip face of a sand dune. (After R. A. Bagnold, Methuen & Co., Ltd., London, 1941.)

Where the prevailing wind direction is relatively uniform and vegetation is scarce, crescent-shaped *barchans* are formed (Figs. 15-6 and 15-7). The ends, or horns, of the barchan point downwind, and the steep slip face of the dune lies in the concave portion in the wind shadow. The crescent shape is maintained as the barchan advances downwind. Barchans are often best developed on barren desert floors, where there is a scanty supply of sand. They may occur singly or in groups, where individual barchans coalesce to form complex dunes. Since the rate of advance of the slip face of a dune is inversely proportional to the dune height, smaller dunes migrate downwind at a somewhat faster rate than larger dunes.

In regions where the prevailing wind direction is variable, dunes may be strung out in long chains, often referred to as *seif dunes*. They are a great deal larger than barchans, some in southern Iran rising as high as 700 feet above their bases. Seif chains are often very straight for long distances, sometimes as long as 60 to 100 kilometers. According to Bagnold (1941), each chain "grows in height and width by trapping sand during the periods of strong cross-winds, and extends lengthwise during the longer periods of more settled conditions when the wind blows down the chain." (p. 225)

Parabolic dunes have crescent-shaped forms similar to barchans, except that the horns, or points, face in the opposite direction, i.e., upwind rather than downwind (Fig. 15-10), as a result of blowouts of sand which has been partly stabilized by vegetation. The wind blows out the center of the dune, leaving behind the sides which are anchored by vegetation. In some cases

Fig. 15-6   Barchan sand dunes, Sherman County, Oregon. *(G. K. Gilbert, U.S. Geological Survey.)*

the points of the dune form may be left far behind, giving the dune shape a hairpinlike form.

Longitudinal dunes, consisting of long sand ridges parallel to the direction of prevailing winds, may be formed where the wind funnels sand through a gap or notch in a rock ridge. As the wind blows through a gap, the velocity increases, but once through the gap, the velocity drops to that of the general flow, and in so doing a growing deposit of sand forms. Longitudinal dunes are also common below notches in the lee of cliffs, where sand-bearing winds converge through gaps and deposit sand in the shadow zone behind the brink

Fig. 15-7   Air view of a barchan. *(Wards Natural Science Establishment.)*

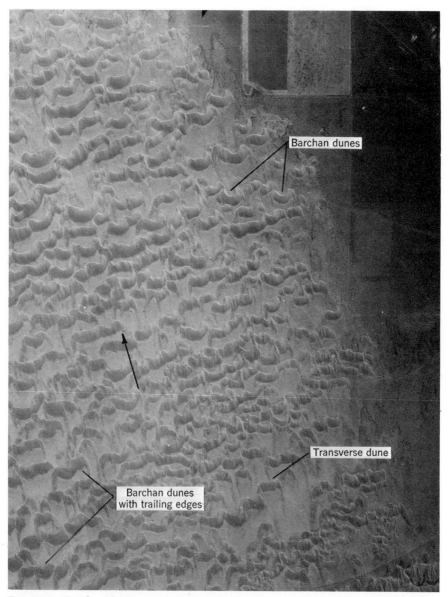

Fig. 15-8 Barchan dunes near Moses Lake, Washington. Arrow indicates direction of prevailing wind. *(U.S. Department of Agriculture.)*

of the cliff. The same relationships also hold for winds blowing through notches up onto a plateau (Fig. 15-12).

Transverse dunes formed at right angles to prevailing winds may develop under certain conditions, but long transverse dune ridges are usually not stable in open country because the wind breaks up the ridges. Occasionally, because of vegetation or other contributing factors, transverse dunes up to several thousand feet in length may form at right angles to the wind.

Fig. 15-9   Modified barchan dunes and parabolic dunes, Moses Lake, Washington. Arrow indicates direction of prevailing wind. *(U.S. Department of Agriculture.)*

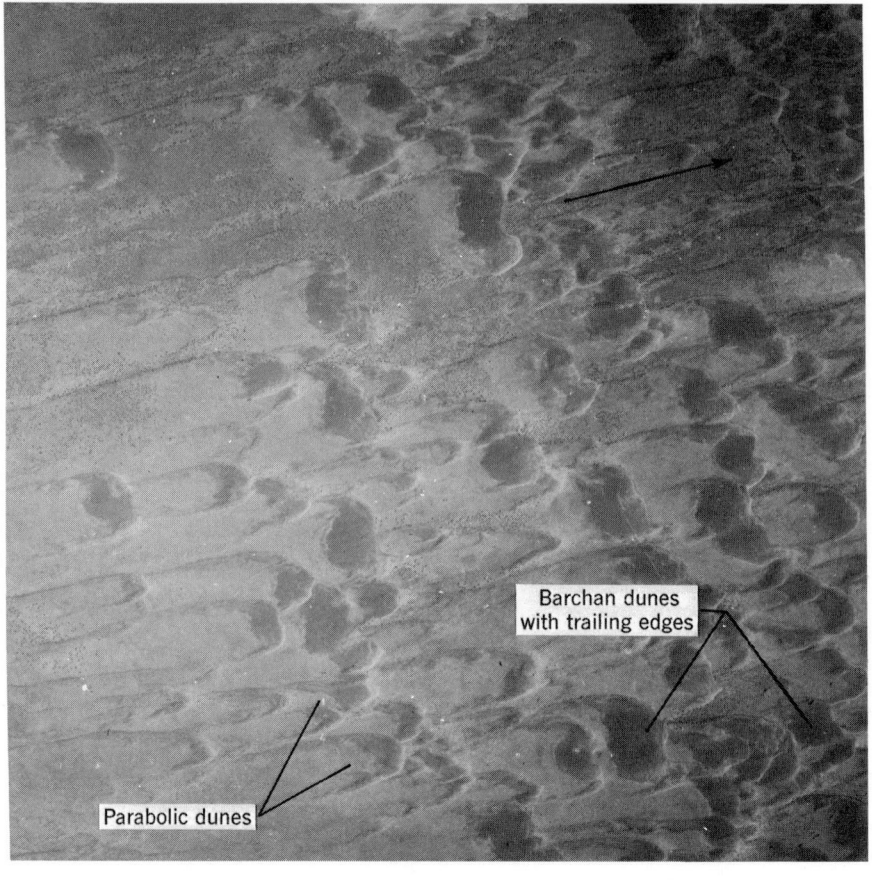

loess

Fine-grained silt winnowed out by wind erosion is sometimes deposited as a blanket of *loess*. This is especially true in areas downwind from glacial outwash plains or alluvial valleys. Extensive loess deposits occur in the Palouse country on the Columbia Plateau in eastern Washington, parts of the Mississippi Valley region, and portions of Europe, China, and Russia.

Fig. 15-10 Parabolic and barchan dunes, White Sands National Monument, New Mexico. Arrow indicates direction of prevailing wind. *(U.S. Department of Agriculture.)*

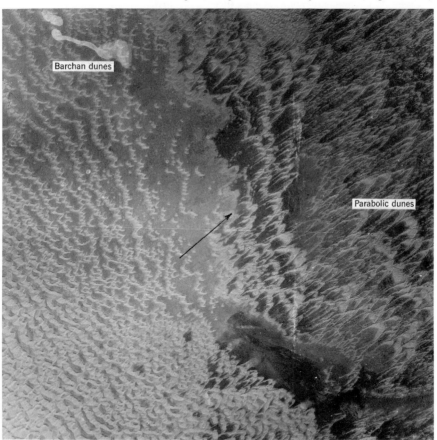

Loess is usually only a few feet thick, and mantles whatever topography it happens to be deposited upon. In some places where the loess is fairly thick and has been deposited by strong unidirectional winds, linear ridges are formed parallel to the wind direction. In parts of China, loess, deposited by silt-bearing winds from the Gobi Desert region, has accumulated to thicknesses of hundreds of feet.

Dunes are not normally formed in loess deposits because the fine-grained nature of the material presents a smooth surface texture and there is a much greater cohesion between particles. Wind-tunnel experiments have shown

Fig. 15-11   Parabolic dunes, Columbia River, Washington.

that wind velocities in excess of those which will move sand often will not move fine dust.

Loess which originates from the winnowing of glacial outwash often contains much rock flour produced by glacial abrasion. The milky water that issues from melting glaciers is highly charged with such material, and as the wind flows across temporarily deposited outwash sediments, such fine material is winnowed out. Loess which originates from nonglacial sources usually consists of fine particles of quartz, feldspar, clay, and other minerals.

Fig. 15-12 Longitudinal dunes, San Juan County, Utah. Arrow indicates direction of prevailing wind. *(U.S. Department of Agriculture.)*

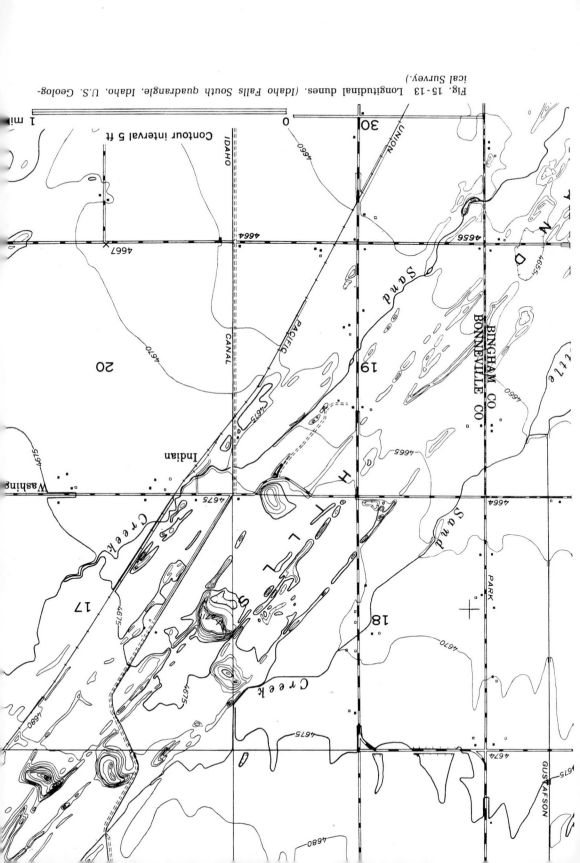

Fig. 15-13 Longitudinal dunes. (Idaho Falls South quadrangle, Idaho. U.S. Geological Survey.)

Contour interval 5 ft

references

Bagnold, R. A.: 1941. The Physics of Blown Sand and Desert Dunes, Methuen & Co., Ltd., London.

Beadnell, H. J. L.: 1910. The sand dunes of the Libyan Desert, *Geog. Jour.*, vol. 35, pp. 379-392.

Berkey, C. P., and F. K. Morris: 1927. Geology of Mongolia, Museum of Natural History, New York.

Cooper, W. S.: 1958. Coastal sand dunes of Oregon and Washington, *Geol. Soc. America Mem. 72.*

Finkel, H. J.: 1959. The Barcans of southern Peru, *Jour. Geol.*, vol. 67, pp. 614-647.

Fisk, H. N.: 1951. Loess and Quaternary geology of the lower Mississippi Valley, *Jour. Geol.*, vol. 59, pp. 333-356.

Hack, J. T.: 1941. Dunes of the western Navajo country, *Geog. Rev.*, vol. 31, pp. 240-263.

Leighton, M. M., and H. B. Willman: 1950. Loess formations of the Mississippi Valley, *Jour. Geol.*, vol. 58, pp. 599-623.

Long, J. T., and R. P. Sharp: 1964. Barchan-dune movement in Imperial Valley, Calif., *Geol. Soc. America Bull.*, vol. 75, pp. 149-156.

7

coastal morphology

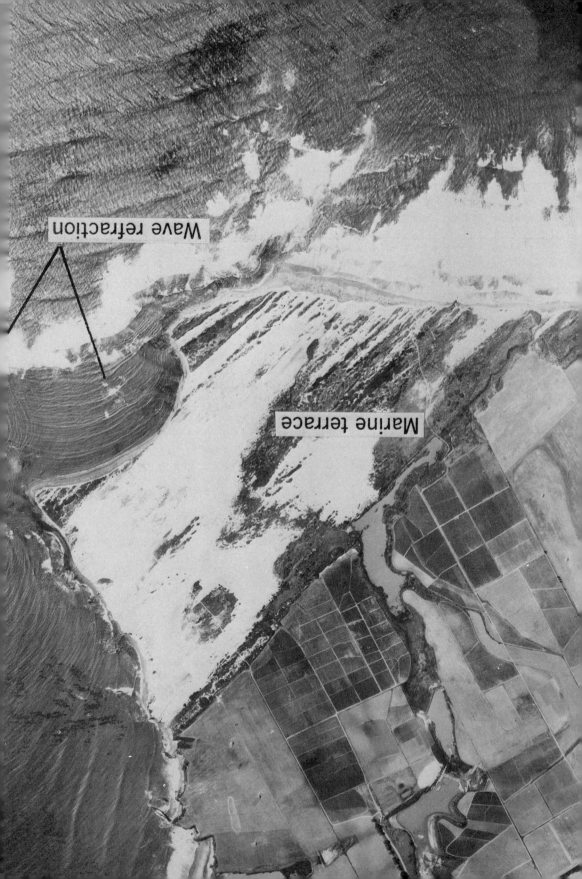
Wave refraction

Marine terrace

16

shorelines

wave motion

The relationship between wind velocity and wave motion is observable along any coastline. As wind passes over the water, drag, produced by friction, disturbs the smoothness of the surface, and ripples are developed. Once the flatness is disturbed, the wind pushes against the upwind side of the wave and imparts additional energy to it. The size of the waves produced depends on (1) the velocity of the wind, (2) the duration of the wind, and (3) fetch (the distance of open water over which the waves may build up without interference).

In the early stages of development the waves tend to be somewhat choppy, relatively small, and irregular. The small waves reach a maximum height and then are destroyed, while the larger ones continue to grow. When a wave crest becomes steeper than about 60° and the wave height becomes about one-seventh of its length, the top of the wave is usually blown off by the wind (Bascom, 1959).

Wave refraction and marine terrace, Santa Cruz, California. (U.S. Department of Agriculture.)

Wave motion is oscillatory. A cork floating in water through which waves are passing does not travel with the wave, but instead describes a nearly circular path, returning approximately, but not quite, to the same point after each wave passes. The diameter of the orbit (Fig. 16-1) is equal to the wave height, the vertical distance from the crest of a wave to the trough (Fig. 16-2). The form of the wave resembles a trochoid, a curve described by a point on a wheel rolled along a flat surface. Figure 16-1 illustrates the rapid decrease in orbital motion with depth and indicates motion of a particle on the water surface. As a wave approaches, a particle in a trough moves upward and toward the crest, and then, as the crest passes, moves downward and back toward its original position in the trough (compare the two positions on the orbital path shown in Fig. 16-1). Below the water surface the orbital motion of water particles is much reduced, and movement occurs in circles of decreasing diameter. The amount of motion at depth depends on the wave length and ratio of wave height to wave length. At a depth of one-half the wave length, the orbital diameters are about 4 percent as large as at the surface (Bascom, 1959), and at depths of one wave length the orbital diameters are about 1/535 of the surface diameter.

Direction of wave movement

Wave crest

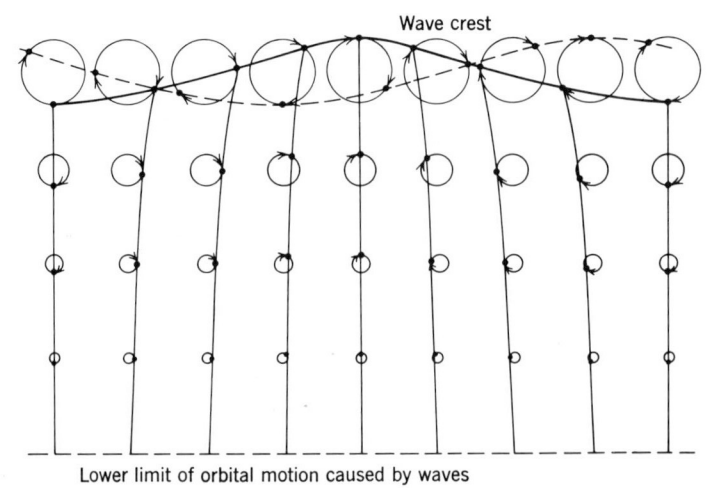

Lower limit of orbital motion caused by waves

Fig. 16-1 Relative movement of water particles caused by waves. Dashed line indicates the position of the water an instant later. (After U.S. Hydrographic Office Publ. 11.)

Waves developed during a storm are usually not simple, but are of different lengths and heights, some of which are superimposed upon one another. Interference of these waves with one another produces compound waves.

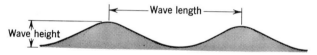

Fig. 16-2   Wave length and wave height.

Where the crests of two wave trains coincide, an exceptionally large wave is developed, but where the crest of one wave train coincides with the trough of another, the two cancel each other. The result is that some waves approaching a shore are larger than others, giving rise to the apparently unfounded popular belief that every seventh wave is larger than the others.

Storm waves over 45 feet high are not uncommon in the open sea, where waves may be built up over long distances. One of the highest waves ever observed, at least 112 feet high, was reported by the *U.S.S. Ramapo* in 1933. This wave was one of many large waves produced by a 30- to 60-knot gale which blew in the North Pacific for several days. Along the Oregon Coast, where severe storms are common, rocks are occasionally thrown through lighthouse windows, which are well above sea level. In one instance a 135-pound rock was hurled 100 feet above sea level, where it fell through the roof of the light-keeper's house.

When waves travel away from the storm that caused them, their crests are usually lower and more smoothly rounded than typical waves near the wind source, and they characteristically have relatively long wave lengths. Such waves, known as *swells*, may travel long distances, far beyond the storm that created them. The average period of swells reaching the shores of the United States is about 5 to 10 seconds. The longest reported swell had a period of 22.5 seconds, which corresponds to a wave length of about 2,600 feet and a velocity of about 78 miles per hour. The height of a swell decreases, and the period and wave length increase, with distance away from the storm area.

tsunamis

A type of ocean wave often of unusual size is the *tsunami*, or seismic sea wave, which is sometimes associated with earthquakes and related landslides. It has a very large wave length, generally about 100 miles or more, and low wave height until the wave approaches shallow water near shore. In the open ocean the waves may travel with velocities of 300 to 400 miles per hour, yet would not be felt by a ship because of the low wave height. When they arrive at a coast, however, the wave height may grow 30 to 40 feet and cause tremendous havoc when they break on the shore. The tsunami which struck

Hilo, Hawaii, on May 23, 1960, brought waves of 30 to 35 feet, which caused heavy damage and took 16 lives.

A wave of exceptional height was generated in Lituya Bay, Alaska, in 1958 by a landslide triggered by an earthquake. On July 9, 1958, an earthquake, set off by fault movement along the Fairweather fault, caused a landslide involving some 40 million cubic yards of material at the upper end of Lituya Bay (Fig. 16-3). The landslide fell into the head of Lituya Bay, creating a wave 1,740 feet high on the opposite shore and sending a wave down the bay at a velocity of about 100 miles an hour (Miller, 1960). Three fishing boats were anchored in the bay at the time. The wave carried one boat like a surfboard over a spit at the lower end of the bay, some 80 feet above trees growing on the spit. The second boat rode out the wave inside the bay, and the third boat was swamped and never seen again. The force of the wave stripped off trees and other vegetation down to bare rock, including some measuring 4 feet in diameter (Fig. 16-4).

Although tsunamis occur only rarely, they are capable of great destruction and can produce significant modifications of shorelines in a very short period of time.

shallow water waves and wave refraction

As waves approach a shoreline and encounter shallower water, the orbital motion of the water particles encounters interference with the sea floor, causing the wave to change velocity and form. The wave length and velocity are decreased, the wave height is increased, and the wave form becomes modified. The wave period, however, remains the same. Usually, these changes begin to take place when the water depth is approximately one-half that of the wave length. In shallow water, orbital motions, which may have been nearly circular in deep water, become elliptical, with progressively flatter orbits near the bottom, until particle movement on the sea floor consists of a back-and-forth motion, seaward under passing wave troughs and landward under passing crests.

The wave length decreases as waves travel shoreward into progressively shallower water as a result of the decrease in velocity with a constant wave period; i.e., the same number of waves must pass a given point each unit of time even though the velocity is less than in deep water. Wave height also undergoes a change. When the water depth is approximately one-twentieth that of the deepwater wave length, the height of the wave begins to increase sharply, and the crest of the wave begins to move ahead of the lower part of the wave. Eventually, the crest moves too far ahead of the rest of the wave, and the unsupported water in the crest curls over, and a *breaker* is formed.

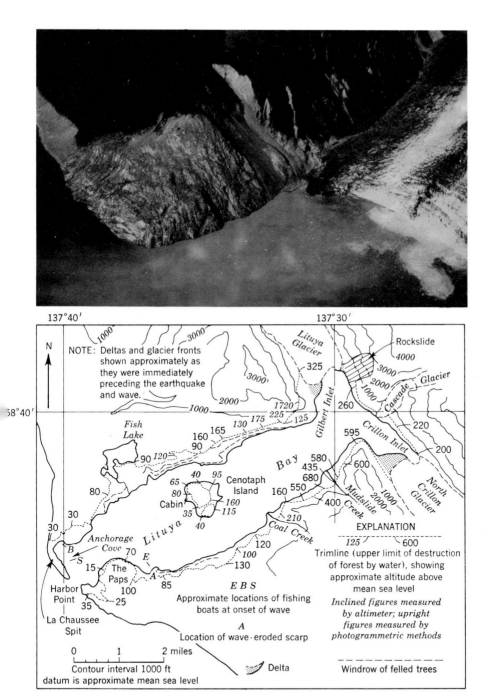

Fig. 16-3 Damage by a giant wave at Lituya Bay, Alaska, following the July 9, 1958, earthquake. *(Top)* Scouring action of the wave at the head of Lituya Bay. Elevation of the top of the scoured area is 1,720 feet. *(D. J. Miller, U.S. Geological Survey.) (Bottom)* Map of Lituya Bay showing effects of 1958 giant wave. *(D. J. Miller, U.S. Geological Survey.)*

The stillwater depth beneath a breaker is generally about one-third times the height of the wave. If the slope of the bottom is relatively steep and a large swell impinges on the sea floor abruptly, the water necessary to fill the wave form decreases rapidly, and a *plunging breaker* forms in which the wave crest curls over in a half-cylinder form, flinging the crest into the preceding trough with a great roar. If the sea floor is more gentle, a *spilling breaker* forms, in which the crest of the wave spills over the advancing front of the wave without completely destroying the wave.

If waves approach a uniformly sloping shoreline at right angles to the shore, each wave will break at approximately the same time, and all the waves will

Fig. 16-4   Scoured area along the margins of Lituya Bay. *(D. J. Miller, U.S. Geological Survey.)*

strike the shore in straight, parallel lines. However, if the waves approach the shoreline at an angle, part of each wave will impinge on the shallow sea floor before the rest of the wave, and that part will be slowed down, resulting in a bending, or *refraction*, of the wave (Fig. 16-5). A series of waves approaching an irregular shoreline, as in Fig. 16-6, will first encounter shallow water in the vicinity of the headlands or peninsulas, causing the waves to become refracted as shown, until the waves begin to parallel the coastline approximately. Lines drawn at right angles to the wave crests converge on the headlands and diverge in the bays, thus concentrating wave energy on the headlands. In Fig. 16-6 the wave energy at *A* is equal to the wave energy

Fig. 16-5   Wave refraction around a small island, Santa Barbara County, California. *(U.S. Department of Agriculture.)*

at B for a given wave, but as the wave becomes refracted, the energy of A becomes concentrated over a small area C on the headland, whereas the same energy of B is spread over a much larger area, D, in the bay. Thus headlands are characterized by high breaking waves, rocky shores, and intense erosion, while bays are typically quiet, with sandy beaches.

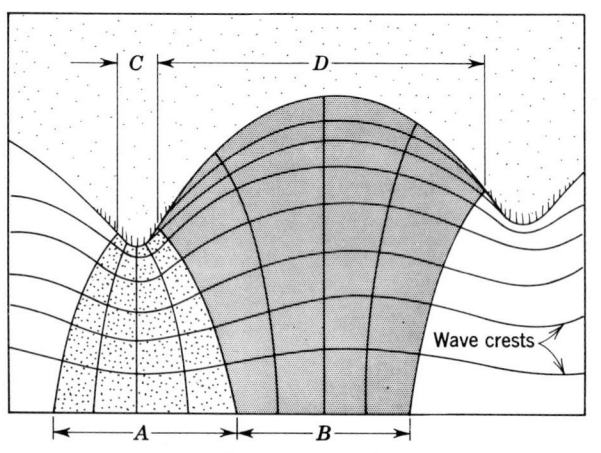

Fig. 16-6 Wave refraction and distribution of wave energy along an irregular coastline. The two wave segments, A and B, have equal energy, but because of wave refraction, the energy of A is concentrated on C, whereas the energy of B is distributed over a much larger area, D.

wave erosion

The energy expended by waves breaking on a shoreline exerts a profound effect on a coastline as a result of mechanical erosion. Erosion is accomplished primarily by the hydraulic pressure and impact of waves striking the shore and by abrasion of sand and pebbles moved incessantly by the water.

The effect of impact and hydraulic pressure is often most obvious where it has affected man-made emplacements. Along the coast of Scotland, where very large storm waves are common, a breakwater weighing 2,600 tons was destroyed in a single storm.

Storm waves are especially effective where rocks along the shore are highly jointed or bedded, and are thus vulnerable to quarrying. Such rocks are eroded block by block as hydraulic pressure and the impact of waves progressively pry loose masses of rock.

Anyone standing on a pebbly beach as waves are breaking on the shore is aware of movement of particles up the beach with each wave and back again seaward as the water runs back. As the particles are dragged back and forth with each wave, they abrade the bedrock along the shore, and also abrade each other, gradually wearing pebbles into sand-size particles and smaller. Wave action thus takes on the characteristics of a great horizontal saw, the teeth of which are the loose particles moved by the waves. Abrasion is not confined to the beach, because each wave in shallow water induces a back-and-forth movement of particles on the sea floor beneath it. The limiting depth at which movement of sediment on the bottom is produced by waves is known as the *wave base*. Small particles may be agitated at depths up to 600 feet, but below that, movement is negligible.

Wave erosion undercuts slopes at the shoreline, and *sea cliffs* are formed, which retreat landward under wave attack. Such cliffs commonly show a notch cut at the base, where wave action is most vigorous. Prolonged erosion and retreat of the sea cliff produce a *wave-cut platform*, or bench (Fig. 16-7), beveling the rocks on the sea floor near the shoreline. Masses of in-

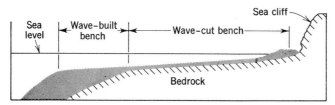

Fig. 16-7   Diagrammatic cross section of near-shore features.

completely beveled rock may remain above the general level of planation as *sea stacks* (Fig. 16-8), and along the sea cliffs selective erosion of weak zones or highly jointed rock may etch out caves. Erosion of caves on a narrow promontory may result in breaking completely through to form an arch, or *natural bridge*.

Rock particles broken from the bedrock do not accumulate in ever-thickening piles, but are transported laterally along the shore or seaward until the motion of the water is no longer able to carry the material farther. Generally, as wave motion on the sea floor becomes reduced in the deeper water seaward from the shoreline, the coarser particles are deposited, whereas the smaller ones continue in motion until they too can no longer be transported. A profile of equilibrium is thus developed, consisting of the rock-planed, wave-cut platform and a wave-built platform which lies seaward. The profile is usually concave upward, reflecting the effect of wave vigor and the size of particles in transport. Near the shore, where wave action is most effective, the slope is relatively steeper than in deeper water, where the slope of the sea floor flattens. On a sea coast where waves are particularly powerful, the profile is steeper than on a coast where waves are less strong or where the coast is protected. Since changes in wave activity and in the level of the sea tend to change the profile of equilibrium, the form of an offshore profile is probably adjusted to long-term conditions or to a given set of conditions, rather than to short-term fluctuations.

transportation and deposition

Waves impinge against most coastlines at an angle other than 90°, and as a

Fig. 16-8   Wave-cut bench and sea stacks near San Luis, California. Note the marine terrace with former sea stack in the left-hand corner. *(Wards Natural Science Establishment.)*

result, each wave which breaks on the shore travels obliquely up the beach. As the water runs back down the beach, it travels directly down the slope, rather than obliquely. Particles carried by each wave are thus carried up the beach at an angle to the slope, and then return downslope in the backwash (Fig. 16-9). Each wave imparts a lateral component of movement to the transported particles and results in *beach drifting*, which moves sediment progressively along the shore.

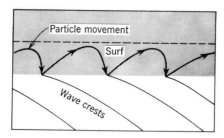

Fig. 16-9 Beach drifting caused by waves approaching a beach at an angle.

Sediment may also be transported in the shallow water just offshore by *longshore currents* flowing parallel to the shore. Both beach drifting and longshore drifting combine to move substantial amounts of sediment laterally along the coastline. The particles transported along a coast may be derived from the erosion along the shoreline or from streams bringing sediment to the sea.

Sediment transported by longshore and beach drifting moves parallel to the shoreline, but when deepwater bays are encountered along an irregular shoreline, sediment carried into the deeper water comes to rest, resulting in the development of *spits* and *bars*. A spit is a ridge of sediment connected at one end to land and terminating in open water at the other end (Fig. 16-10). Once formed, a spit is enlarged in the direction of transport by continued distribution of sediment along the spit. Often the end of the spit curves landward (Fig. 16-11). Eventually, a spit may extend all the way across a bay, forming a *baymouth bar*, enclosing a *lagoon*. A narrow tidal channel may persist through the bar until the tidal currents are no longer able to sweep out the influx of sediment. Occasionally, a spit will connect an island to the mainland, forming a *tombolo*.

The zone along the shore between high and low tide is commonly referred to as the *beach*. It is usually characterized by sand or gravel deposited by wave activity, the coarseness of material depending in part on the vigor of wave attack and the nature of the source of the deposits. Some beaches

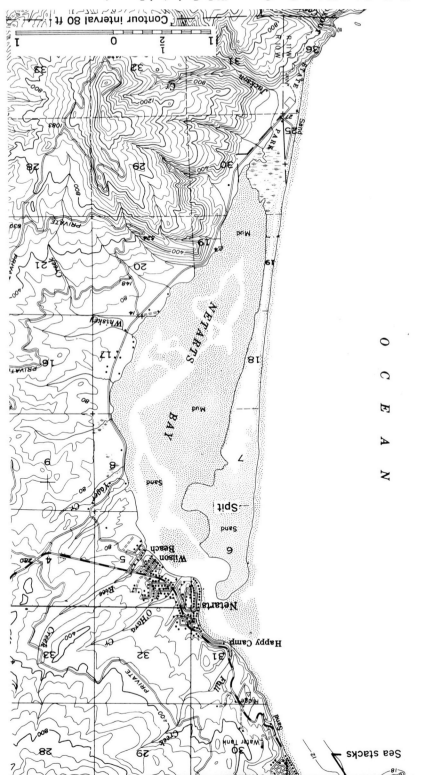

Fig. 16-10 Spit at Netarts Bay, Oregon. (U.S. Geological Survey.)

consist of sand during the summer months, but during the winter the sand is swept away by storm waves, leaving only gravel. The effects of wave refraction often result in development of *bayhead beaches* in the upper portions of bays. High storm waves drive beach sand landward, where the sand often accumulates to form a *storm beach* above normal high tide.

Along shorelines where there is an unusual abundance of sediment, as where streams bring much material to the ocean, sand may accumulate more rapidly than it can be swept away by waves and currents, progressively shifting the shoreline seaward. Such shorelines are said to be *prograding*.

Offshore bars, also known as *barrier beaches* (Figs. 16-12 and 16-13), form some distance from the coast, and are separated from the mainland by a lagoon. Most such bars are formed on gently sloping sea floors and are usually composed of sand.

The origin of offshore bars is not fully understood beyond the fact that they are formed by wave and current actions. Johnson (1919) thought that they were made by breakers offshore throwing up sand into a submarine ridge, which eventually evolves into a chain of islands, and finally into a continuous bar above sea level.

Man-made groins, built to trap sand along a shore, may create local beaches, but groins interfere with the normal lateral movement of sand along the beach, and consequently areas down the beach are "starved" by the cutting off of the sand supply. Along the New Jersey coast sand is dumped artificially to act as a source for distribution along the resort beaches by wave action.

classification of shorelines

The complexity of most shorelines makes them difficult to fit into a unified scheme of classification. A number of classifications which would try to include all major types of shorelines have been attempted, but none have proved to be entirely adequate, largely because of the diversity and compound nature of many shorelines.

A widely used classification is that of Johnson (1919), which separates shorelines into four main categories:

1. *Shorelines of submergence,* formed by partial submergence of a land mass as a result of rise of sea level or subsidence of the land. Such shorelines are characterized by drowned valleys, deep embayments, and a very irregular configuration (Figs. 16-14 and 16-15).

2. *Shorelines of emergence,* formed by uplift of the land or lowering of sea level. Such shorelines are characterized by relatively straight coasts of low relief, marine terraces, and offshore bars.

3. *Neutral shorelines*, whose essential characteristics are independent of either submergence or emergence. Among the various examples of these shorelines are

   a. Deltas, alluvial plains, outwash plains
   b. Volcanic shorelines
   c. Coral reefs
   d. Faulted shorelines

4. *Compound shorelines*, characterized by features of more than one of the categories listed above. For example, a shoreline which exhibits both marine terraces and drowned valleys, as a result of fluctuations in relative sea level, is called *compound*.

Fig. 16-11 Spit at Monomoy Point, Massachusetts. Note the former position of the

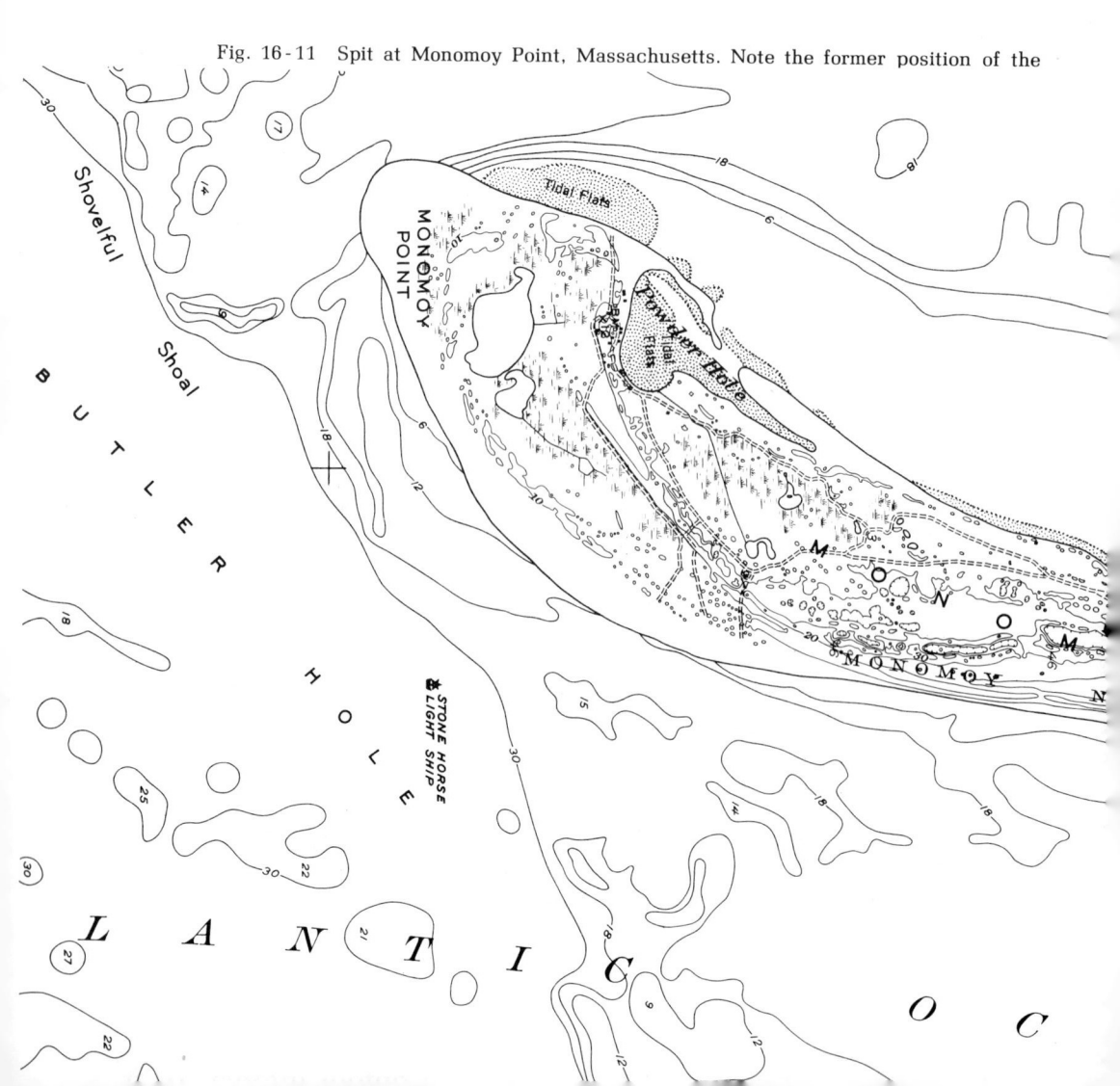

Johnson's classification has the advantage of being simple and genetic, and it includes all the major types of recognized shorelines. One of the principal disadvantages of the system is that, because of changes of sea level during the Pleistocene glaciations and interglaciations, many shorelines have undergone both submergence and emergence, including some shorelines best classified as neutral on the basis of other factors. However, the classification remains very useful, because it focuses attention on factors most likely to exert profound influences on a shoreline.

Shepard (1948) proposed the following classification, which places a somewhat different emphasis on the various factors related to shoreline development.

curved end of the spit at Inward Pt. *(U.S. Geological Survey.)*

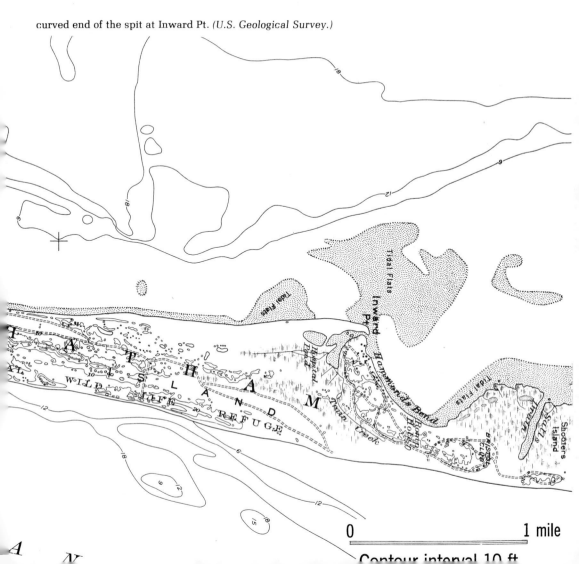

0   1 mile

Contour interval 10 ft

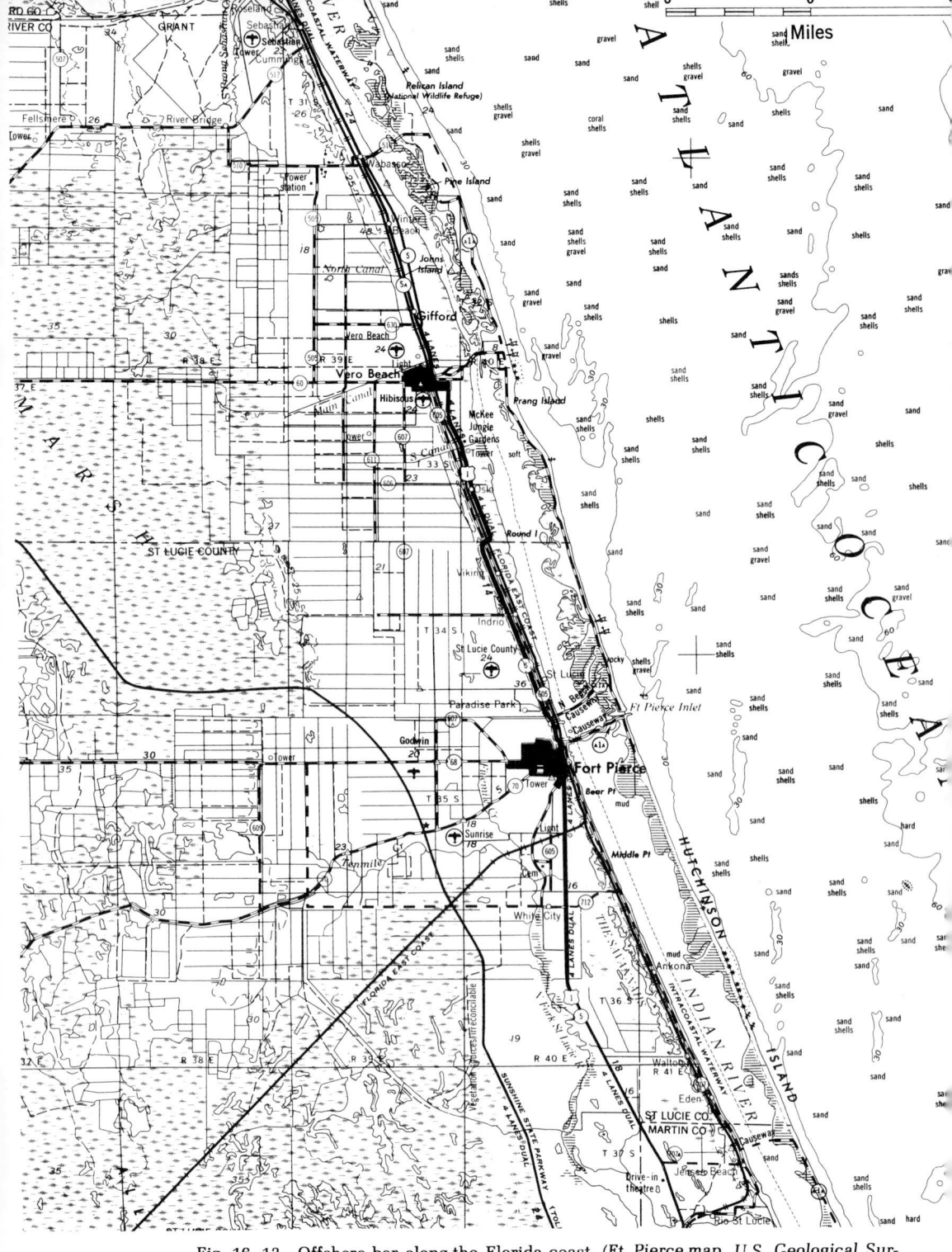

Fig. 16-12 Offshore bar along the Florida coast. (*Ft. Pierce map, U.S. Geological Survey.*)

I. Primary, or youthful, coasts and shorelines, produced chiefly by nonmarine agencies

    A. Shaped by erosion on land and drowned as a result of deglaciation or downwarping

        1. Drowned river coasts (ria coasts)

        2. Drowned glaciated coasts

    B. Shaped by deposits made on land

        1. River-deposition coasts

            a. Deltaic coasts

            b. Drowned alluvial plains

Fig. 16-13 Offshore bar along the Louisiana coast. (*Wards Natural Science Establishment.*)

2. Glacial-deposition coasts

   a. Partially submerged moraines

   b. Partially submerged drumlins

3. Wind-deposition coasts (prograding sand dunes)

4. Vegetation-extended coasts (mangrove swamps)

C. Shaped by volcanic activity

   1. Recent lava flows

   2. Volcanic collapse or explosion (calderas)

D. Shaped by diastrophism

   1. Fault-scarp coasts

   2. Coasts related to folding

II. Secondary, or mature, coasts and shorelines shaped primarily by marine agencies

A. Shaped by marine erosion

   1. Sea cliffs straightened by wave erosion

   2. Sea cliffs made irregular by wave erosion

B. Shaped by marine deposition

   1. Coasts straightened by deposition of bars across estuaries

   2. Coasts prograded by deposits

   3. Shorelines with offshore bars and longshore spits

   4. Coral reefs

Shepard's classification is also quite useful, but, as with other classifications, it has a number of drawbacks. Erosion and deposition are likely to be occurring on a coast simultaneously, eroding headlands and depositing the material on spits and bars, so that it is often quite difficult to describe a coast as erosional or depositional. In addition, the classification suffers from the same difficulties as Johnson's as a result of fluctuations of Pleistocene sea levels.

evolution of shorelines

Shorelines subjected to long-continued wave erosion and deposition undergo progressive change, and will evolve through a sequence of forms. At each stage of development certain characteristic forms are typically found.

A shoreline recently submerged is quite irregular in configuration, especially if there was appreciable relief of the area prior to submergence. Stream
valleys are drowned, forming deep estuaries; previous stream divides become
peninsulas; and small islands are numerous. Soon after a coast is submerged,
wave erosion begins to attack headlands and the seaward margins of islands,
developing sea cliffs and initiating the development of wave-cut benches
(Fig. 16-16). Material eroded from the headlands is distributed laterally by
beach and longshore drifting, and spits are constructed from the headlands
into the adjacent bays and estuaries. Bayhead beaches begin to form at the
heads of the bays, and deltas begin to form where streams enter the bays.

Fig. 16-14 Submergent shoreline along the coast of Maine near Brunswick. *(John
Shelton.)*

Fig. 16-15 Submergent shoreline near Boothbay Harbor, Maine. (*Boothbay, Maine, quadrangle, U.S. Geological Survey.*)

Some of the islands become tombolos, attached to the mainland by spits. With continued erosion of the headlands and distribution of sediment, baymouth bars extend across the estuaries, enclosing lagoons which gradually become reduced in size by delta building at the head of the lagoon. As the lagoons are filled with sediment, they first become swamps, which are themselves destroyed by sedimentation. As the shoreline approaches maturity, it is considerably straighter than its youthful form. With further wave erosion, the shoreline retreats landward in a more or less straight line. As the wave-cut bench becomes more and more extensive by beveling of the land, waves break

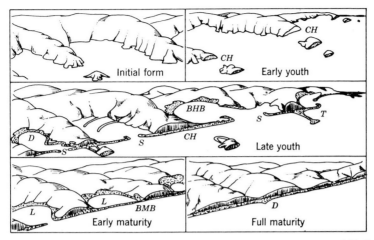

Fig. 16-16   Evolution of shorelines along a submergent coast. *CH*, cliffed headland; *S*, spit; *BHB*, bayhead beach; *T*, tombolo; *D*, delta; *L*, lagoon; *BMB*, baymouth bar. *(A. N. Strahler, John Wiley & Sons, Inc., New York, 1960.)*

initially farther and farther from shore on the shallow water of the bench, thus reducing the vigor of wave attack on sea cliffs, and as old age is approached, an extensive marine plain exists, and sea-cliff recession takes place at an ever-decreasing rate.

Recently emergent shorelines would present a much different assemblage of shoreline features. Initially, if the former sea floor was relatively flat, the emergent shoreline is quite straight, and waves breaking in the shallow water offshore create an offshore bar, separated from the mainland by a lagoon. Marine terraces mark the position of wave-cut benches formed prior to emergence (Figs. 16-19 and 16-20). Wave attack on the offshore bar and regrading of the sea floor cause the bar to shift landward, and swamps begin to fill in the lagoon (Fig. 16-18). Continued migration of the offshore bar toward the land until the swamps in the lagoon have been eliminated marks the attainment of maturity. Thereafter, the shoreline retreats in a more or less straight line under wave erosion, and an extensive wave-cut bench is developed as old age is approached.

coral reefs

Shorelines in tropical climates are likely to be strongly affected by the growth of reef-forming corals and algae. The environment in which coral and algae

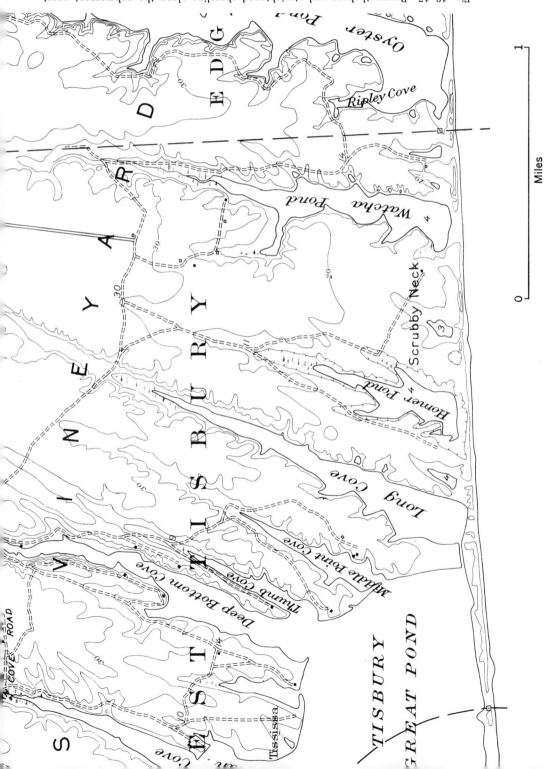

Fig. 16-17 Baymouth bars and straightened shoreline along the submergent coast of Martha's Vineyard, Massachusetts. (Tisbury Great Pond quadrangle, U.S. Geological Survey.)

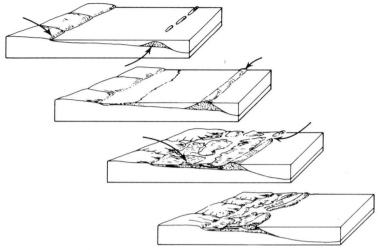

Fig. 16-18 Evolution of shorelines along an emergent coast. (A. N. Strahler, John Wiley & Sons, Inc., New York, 1960.)

can flourish is restricted to tropical ocean water, largely as a result of temperature conditions necessary for growth. Corals can prosper in water temperatures around 68 °F, but do best in water between 77 and 86 °F. Thus coral reefs are limited to latitudes approximately between 30N and 25S. They also require shallow water, usually not more than about 200 feet deep, partly because algae require sunlight for growth, and sunlight generally does not penetrate much below that depth. The water must be relatively clear and free of sediment and of normal salinity. Where seawater is diluted by freshwater streams, coral growth is inhibited. Since corals are attached to the sea floor and cannot move about, they are dependent on agitation of the water for bringing them food and oxygen. They may obtain some oxygen from algae, but appear to thrive on the windward side of islands, where wave action is greater.

Charles Darwin, on the famous voyage of the *Beagle,* visited only a few reefs in the Pacific, but recognized three different types, and postulated a theory of their origin which appears to be essentially correct in the light of present knowledge. *Fringing reefs* occur on the coastline, and are attached to the land in a belt varying in width from a few hundred feet to half a mile. They are often absent where streams enter from the land, because of dilution by the freshwater, and in muddy bays. The reef may be exposed at low tide, but since corals cannot live for long out of the water, they do not persist very much above the low-tide zone. *Barrier reefs* occur just offshore, and are separated from the mainland by a lagoon which may vary in width from ½ mile or less to more than 10 miles. Coral fragments, mollusc shells,

foraminifera, and other calcareous organisms collect on the seaward side
of the reef as a debris pile and contribute to the formation of the reef mass.
One of the largest barrier reefs in the world is the Great Barrier Reef, which
extends for more than 1,200 miles along the north coast of Australia. A third
type of reef is the *atoll*, a ringlike reef enclosing a lagoon without a central
island (Fig. 16-22).

The origin of barrier reefs and atolls has provoked much discussion since
the publication of Darwin's reef studies in 1842. The principal problems are
the origin of the lagoon behind barrier reefs and atolls and the origin of the
ringlike form of atolls. Darwin proposed that coral growth originated around
the margins of island (usually volcanic cones) as a fringing reef and that, as

Fig. 16-19   Marine terraces on the Palos Verdes Hills near Los Angeles, California.
(John Shelton.)

slow subsidence of the island took place, the coral was able to grow rapidly enough to maintain shallow-water conditions necessary for survival. The growth of the reef is mostly upward, and perhaps a bit seaward, where wave action keeps the water agitated, so that, as the central island subsides, a lagoon is formed, which gradually enlarges with continued subsidence and reef growth. Eventually, if the central island subsides below sea level, only the reef is left, and an atoll is formed. Because of the important implications of subsidence involved, this theory has become known as the *subsidence theory*.

A different theory of origin, related to Pleistocene changes in sea level, was proposed by Daly (1934, 1942), who noted that the lagoons behind many barrier reefs and atolls have very similar depths, and postulated that present

Fig. 16-20 Marine terrace at Palos Verdes, California. Note how the former wave-cut bench bevels the rocks. *(Aero Service Division, Litton Industries.)*

reefs were formed during postglacial rise of sea level, following a period of erosion, during which wave erosion beveled broad rock platforms. Daly's theory, known as the *glacial control theory*, called for wide destruction of coral reefs during a low stand of sea level in glacial episodes, partly by reduction of the rate of growth of corals in lower water temperatures and partly by wave erosion, resulting in development of broad marine benches. Then, with the return of warmer water and rise of sea level in postglacial times, reef building increased rapidly, and reefs were formed on the beveled platforms.

Fig. 16-21 Ancient shorelines along the east end of Great Bear Lake, Northwest Territories. (*National Air Photo Library, Surveys and Mapping Branch, Department of Energy, Mines, and Resources.*)

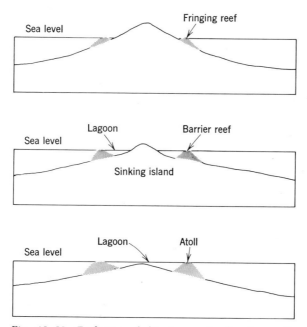

Fig. 16-22 Evolution of fringing reefs, barrier reefs, and atolls by subsidence.

Although water was almost certainly lowered somewhat during the Pleistocene, and sea level did fluctuate several hundred feet, there is considerable doubt that these changes were of sufficient magnitude to produce the effects postulated in Daly's theory.

Of critical importance in any theory of origin of reefs is the depth to which reef deposits are found. According to the glacial control theory, reef deposits should be relatively thin, but according to the subsidence theory of Darwin, the deposits should be much thicker. Test holes drilled on Bikini in 1947 penetrated 2,558 feet of coral material, and two holes drilled on Eniwetok in 1951 and 1952 went through 4,000 feet of shallow-water reef deposits before encountering basalt, conclusively demonstrating that subsidence has played a major role in the origin of these islands. Although not all reefs necessarily show such profound subsidence, the evidence from drilling on Bikini and Eniwetok suggests that Darwin's subsidence theory of atoll formation is correct. However, the rise and fall of sea level during the glaciations of the Pleistocene undoubtedly had some effect on coral-reef development.

references

Bascom, Willard: 1959. Ocean waves, *Sci. American,* vol. 201, pp. 74-84.

Cloud, P. E.: 1958. Nature and origin of atolls, *Eighth Pacific Sci. Cong.,* vol. IIIA, pp. 1009-1035.

Daly, R. A.: 1934. The Changing World of the Ice Age, Yale University Press, New Haven, Conn.

――――: 1942. The Floor of the Ocean, The University of North Carolina Press, Chapel Hill, N. C.

Dana, J. D.: 1885. The origin of coral reefs and islands, *Am. Jour. Sci.,* vol. 30, pp. 89-105, 169-191.

Darwin, C.: 1898. The Structure and Distribution of Coral Reefs, 3d ed., D. Appleton & Company, Inc., New York.

Davis, W. M: 1928. The coral reef problem, *Amer. Geog. Soc. Spec. Pub. 9.*

Dobrin, M. B., B. Perkins, and B. L. Snavely: 1949. Subsurface constitution of Bikini Atoll as indicated by seismic refraction survey, *Geol. Soc. America Bull.,* vol. 60, pp. 807-828.

Eaton, J. P., D. H. Richter, and W. V. Ault: 1961. The tsunami of May 23, 1960, on the island of Hawaii, *Seismol. Soc. America Bull.,* vol. 51, pp. 135-137.

Emery, K. O.: 1948. Submarine geology of Bikini Atoll, *Geol. Soc. America Bull.,* vol. 59, pp. 855-860.

――――, J. I. Tracey, and H. S. Ladd: 1954. Geology of Bikini and nearby atolls, *U. S. Geol. Survey Prof. Paper 260-A.*

Gilbert, G. K.: 1890. Lake Bonneville, *U.S. Geol. Survey Mon. 1.*

Guilcher, André: 1958. Coastal and Submarine Morphology, Methuen & Co., Ltd, London.

Johnson, D. W.: 1919. Shore Processes and Shoreline Development, John Wiley & Sons, Inc., New York.

――――: 1925. The New England-Acadian Shoreline, John Wiley & Sons, Inc., New York.

King, C. A. M.: 1959. Beaches and Coasts, Edward Arnold (Publishers) Ltd., London.

Kuenen, P. H.: 1950. Marine Geology, John Wiley & Sons, Inc., New York.

Ladd, H. S.: 1961. Reef building, *Science*, vol. 134, pp. 703-715.

Miller, D. J.: 1960. Giant waves in Lituya Bay, Alaska, *U.S. Geol. Survey Prof. Paper 354-C.*

Shepard, F. P.: 1948. Submarine Geology, Harper & Row, Publishers, Incorporated, New York.

Thornbury, W. D.: 1954. Principles of Geomorphology, John Wiley & Sons, Inc., New York.

Wiens, H. J.: 1962. Atoll Environment and Ecology, Yale University Press, New Haven, Conn.

(8)

relationship
of geologic structure
to topography

folded
sedimentary rocks

evolution of topography
developed on folded sedimentary rocks

The effect of folded rocks on the development of landforms is in part related to initial forms developed by warping of the earth's crust and in part to differential erosion of rocks of folded structures. In view of rates of weathering and erosion relative to rates of folding, probably far more landforms developed on folded structures are related to differential erosion than to initial form caused by deformation. Few folds persist long without modification by erosion.

initial forms

Forces beneath the earth's surface, responsible for deformation of the rocks of the crust, produce various types of folds. In some places folding has been geologically recent enough so that weathering and erosion have not yet markedly modified the initial shape of the folds.

Folding of rocks into a series of anticlines and synclines initially results

Sheep Mountain anticline, Wyoming. *(John Shelton.)*

in *anticlinal ridges,* marking the location of each anticline, and *synclinal valleys,* marking the location of each syncline. If the axes of the folds are horizontal, a series of parallel ridges and valleys is formed, but if the fold axes are not horizontal, a series of tapering ridges and valleys is developed. A ridge in which the strata dip in opposite directions away from the crest may be termed an anticlinal ridge, even though erosion may have stripped younger beds from the top.

Examples of anticlinal ridges and synclinal valleys occur in the Columbia Plateau of eastern Washington, where Miocene basalt flows have been folded into a series of anticlines and synclines which have not yet been breached by erosion.

Domes are circular or elliptical folds in which the beds dip away from a central point in all directions (Fig. 17‑1). Bowing up of the beds in domes may be accomplished either by tectonic warping or by intrusion of igneous material from below. In each case the resulting landform in initial stages is an oval-shaped hill or mountain. Until erosion has stripped off the upper beds in a dome, it is not possible to distinquish igneous domes from structural domes without subsurface information such as drill-hole data or magnetic, gravity, or seismic surveys. Intrusive igneous domes are discussed further in the section on igneous landforms. Drainage on newly uplifted domes is consequent in origin and forms a radial pattern.

Structural basins are folds in which beds dip toward a common center, forming initially oval basins or large depressions. Consequent streams flowing down the dip of the beds toward the center of the basin form a centripetal drainage pattern.

Flat-lying beds, or beds gently dipping in one direction, may locally assume steeper dips to form monoclines (Fig. 17‑2). Topographically, monoclines make steplike benches.

differential erosion of folds

Considerable differences in weathering and erosional characteristics of the more common sedimentary rocks lead to the etching out of less resistant rocks, leaving the more resistant ones standing higher. Well-cemented sandstones and conglomerates are relatively resistant and produce prominent ridges, or upland areas, whereas shales are relatively less resistant and produce valleys, or lowlands. The resistance of carbonate rocks, such as limestone and dolomite, varies with climatic conditions. In humid regions where rainfall is abundant, limestones are susceptible to solution, and hence produce valleys, or lowlands, whereas in arid regions where water is much less available for solution, limestone rocks are usually resistant ridge-makers.

In an area of recent folding, increased elevation of anticlinal ridges induces accelerated erosion, which begins to modify the ridges. If the rate of erosion is high relative to the rate of uplift, the initial topographic form is modified. If the axial portion of an anticline consists of resistant rock, a topographic high is maintained, but if the axial portion of an anticlinal ridge has been breached by erosion to expose weak underlying shales (or limestone, in a humid

Fig. 17-1   Circle Ridge dome, a structural dome in Fremont County, Wyoming. *(U.S. Department of Agriculture.)*

climate), an *anticlinal valley* (Figs. 17-3 and 17-4) is developed parallel to the axis of the structure, bounded on both sides by homoclines. *Homoclines* consist of beds dipping in one direction along the limbs of eroded folds. They differ from monoclines in that they do not have the steplike flexure of monoclines.

Probably the most common topography developed on folded sedimentary rocks is alternating *linear homoclinal ridges* and *homoclinal valleys*, formed on tilted beds of differing resistance. Homoclinal ridges are, typically, asymmetric in cross profile (Fig. 17-5), with a steep scarp face and a more gentle dip slope, which corresponds to the dip of the resistant stratum. The scarp face is kept steepened by sapping, which occurs when erosion of weaker

Fig. 17-2 San Raphael swell, Utah, a flat-topped upwarp bounded by monoclinal flexures. *(John Shelton.)*

material beneath a resistant unit undercuts the resistant unit. Homoclinal ridges become progressively less asymmetric with increase in dip, and may become symmetrical when the angle of dip approaches the angle of repose of the material on the scarp face. Ridges developed on steeply tilted resistant beds are known as *hogbacks* (Fig. 17-6). With decrease in dip, hogbacks merge imperceptibly with linear ridges, called *cuestas,* composed of beds having more gentle dips (Fig. 17-7), and these in turn grade into *mesas,* developed on nearly horizontal beds. The dividing line between hogbacks, cuestas, and mesas is arbitrary. All are examples of stripped structural surfaces.

A classic area of topographic adjustment to structures on folded sedimentary rocks is the Valley and Ridge province of the Appalachian Mountains

Fig. 17-3 Gypsum Valley anticline, Colorado. The axial portion of the anticline has been breached by erosion. *(John Shelton.)*

in Pennsylvania, New York, Virginia, and adjacent states. Differential erosion of closely folded, interbedded resistant conglomerates and sandstones and less resistant shales and limestones has produced a series of alternating ridges and valleys. Where folds have been breached by erosion, homoclinal ridges occur. If the axis of a fold is horizontal, homoclinal ridges on a breached structure are parallel to one another, but if the axis is inclined, homoclinal ridges converge or diverge, depending on the kind of fold. *Homoclinal ridges converge in the direction of plunge of anticlines, but diverge in the direction of plunge of synclines* (Fig. 17-8).

Figure 17-9 shows an example of a partially breached anticline in the Appala-

Fig. 17-4   Anticlinal valley in Banff National Park, Alberta, Canada. The axis of the anticline lies along the axis of the valley.

chian Mountains. Nippenose Valley and Mosquito Valley have formed because of breaching of the resistant Tuscorora sandstone and erosion into the underlying weak Ordovician limestone. The area along the axis of the anticline between the two valleys is an unbreached portion where the resistant sandstone has not yet been removed by erosion.

The Virgin anticline in southwestern Utah (Fig. 17-10) is an example of a breached plunging anticline. The prominent homoclinal ridges, composed of the resistant Shinarump conglomerate, converge in the direction of plunge of the anticline. The core of the fold is composed of weaker Moenkopi shales and siltstones.

Fig. 17-5 Homoclinal ridge near the Henry Mountains, Utah. *(C. B. Hunt, U.S. Geological Survey.)*

If resistant beds are interlayered with less resistant beds, prolonged differential erosion may strip away the weaker material and leave remnants of resistant beds preserved along the axis of synclines. Ridges developed in this manner are termed *synclinal ridges* (Fig. 17-11). A structural low (syncline that makes a topographic high is an example of inversion of topography.

Monoclines are characteristically bounded by plateaus (Fig. 17-12). Following erosion, hogbacks mark the steeply dipping units close to the axis,

Fig. 17-6 Dakota hogback west of Denver, Colorado. The ridge is composed of resistant Dakota sandstone. Note the V-shaped notch in the hogback where it is crossed by a stream. *(T. S. Lovering, U. S. Geological Survey.)*

while mesas or cuestas occur where the dips are more gentle away from the
axis.

domes and basins

In its initial stages, drainage consequent on a newly uplifted dome may
be radial, but as erosion progresses, differential erosion of resistant and weak
beds develops a series of concentric ridges and valleys (Fig. 17-13). Subse-

Fig. 17-7   Cuesta at Book Cliffs, Utah. *(J. R. Balsley, U. S. Geological Survey.)*

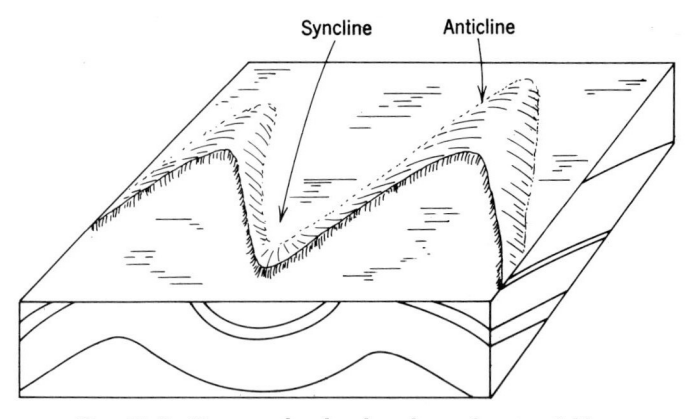

Fig. 17-8   Topography developed on plunging folds.

quent streams, formed in valleys on weak rock, capture the consequent radial streams, thereby converting the drainage to an annular pattern. Resistant beds of conglomerate and sandstone (or limestone if the climate is arid) stand out as encircling hogbacks or cuestas. If the rocks in the core of the dome are resistant, the central area may remain a topographic high. If relatively weak, the rocks may produce a topographic basin in the core.

Breaching of a dome exposes older rocks near the center of the structure. If the sedimentary beds were deposited unconformably on a crystalline basement of igneous or metamorphic rocks, exposure of these older rocks in the core results in a massive central area surrounded by concentric linear ridges of sedimentary rocks. A good example of this type of dome is the Black Hills upwarp, an elliptical, flat-topped dome about 125 miles long and 65 miles wide. Dips around the flanks of the structure are steeper than those near the central area. Erosion of the central and eastern portions has stripped off the overlying Paleozoic and Mesozoic sedimentary beds, to expose pre-Cambrian igneous and metamorphic rocks. On the eastern flank resistant Paleozoic limestones make homoclinal ridges, while to the west of the crystalline core incomplete erosion of the same gently dipping limestones has produced a broad stripped structural surface known as the Limestone Plateau. Encircling the entire dome is the Red Valley, developed by differential erosion of weak Triassic red shales, and the Dakota hogback, developed on the resistant Dakota sandstone.

A series of broad regional upwarps occurs in the Paleozoic rocks of the eastern United States, among which are the Cincinnati arch and the Nashville dome. Dips along the flanks are gentle, and cuestas, rather than hogbacks, encircle the core of Ordovician limestone. The core of each of these domes is a topographic basin. At the Nashville dome in Tennessee, gently

Fig. 17 - 9 Partially breached plunging anticline in the Appalachian Mountains. Note also the plunging syncline between North and South White Deer Ridges. *(Williamsport quadrangle, U.S. Geological Survey.)*

dipping Mississippian sandstone and chert make concentric cuestas around a basin eroded in weaker Ordovician shales.

interpretation of geologic structures from topography

The close relationship of topography to geologic structure allows identification of structures by detailed analysis of topography on maps or air photographs. Such studies have become an integral part of geologic mapping. Several criteria which may be employed to determine the direction of dip of beds, to distinguish anticlinal folds from synclinal folds, and to determine the direc-

Fig. 17-10 The Virgin anticline, a breached plunging anticline. Note the V-shaped notches cut in the prominent hogback. *(U.S. Department of Agriculture.)*

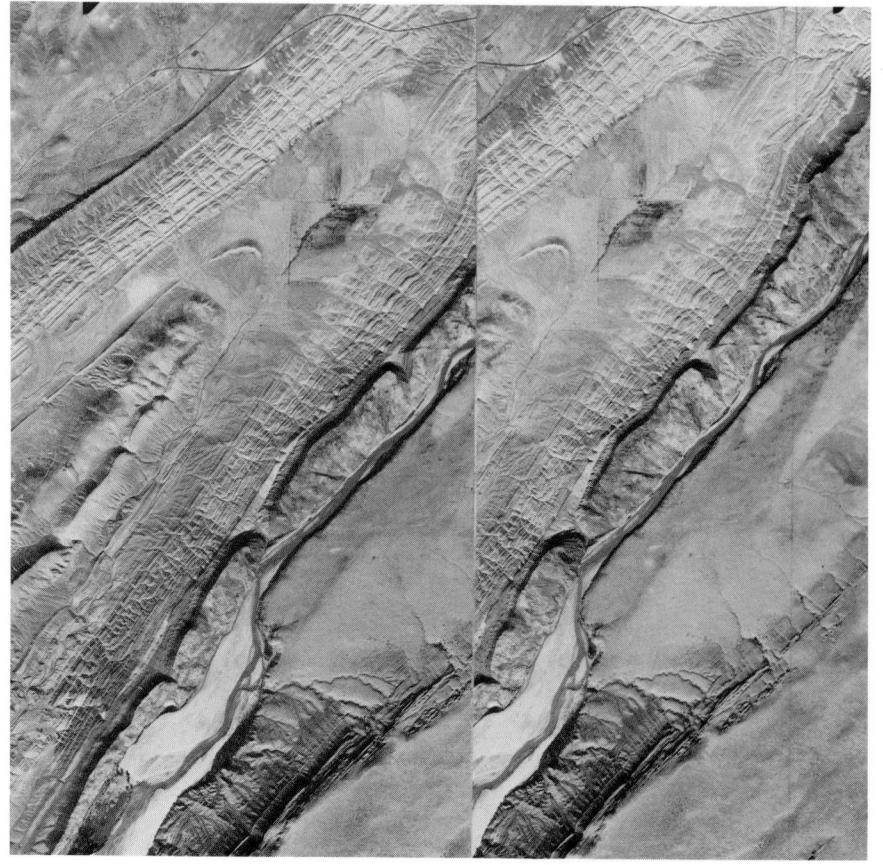

tion of plunge of folds are individually discussed in the sections that follow.

asymmetry of homoclinal ridges

Since homoclinal ridges developed on tilted beds are typically asymmetric, the direction of asymmetry can be used to determine the dip. The dip slope, or gently inclined portion of the ridge, slopes in the direction of the dip of the bed (Fig. 17-5). The scarp on the opposite side of the crest of the ridge faces in the direction opposite to that of the dip of the bed. This method is usable for cuestas and hogbacks developed in strata dipping less than about 35°. The asymmetry of the ridge tends to be obscured on beds with greater

Fig. 17-11 Syncline making up a ridge near Sunwapta Pass, Jasper National Park, Alberta, Canada.

dips because the angle of repose of weathered debris on the scarp face usually approaches 30 to 35°.

The angle of dip of a resistant bed making up a homoclinal ridge may be calculated if the dip slope is well defined. On Fig. 17-14 the angle of dip, α, is the angle between the inclined plane and a horizontal plane, and may be calculated by determining the horizontal distance h relative to the vertical distance v.

$$\tan \alpha = v/h$$

The value of v may be determined from a contour map, and the value of h may be scaled from the same map. If v is 200 feet and h is 1,620 feet, then $v/h =$

Fig. 17-12   Comb Ridge monocline near San Juan River, Utah. *(John Shelton.)*

200/1,620 = 0.123, which corresponds to an angle of $7°$ in a table of trigono-metric values.

rule of V's

In places where streams cross homoclinal ridges, a notch is cut through the resistant unit. Such a notch has a V shape in map view (Fig. 17-15). The reason for the V shape is the decrease in height of the resistant bed above the stream in the direction of dip. Thus the apex of the V points in the direction of dip.

Fig. 17-13 Dome southeast of Lander, Wyoming. *(John Shelton.)*

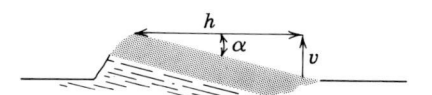

Fig. 17-14 Calculation of angle of dip from a homoclinal
ridge: tan $\alpha = v/h$.

Contour interval 20 ft

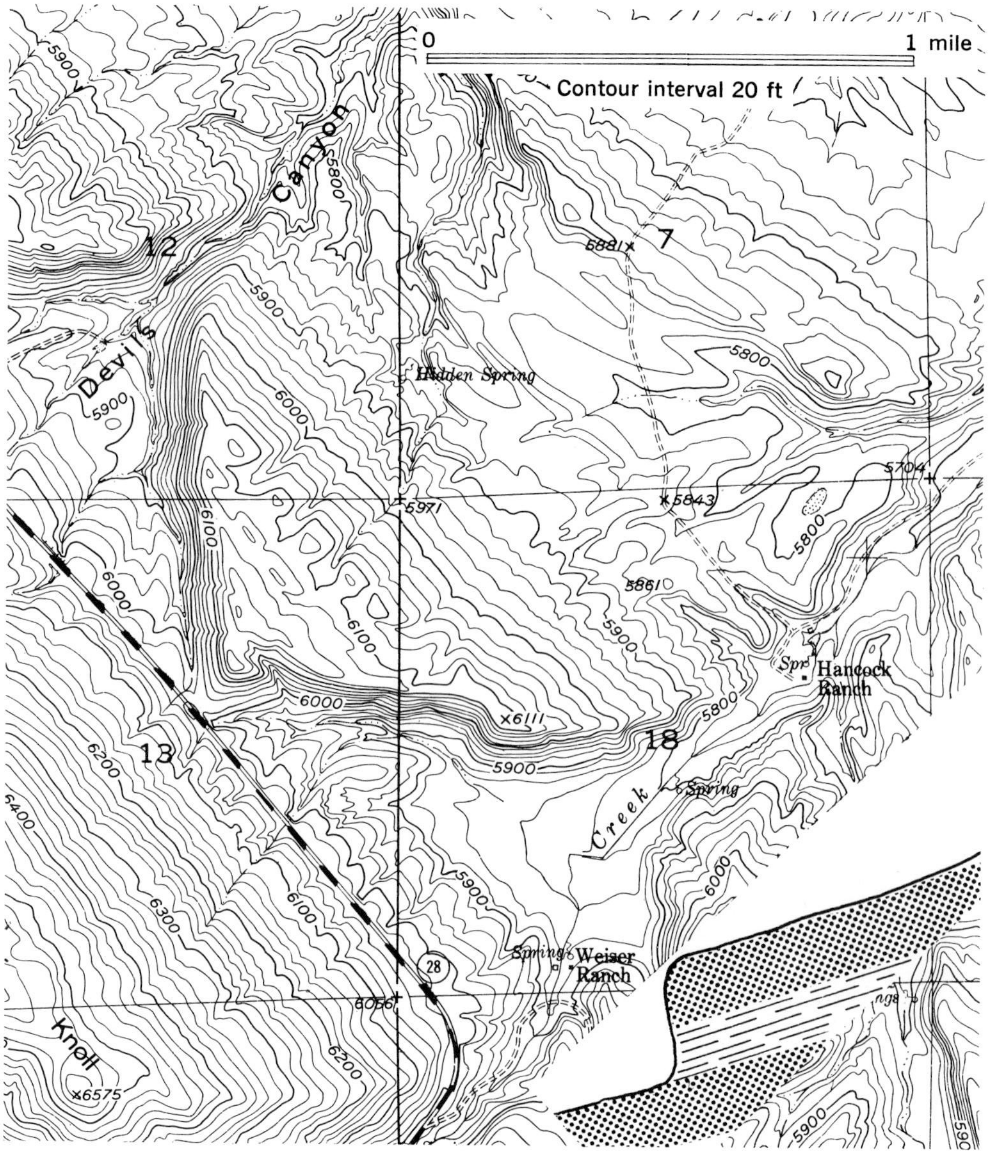

Fig. 17-15 V-shaped notches and asymmetric profile of a homoclinal ridge. (Weiser
Pass quadrangle, Wyoming, U.S. Geological Survey.)

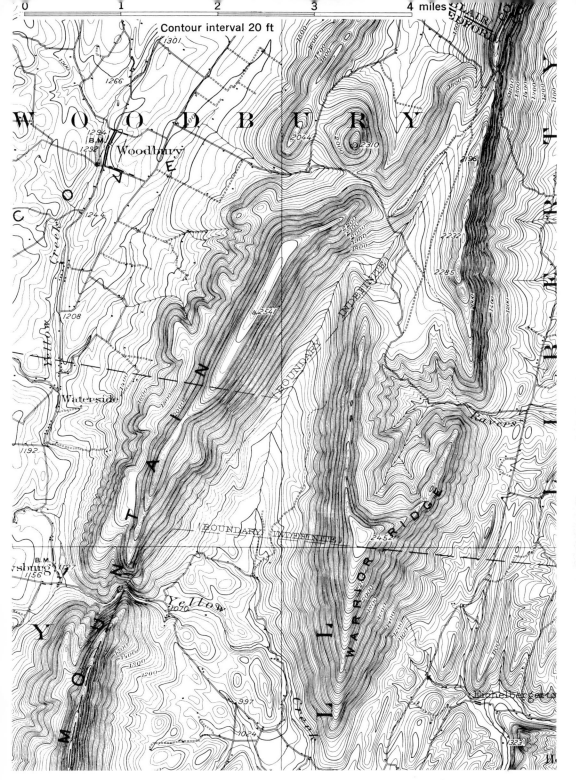

Fig. 17-16 Tapering nose of a plunging anticline. *(Everett quadrangle, Pennsylvania, U.S. Geological Survey.)*

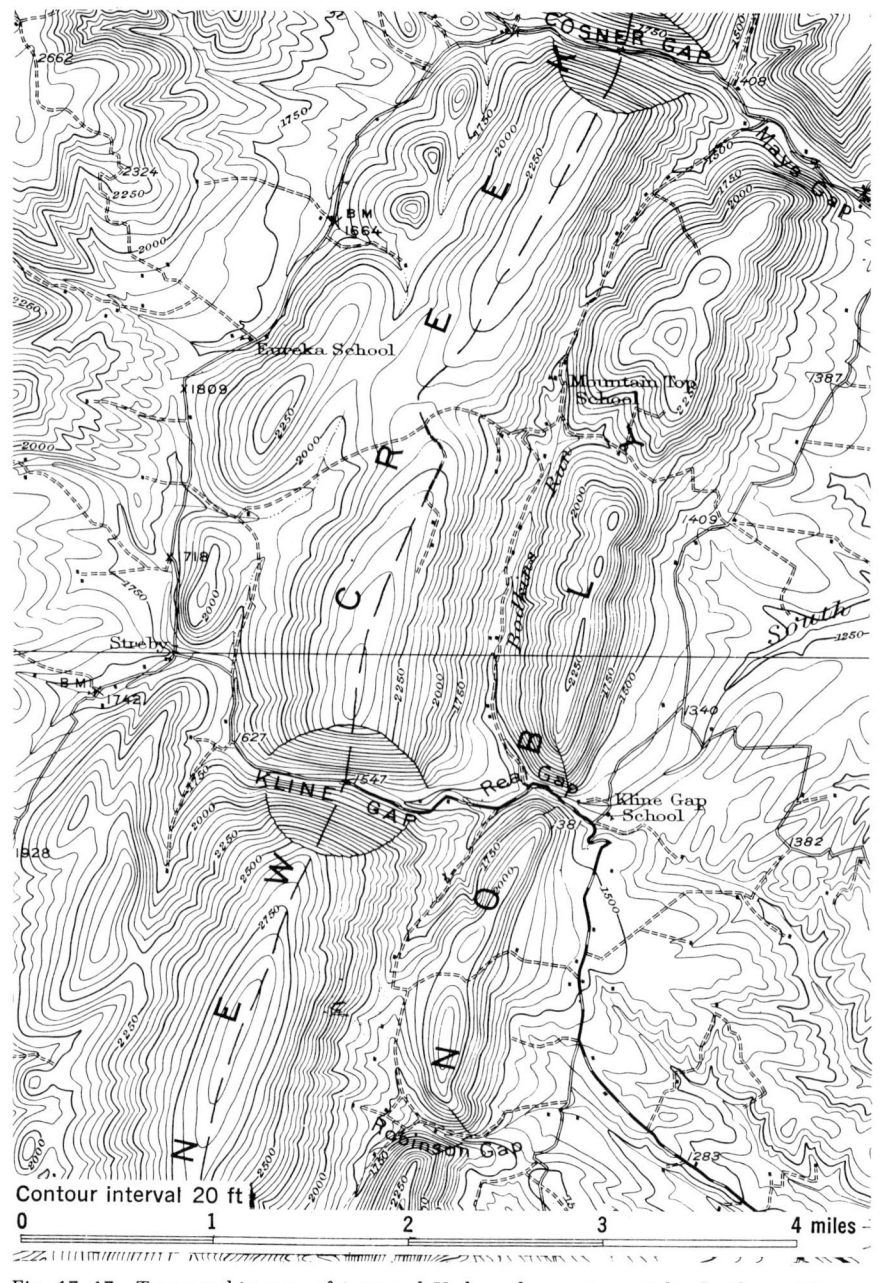

Contour interval 20 ft

0   1   2   3   4 miles

Fig. 17-17 Topographic map of inverted U-shaped gaps in anticlinal ridges. *(Green-land Gap quadrangle, West Virginia, U.S. Geological Survey.)*

apex of plunging structures

The nature of the apex of converging homoclinal ridges of breached plunging folds may be used to determine the type of structure. Because the dip of the beds is the same as the plunge of the fold at the apex of a plunging structure, the ridges at the nose of plunging anticlines will tend to be asymmetric. The dip slope makes a tapering nose on the convex side of the ridge in the direction of plunge (Fig. 17-8). At the nose of a plunging syncline, however, the dip of the beds is away from the direction of convergence of the ridges, and hence the scarp, rather than the dip slope, is on the convex side (Fig. 17-8). Warrior Ridge (Fig. 17-16) illustrates on a topographic map the tapering nose at the apex of a plunging anticline.

inverted U-shaped gaps in anticlinal ridges

Where a stream has cut through an anticlinal ridge, a vertical cliff corresponding to a resistant rock layer will affect the spacing of contours in such a way as to produce a U-shaped line (Fig. 17-17). In crossing the vertical face, contours appear to be offset, and merge with a single line transverse to the contour. The reason for this is that all contours on a vertical face are "stacked up" one upon another when viewed from above, as on a topographic map.

A layer of resistant rock making a vertical cliff is shown on Fig. 17-18. Tracing any contour along the side of the ridge, it becomes apparent that when the contour crosses the cliff, it must make a sharp turn, and if the cliff is vertical, the contours will be superimposed upon one another and merge to form a single line.

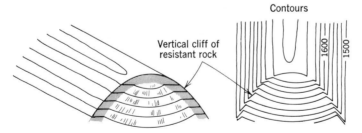

Fig. 17-18   Inverted U-shaped gap in an anticlinal ridge.

projection of structure along the strike

If the structure of one part of a complexly folded area can be determined, often the remainder can be worked out by extrapolation. Figure 17-19 is a

Fig. 17-19   Topographic map of a folded area in the Appalachian Mountains of West Virginia. (Strasburg quadrangle, West Virginia, U.S. Geological Survey.)

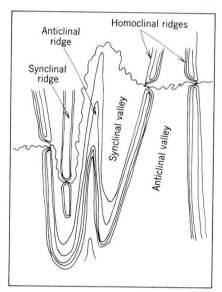

Fig. 17-20 Relationship of topography
to geologic structure in a region of folded
sedimentary rocks.

topographic map representing a series of folds in the Appalachian Mountains
of West Virginia. Analysis of the structure may be made by examining two
critical localities, Little Crease Mountain and Signal Knob. The tapering nose
of Little Crease Mountain indicates that it is an anticline plunging SW. Thus
the dip of the bed making the ridge northeastward toward High Peak must
be to the NW, and a synclinal axis must lie in the valley of Passage Creek.
The short blunt nose of the fold at Signal Gap suggests a syncline, the axis
of which lies in Fort Valley. If synclinal axes occupy Fort Valley and Passage Creek Valley, then the ridge between them must be an anticline.

references

Lobeck, A. K.: 1939. Geomorphology, McGraw-Hill Book Company, New York.

Strahler, A. N.: 1963. The Earth Sciences, Harper & Row, Publishers, Incorporated, New York.

Thornbury, W. D.: 1954. Principles of Geomorphology, John Wiley & Sons,
Inc., New York.

Von Engeln, O. D.: 1942. Geomorphology, The Macmillan Company, New York.

topography associated with faulting

Faulting may be responsible for a variety of topographic features. Some are initial forms produced directly at the time of faulting, and some are indirectly related to faulting as a result of differential erosion of displaced beds. Often the most conspicuous landforms caused by faults are scarps, but they seldom persist for long periods of geologic time since erosion rapidly attacks the upthrown block.

Faults which have a high angle of dip and dominantly vertical movement may be either *normal* or *reverse*, depending on a relative movement on opposite sides of the fault. For a normal fault the block above a fault plane (hanging wall) moves down relative to the block beneath (footwall). For a reverse fault the block above the fault plane moves up relative to the block beneath (Fig. 18-1).

fault and fault-line scarps

Probably the most common landforms produced by faulting are scarps, formed either as a direct result of fault movement or indirectly by differential

West Bay fault, Yellowknife, Northwest Territories, *(National Air Photo Library, Surveys and Mapping Branch, Department of Energy, Mines, and Resources.)*

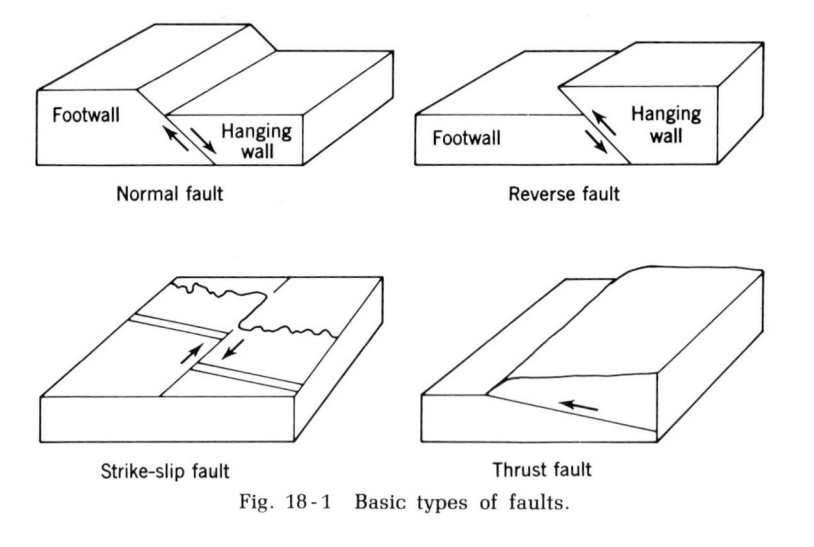

Fig. 18-1   Basic types of faults.

erosion on opposite sides of the fault. If a scarp originates directly from fault
displacement, it is called a *fault scarp* (Fig. 18-2). With time, a fault scarp
loses its sharpness of form, but even if the scarp has been much dissected
by erosion, it may still be considered a fault scarp as long as the relief is related to the original faulting. A *fault-line scarp* is formed when the original
fault scarp has been destroyed but a scarp remains along the line of the fault
plane because rocks of differing resistance occur on opposite sides of the
fault (Fig. 18-3). In such an instance the reason for the scarp is differential
erosion, not displacement of the land surface by fault movement.

Davis (1913) recognized the differences in geomorphic history implied by
interpretation of scarps as fault scarps or fault-line scarps. He also was cogni-
zant of the fact that a fault-line scarp might face either the upthrown or downthrown block of a fault, depending on the resistance of the rock on either
side of the fault. A scarp developed on the downthrown block by differential
erosion is called an *obsequent fault-line scarp* (Fig. 18-3) because the scarp
faces the opposite direction as the original fault scarp. A scarp developed on
the upthrown block by differential erosion (Fig. 18-3) is known as a *resequent
fault-line scarp* because the new scarp faces the same direction as the original fault scarp but is not related to the original displacement of the surface.
Distinguishing a fault scarp from an obsequent fault-line scarp is not as difficult as distinguishing a resequent fault-line scarp from a fault scarp.

Renewed movement of a fault along a fault-line scarp results in a *composite
scarp* in which the upper part is a fault-line scarp but the lower part is a fault

scarp. Such scarps may sometimes be recognized by a fresh scarp at the base of a dissected fault-line scarp.

A number of features associated with faulting may be useful in distinguishing between fault scarps and fault-line scarps. Among the criteria for identifying fault scarps are the following:

Alluvial Scarps. Scarps along faults which cut across unconsolidated sediments such as alluvium are strong evidence of recent faulting since there is no appreciable difference in resistance to erosion on opposite sides of the fault. Such scarps do not persist long, attesting to recency of fault movement.

Fig. 18-2   Fault scarp produced during the Hebgen Lake earthquake, 1959. *(I. J. Witkind, U.S. Geological Survey.)*

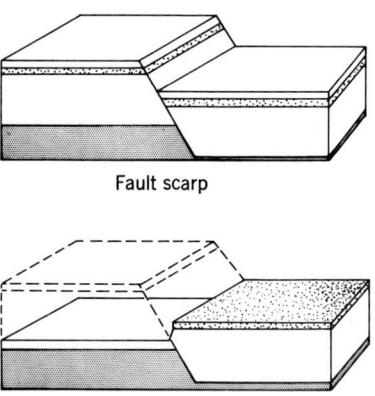

Fault scarp

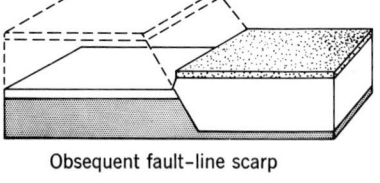

Obsequent fault–line scarp

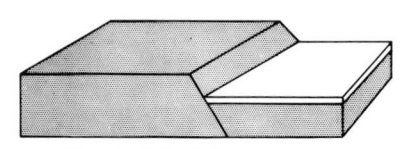

Resequent fault–line scarp

Fig. 18-3 Evolution of fault-line scarps.

Little Correlation between a Scarp and Rock Resistance on Opposite Sides of a Fault. Since a fault-line scarp is a direct result of differential erosion, if the material on opposite sides of a fault shows little difference in resistance to erosion, it is probable that the scarp is related to surface displacement.

Ponded Drainage. Fault scarps interrupt normal stream drainage, often resulting in ponds or small lakes being formed at the base of such scarps (Fig. 18-4).

Displacement of Recent Features. Movement of recent features whose original position may be inferred strongly indicates recent faulting. Obvious examples include roads, houses, orchards, and other man-made features (Fig. 18-5).

Disruption of Once-continuous Topographic Features. Scarps across lava flows or stream terraces whose original continuity can be determined may be considered fault scarps.

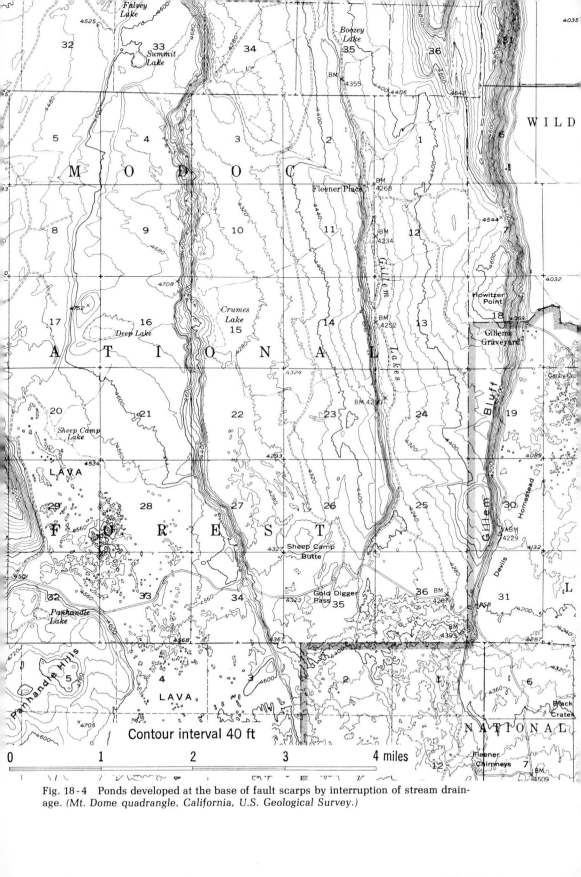

Fig. 18-4 Ponds developed at the base of fault scarps by interruption of stream drainage. (Mt. Dome quadrangle, California, U.S. Geological Survey.)

Evidence that a scarp is a fault-line rather than a true fault scarp includes the following:

Close Correlation between a Scarp and Rock Resistance. If a scarp corresponds to a resistant bed with weaker beds on the lower block, differential erosion may be suspected of causing the scarp. However, such evidence is not necessarily always conclusive.

A Scarp Located on the Downthrown Block. A scarp on the downthrown block of a fault could not have formed by direct surface displacement, and thus must be a fault-line scarp caused by differential erosion. Such a scarp would have to be an obsequent fault-line scarp.

Scarps Which Pass beneath Lava Flows without Disrupting Them. Figure 18-8 shows a scarp intersected by a lava flow. Since neither the surface of the flow nor the lower contact is affected by the fault trace, the lava must have crossed the fault after it was formed but at a time when there was no relief. Thus the scarp must have been formed by differential erosion after the flow crossed the fault.

Fig. 18-5 Fault scarp passing through the dwellings of a ranch, Hebgen Lake earthquake, 1959. *(I. J. Witkind, U.S. Geological Survey.)*

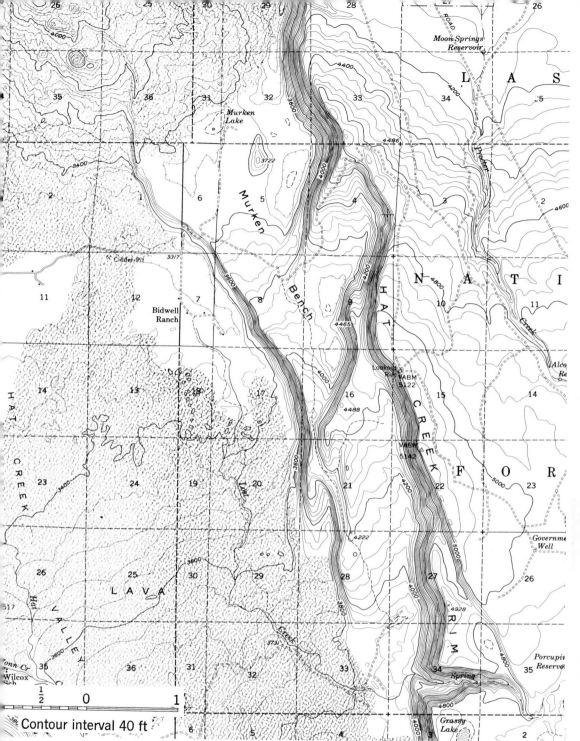

Fig. 18-6 Fault scarps in northern California. The scarps differ from those made by resistant sedimentary beds in that they vary in height, and "splinter" scarps extend from the main scarp. Note the cinder cones in the upper left corner which are aligned with the scarp at Murken Bench. *(Jellico quadrangle, U.S. Geological Survey.)*

other topographic features of faulting

Among a number of other topographic features related to faulting are *tri-angular facets* developed on spurs between valleys. These are formed when V-shaped valleys cut into a scarp, and the interstream divide becomes a sharp-crested ridge. The area on the fault face between streams thus assumes a triangular shape, with the apex upward (Fig. 18-10). Although triangular facets are commonly associated with fault scarps, they may also be formed by glaciation and other surface processes.

Breaks in stream profiles may occur where streams cross faults. This is

Fig. 18-7 Fault plane near Klamath Falls, Oregon. (*Wards Natural Science Estab-lishment.*)

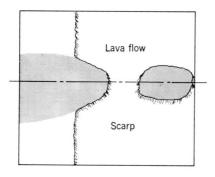

Fig. 18-8 Fault-line scarp formed by differential erosion after a lava flow covered the fault at a time when there was no relief across the fault.

a common occurrence where fault movement has taken place rapidly, sometimes 10 to 25 feet in a single jog, so that the stream has not yet been able to cut down enough to reestablish its profile across the fault. In some cases *hanging valleys* may be produced, and a waterfall results. If a scarp is formed facing upstream, the drainage becomes ponded.

Alluvial fans along a fault at a mountain front are common, but do not in themselves prove faulting. If the fans are related to faulting, they may be associated with hanging valleys, breaks in stream profiles, and triangular facets.

Displacement of sedimentary beds by faulting leads to topographic displacement of ridges. The bed of resistant rock in Fig. 18-11 has been offset by a fault, and as a result the ridge terminates abruptly at the fault, and is offset. Bending of the ridge near the fault is caused by frictional *drag*. A subsequent stream now follows the fault at its northern end and clearly outlines the trace of the fault.

Faulting at the base of a mountain range not only creates scarps, but also causes truncation of beds transverse to the strike. A feature commonly observed in the tilted-fault-block ranges of Nevada, Arizona, eastern California, and western Utah is the absence of long spurs extending from a range as a result of differential erosion of weak and resistant beds. Instead, resistant and weak beds alike terminate abruptly at the range front.

relationship of geologic structure to topography

grabens and horsts

Fault blocks bounded on both sides by high-angle faults may be raised or lowered by parallel movement along the faults. A block downdropped relative to the rocks on either side is a *graben*, and a block uplifted relative to the adjacent rocks is a *horst* (Fig. 18-12). The absolute movement of a graben is not necessarily downward since the same structural relationships could be produced by uplift of horsts on either side. The size of a graben or horst varies from a few tens of feet across to many miles (Figs. 18-13 and 18-14). Initial displacement of the land surface by a graben produces a topographic valley or basin. With time, erosion may considerably alter the initial topog-

Fig. 18-9 East front of the Sierra Nevada, California, marking a fault zone. (*F. E. Matthes, U.S. Geological Survey.*)

raphy, however, and grabens are not necessarily always topographic lows, nor are horsts necessarily always topographic highs.

strike-slip faults

Not all faults have vertical components of movement. Faults which have a dominance of horizontal over vertical movement are known as *strike-slip faults* (Fig. 18-1). They are usually characterized by a steeply dipping fault plane in which the rocks are intensely ground up. If the fault is extensive, it often consists of a fracture zone of numerous parallel faults.

Much of the Coast Range of California is cut by strike-slip faults, the larg-

Fig. 18-10 Triangular facets along a fault scarp at the western margin of the Wasatch Range near Salt Lake City, Utah.

Fig. 18-11 Offsetting of a hogback by faulting. The bending of the ridge is a result of drag along the fault. A small subsequent stream has worked its way headward along the fault. (U.S. Department of Agriculture.)

relationship of geologic structure to topography

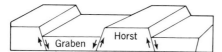

Fig. 18-12   Grabens and horsts.

est of which is the San Andreas fault extending from Mexico to the Pacific Ocean north of San Francisco. The total known length of San Andreas is some 600 miles before it disappears into the Pacific Ocean at its northern end and beneath the Salton Basin at its southern end. Along much of its extent the

Fig. 18-13   Graben cutting a lateral moraine at Bells Canyon, near Salt Lake City, Utah.

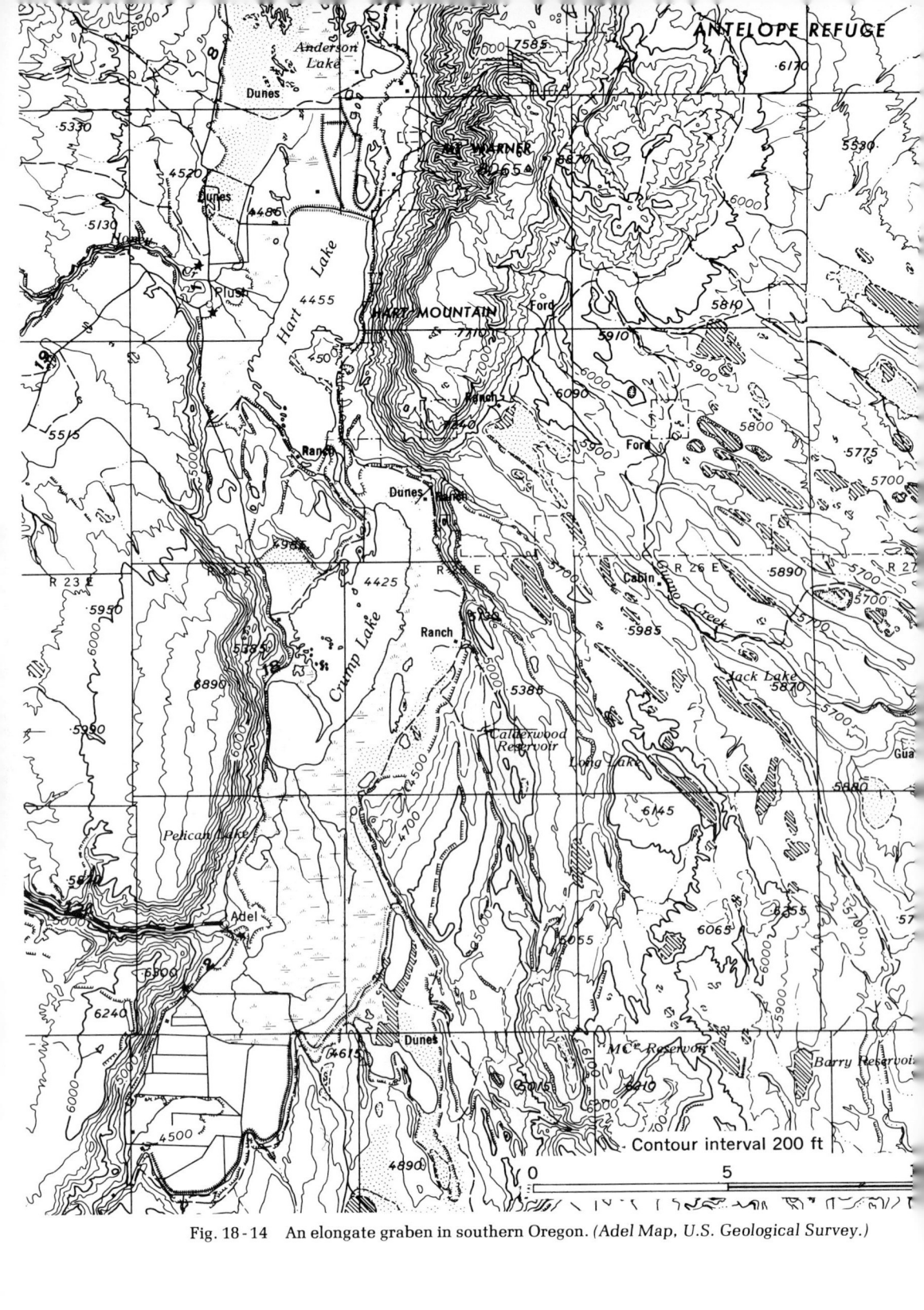

Fig. 18-14   An elongate graben in southern Oregon. *(Adel Map, U.S. Geological Survey.)*

San Andreas consists of a wide zone of broken and ground-up rock, and is marked by a valley bounded by scarps, depressions, and offset streams and ridges.

The famous San Francisco earthquake of 1906 was caused by lateral movement along the San Andreas of as much as 21 feet. Offsetting of roads, fences, orchards, and other surface features indicates that the movement was *right-lateral;* i.e., when looking across the fault, objects are seen displaced to the right. Streams crossing the fault in many places are offset several hundred feet or more (Figs. 18-15 and 18-16). In some instances ridges have been offset just enough to block valleys on the opposite side of the fault. Such ridges are referred to as *shutter ridges.*

Fig. 18-15 San Andreas fault, California, a right-lateral strike-slip fault. Note the offset drainage. *(J. R. Balsley, U.S. Geological Survey.)*

thrust faults

Thrust faults are low-angle reverse faults. They are formed as a result of compressive forces which push rocks along a gently inclined fault plane, and may achieve displacements measured in many miles. Generally, they do not produce scarps as imposing as high-angle faults, but if resistant rocks are faulted over less resistant units, differential erosion develops steep cliffs along the fault trace (Fig. 18-17). With prolonged erosion, outliers consisting of erosional remnants isolated from the main thrust sheet develop into *klippes*.

determination of direction of fault movement

Displacement of tilted resistant sedimentary beds by faulting results in offsetting of ridges. Analysis of geometric relationships of displaced ridges

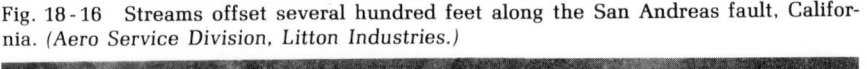

Fig. 18-16 Streams offset several hundred feet along the San Andreas fault, California. *(Aero Service Division, Litton Industries.)*

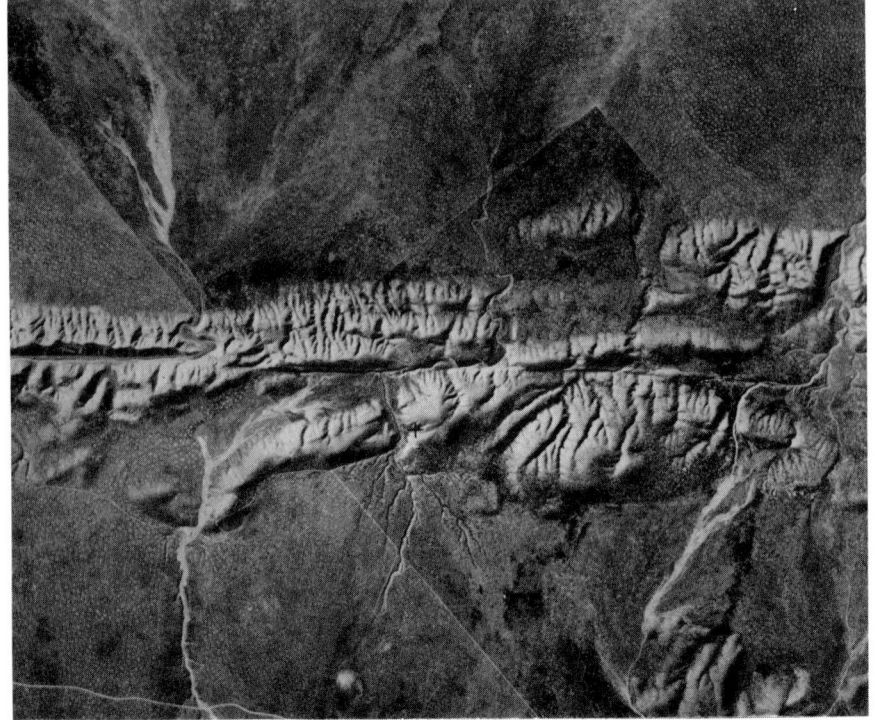

observable on topographic maps or air photographs allows determination of upthrown and downthrown blocks.

Figure 18-18 is a topographic map of homoclinal ridges along the limbs of a breached fold. The asymmetry of the ridges and the V-shaped notches indicate that the bed making the ridge at A dips to the left, and the bed at B dips to the right. The fold is thus a breached anticline. The beds have been displaced along a fault f-f, and the land surface has been eroded to a common level on both sides of the fault. The wider spacing between the ridges on the north side of the fault results from the fact that the structure has been interrupted at the fault. Since the axis of the fold has not been shifted horizontally, movement along the fault must have been up or down the dip of the fault, without a strike-slip component. Thus the block north of the fault must have moved up or down relative to the block south of the fault. A simple rule

Fig. 18-17   Lewis thrust fault, Glacier National Park, Montana. The lower gentle slopes consist of Cretaceous shales and sandstone which have been overridden by pre-Cambrian sedimentary rocks.

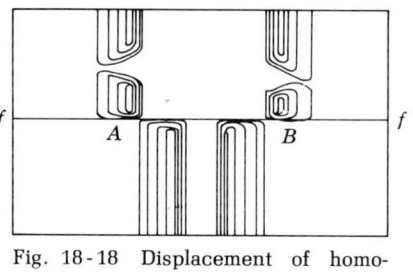

Fig. 18-18 Displacement of homoclinal ridges by faulting.

may be used to determine which is the downthrown and which the upthrown block. Consider the displaced bed on the east side of the axis of the fold. Rotate the map counterclockwise 90° and cover the lower half of the map. You are now looking down the projected dip of the displaced bed; i.e., the direction of dip is away from you toward the top of the page. The bed on the left lies higher on the page than the displaced bed on the right. The block on the left of the fault is the upthrown block, and the block on the right is the downthrown block. Orienting the map of any displaced homoclinal ridge in this manner will always result in the bed higher on the page (toward the top) being on the upthrown block. This method of determining geologic structures is known as the *downdip method* (Mackin, 1950).

references

Billings, M. P.: 1938. Physiographic relations of the Lewis overthrust in northern Montana, *Am. Jour. Sci.,* vol. 35, pp. 260-272.

———: 1942. Structural Geology, Prentice-Hall, Inc., Englewood Cliffs, N. J., pp. 124-225.

Blackwelder, E.: 1928. The recognition of fault scarps, *Jour. Geology,* vol. 36, pp. 289-311.

Cotton, C. A.: 1950. Tectonic scarps and fault valleys, *Geol. Soc. America Bull.,* vol. 61, pp. 717-758.

Davis, W. M.: 1913. Nomenclature of surface forms on faulted structures, *Geol. Soc. America Bull.,* vol. 24, pp. 187-216.

Fuller, R. E., and A. C. Waters: 1929. The nature and origin of the horst and graben structure of southern Oregon, *Jour. Geology,* vol. 37, pp. 204-238.

Johnson, D.: 1939. Fault scarps and fault-line scarps, *Jour. Geomorphology,* vol. 2, pp. 174-177.

Mackin, J. H.: 1950. The down-structure method of viewing geologic maps, *Jour. Geology,* vol. 58, pp. 55 - 72.

Willis, B.: 1938. San Andreas Rift, California, *Jour. Geology,* vol. 46, pp. 793 - 827.

———— : 1938. San Andreas Rift in southwestern California, *Jour. Geology,* vol. 46, pp. 1017 - 1057.

Witkind, I. J.: 1959. The Hebgen Lake, Montana, earthquake of Aug. 17, 1959, *U. S. Geol. Survey Prof. Papers* 435 - A, 435 - B, 435 - G.

igneous landforms

volcanic landforms

Eruption of volcanic material on the earth's surface substantially alters the preeruption topography. Most volcanic landforms are constructional forms produced by accumulation of igneous material on the surface, but secondary features are also formed by differential erosion of rocks of differing resistance.

Volcanic landforms are relatively easy to identify if they are recently formed. Each type of volcanic landform possesses certain characteristic features, but these become modified or obliterated by weathering and erosion with time. Evidence of volcanism in the geologic past is found in many areas of the world, but only relatively recent eruptions have left landforms which still persist today.

Not all parts of the world are subject to volcanic activity at present. Volcanos are not randomly scattered, but are restricted to certain fairly well defined zones. The distribution of recently active volcanos is concentrated in a series of chains, the most impressive of which is the circum-Pacific-zone encircling most of the Pacific Ocean. This zone extends northward from the

Lava fields, Craters of the Moon National Monument, Idaho. *(John Shelton.)*

southern tip of Chile, through the Andes Mountains, along the west coast of South America, through Central America and the Cascade Mountains of the western United States, into Canada. The belt continues from Alaska, across the Aleutian Islands to Japan and the Philippines, and then makes a broad arc through the East Indies.

volcanic cones

Volcanic cones are easily recognized by their topographic form, which usually consists of a conical hill, often having a circular depression, or crater, developed on the top. Volcanos erupt three kinds of material: (1) gaseous material, consisting predominantly of steam, (2) molten material in the form of lava flows, and (3) pyroclastic material composed of solid fragments ripped from the sides of the vent during eruption and of incandescent material which has solidified in flight (Fig. 19-1). The topographic expression of a volcanic cone depends largely on the nature of the erupted material, whether dominantly lava flows or dominantly fragmental material.

If the material consists almost entirely of fragments, a *cinder cone* is developed, with fairly steep sides lying at the angle of repose of the loose fragments, usually about 30 to 35° (Figs. 19-2 and 19-3). Cinder cones often occur in clusters (Fig. 19-4), and are usually readily identified on topographic maps or air photos by their shape and the presence of a crater at the top.

Volcanic cones made almost entirely from the eruption of lavas are termed *shield volcanos*. They tend to have more gently sloping sides than cinder cones. Among the best examples of shield volcanos are the Hawaiian Islands, several very broad-based volcanic cones rising approximately 30,000 feet above the ocean floor (Fig. 19-5). Nearly all the material in these cones consists of basalt flows that emanated not only from the central portion of the cone, but also from fissure eruptions along the flanks of the volcanos. These eruptions are characterized by relatively quiet outpouring of molten material, which is dominantly basaltic in composition. Basaltic lavas tend to be more fluid than silicic lavas, and explosive activity and ejection of fragmental material are comparatively mild relative to more silicic volcanos.

The third type of volcanic cone is the *composite cone*, composed of interlayered lava flows and fragmental material having slopes intermediate between cinder cones and the shield volcanos (Fig. 19-6). Composition of the material may range from rhyolite to basalt, although andesites are more common than either of these two.

calderas

Many volcanic cones are characterized by craters in their centers, measuring hundreds or a few thousands of feet in diameter. However, some volcanic depressions have diameters measured in miles, several times that of a normal crater. Such depressions, known as *calderas*, are produced by explosion of the summit area or by subsidence and collapse of the inner part of a volcanic cone.

A typical example of a caldera is the one occupied by Crater Lake in Oregon

Fig. 19-1 Volcanic eruption in Hawaii. Note the dark cloud of pumice and ash to the right of the main eruption. *(Wards Natural Science Establishment.)*

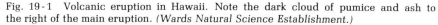

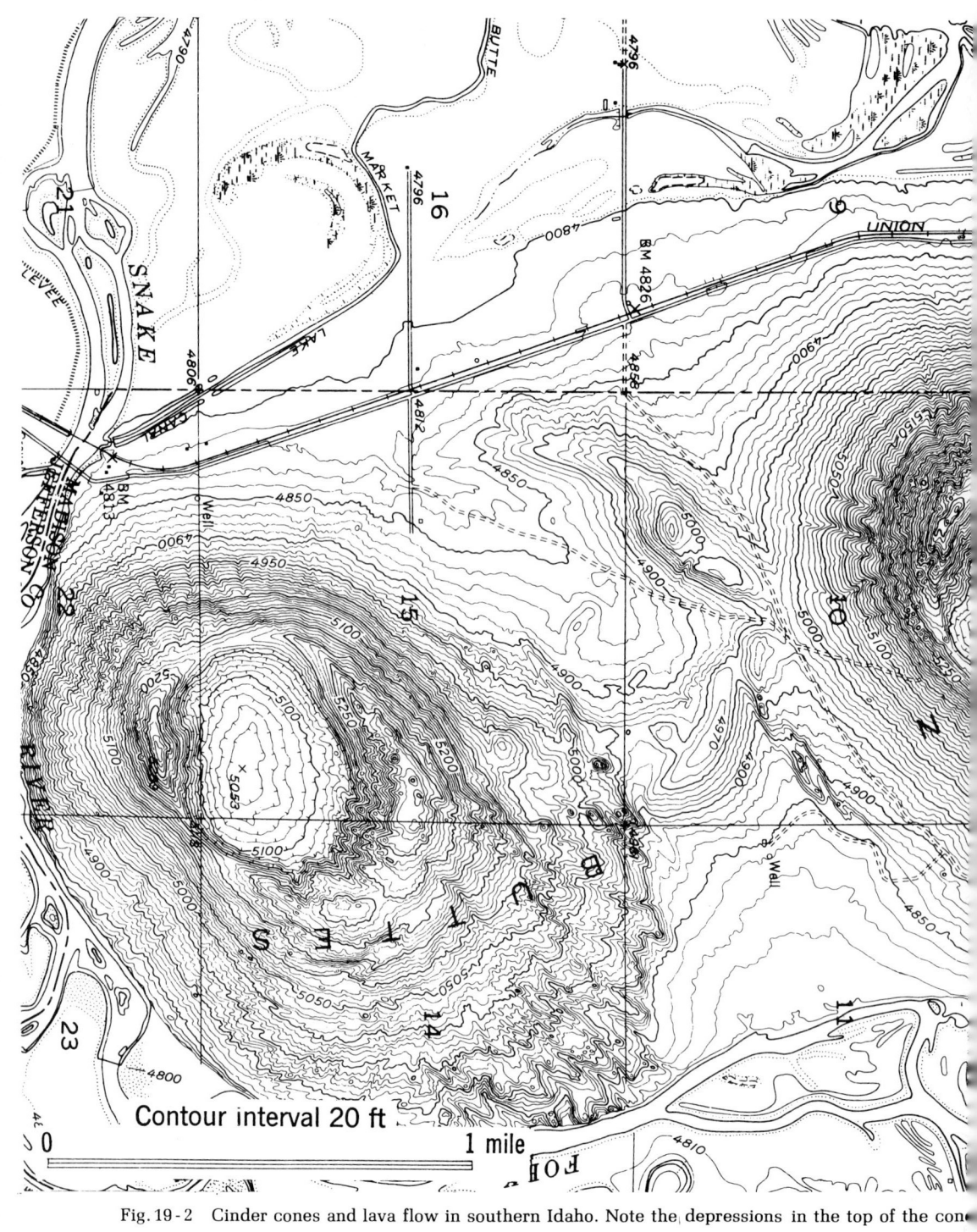

Fig. 19-2 Cinder cones and lava flow in southern Idaho. Note the depressions in the top of the con

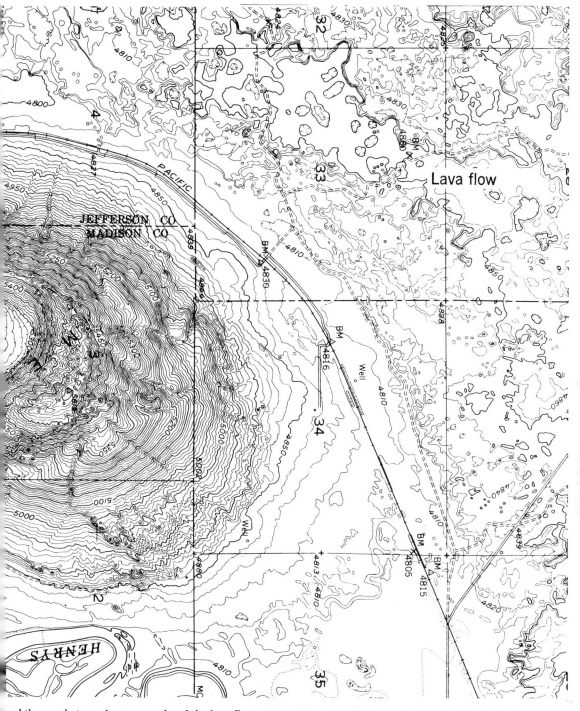

and the rough, irregular topography of the lava flow. *(Menan Buttes quadrangle, Idaho, U.S. Geological Survey.)*

(Fig. 19-7). The caldera is about 5½ miles in diameter and nearly circular in outline. The lake, as much as 2,000 feet deep in places, lies at an elevation of 6,160 feet, and is surrounded by nearly vertical cliffs which rise to altitudes of approximately 8,000 feet around the periphery of the caldera. There is considerable evidence that the present crater is only the stump of a former volcanic cone and that at one time the former summit of the volcano extended more than 6,000 feet above the present rim. Diller (1902), who was among the early workers to study the caldera, recognized that the summit must have been considerably higher in order to support the glaciers required for the deep glacial troughs which presently furrow the side of the mountains, and also to provide the glacial debris interbedded with volcanic debris at

Fig. 19-3   Capulin Mountain, New Mexico, an extinct cinder cone. (Wards Natural Science Establishment.)

the rim of the present caldera. Diller thought that the destruction of the former summit could not have been accomplished by a gigantic explosion. He calculated that about 17 cubic miles of material was gone from the former summit, and argued that if all this had been blown out by explosions, a tremendously thick layer of fragmental debris should have been found around the base of the volcano. Since only a small fraction of that amount can be found nearby, he concluded that the crater was formed primarily by collapse of the volcano's sides. Williams (1941), who studied the Crater Lake caldera in great detail, found that approximately 1.5 cubic miles of fragmental material could have been removed from the summit of the volcano by explosion,

Fig. 19-4 Cinder cones and lava flows at Craters of the Moon National Monument, Idaho. Note the rift in the cone from which the dark lava flow emanated. *(John Shelton.)*

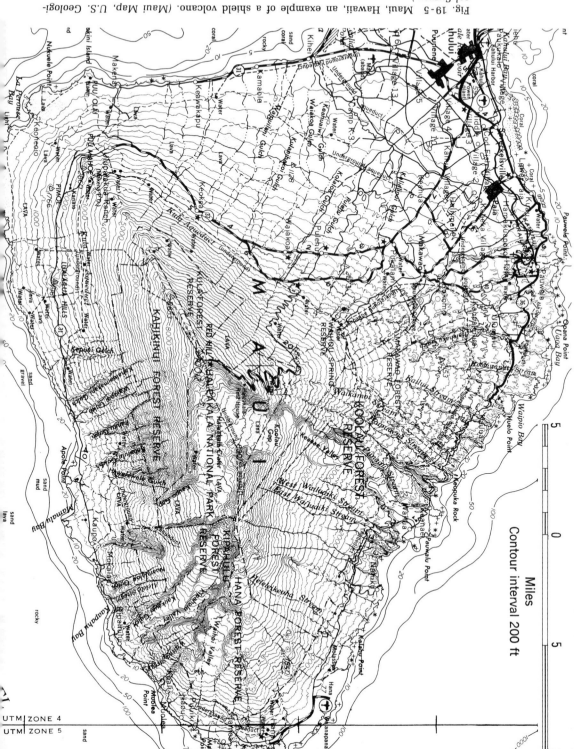

Fig. 19-5 Maui, Hawaii, an example of a shield volcano. (Maui Map, U.S. Geological Survey.)

Miles

Contour interval 200 ft

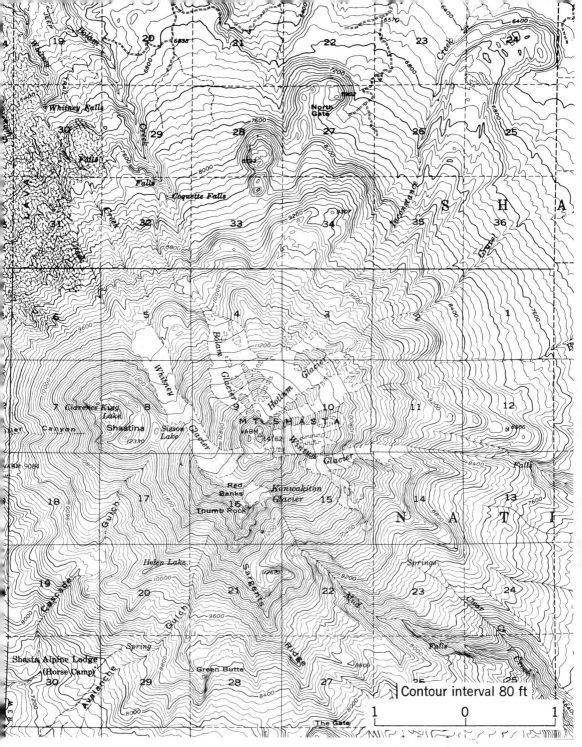

Fig. 19-6 Mt. Shasta, California, a composite volcanic cone. Note the lobate form of the lava flow in the northeast corner. *(Shasta quadrangle, California, U.S. Geological Survey.)*

but that the rest must have been removed by some other process. He calculated that another 5 cubic miles may have caved into the magma chamber to replace magma which had erupted during the last stages of eruptive activity. This left approximately 10.5 cubic miles of material to account for in the former summit area. To explain this, Williams suggested that the 10.5 cubic miles of material could be accounted for by withdrawal of magma from below as a result of the intrusion of the magma into adjacent country rock. Williams felt that the collapse probably followed shortly after a series of very violent eruptions which spread as nuées ardentes down the side of the volcano. Radiocarbon dating of volcanic ash ejected during the final eruptive phases of Mt. Mazama indicates an age of about 6,700 years.

Fig. 19-7   Crater Lake caldera, Oregon. *(John Shelton.)*

fissure eruptions

flood basalts

Not all volcanic eruptions produce cones. Eruptions of large amounts of lava from fissures, such as those which have occurred in recent time in Iceland, may form plateaus of *flood basalts.*

Examples of this type of eruption occur in the Columbia Plateau of Washington, Oregon, and Idaho and the Deccan area of western India. The lavas of the Columbia Plateau cover approximately 200,000 square miles and are, in places, more than 10,000 feet thick. Eruption of 60,000 to 75,000 cubic miles of basaltic lava through many fissures occurred during the Miocene. The lava was not erupted all at once, but accumulated by the piling up of many individual flows, varying in thickness from a few feet to several hundred feet (Fig. 19-8). Many of the flows cover very large areas, indicating that the basalts were quite fluid. As the flows advanced into an area of moderate relief, streams were ponded by the flows to form lakes into which subsequent lavas poured. Lake sediments and pillow lavas which formed in the lakes were overridden by successive flows, and the preeruption topography was covered by basalt flows until only a few "islands" of granite rose above the level of the basalt plateau.

In western India similar eruptions of basalt from fissures have produced a vast basalt plateau covering approximately 250,000 square miles.

plug domes

A somewhat peculiar type of eruption is the *plug dome,* in which very viscous lava rises to the surface as a pasty mass but does not spread far from the vent because of the high viscosity (Fig. 19-9). Many such domes are composed of obsidian and pumice. Flows associated with the domes usually have steep margins (Fig. 19-10)..

lava flows

The constructional surfaces of lava flows may make conspicuous topographic features readily observed on aerial photographs or topographic maps. As a flow advances, the top solidifies more rapidly than the interior, and forms a crust which breaks up into blocks as the still-molten interior of the flow continues to move. At the front of the flow solidified blocks of lava tumble over the edge and are overridden as the flow advances (Fig. 19-11). The surface of such a flow is characterized by a very rough, blocky appearance.

Topographic forms of fairly recent lava flows are usually well preserved. They may be identified on maps and air photos by various surface features and by a general lobate form extending from the eruptive center, which often consists of a volcanic cone (Figs. 19-12 and 19-13). Among the surface features commonly found are *pressure ridges* formed by movement of the congealing flow while the interior is still mobile (Figs. 19-14 and 19-15). Such ridges are usually transverse to the direction of flow and are convex downslope. Many have conspicuous cracks along the crest of the ridge. A related

Fig. 19-8   Columbia River basalt near Palouse Falls, Washington. *(F. O. Jones, U.S. Geological Survey.)*

form is a mound known as a *squeeze-up,* formed by lava pushing up through the earlier-formed crust.

Occasionally, the molten interior of a flow will break out and leave the earlier solidified sides and top, producing *lava tunnels.* Collapse of portions of the roofs of such tunnels leaves elongate depressions on the surface of the flow, and occasionally natural bridges are formed where all but a small part of the tunnel has collapsed. With time, the constructional effects of lava flows are obliterated by erosion, and differential erosion may produce new forms.

Lava flowing down a valley has an elongate form, and as it fills the valley, it destroys the preflow drainage. Later, as local tributary streams enter from the valley sides, drainage is reestablished along the margins between the flow and the valley sides. This happens largely as a result of the fact that

Fig. 19-9   Mono Craters plug domes, south of Mono Lake, California. *(John Shelton.)*

the surface of the flow is very rough and contains many depressions which inhibit flow of water on the surface. Thus two streams are developed along the sides of the flow, and as these streams cut their channels downward, the lava flow, which formerly occupied the topographic low in the valley, eventually becomes a stream divide (Fig. 19-18). This process results in inversion of topography. The ridge which is thus formed has a sinuous pattern in map view.

topographic effects of igneous intrusions

Since igneous intrusions form beneath the surface of the earth, no constructional forms are produced directly by them. Topographic effects stem pri-

Fig. 19-10   Inyo Craters plug domes and obsidian flows, California. *(John Shelton.)*

marily from differential erosion after they are exposed at the surface and from the bowing up of overlying rocks.

Erosion of a volcanic cone may expose a *volcanic neck* (Fig. 19-19), a cylindrical to spire-shaped form consisting of magma solidified in the vent, or conduit, of a former volcano. They stand above the surrounding area because the igneous rock is more resistant than the country rock. Volcanic necks differ from cones in lacking a crater at their top and in having steep-sided, spire-shaped forms.

Magma which has solidified below the surface in fissures cutting across preexisting rocks forms *dikes*. When exposed at the surface, they often develop vertical tabular ridges. Dikes occasionally radiate from volcanic necks or

Fig. 19-11   Front of an advancing lava flow. *(Wards Natural Science Establishment.)*

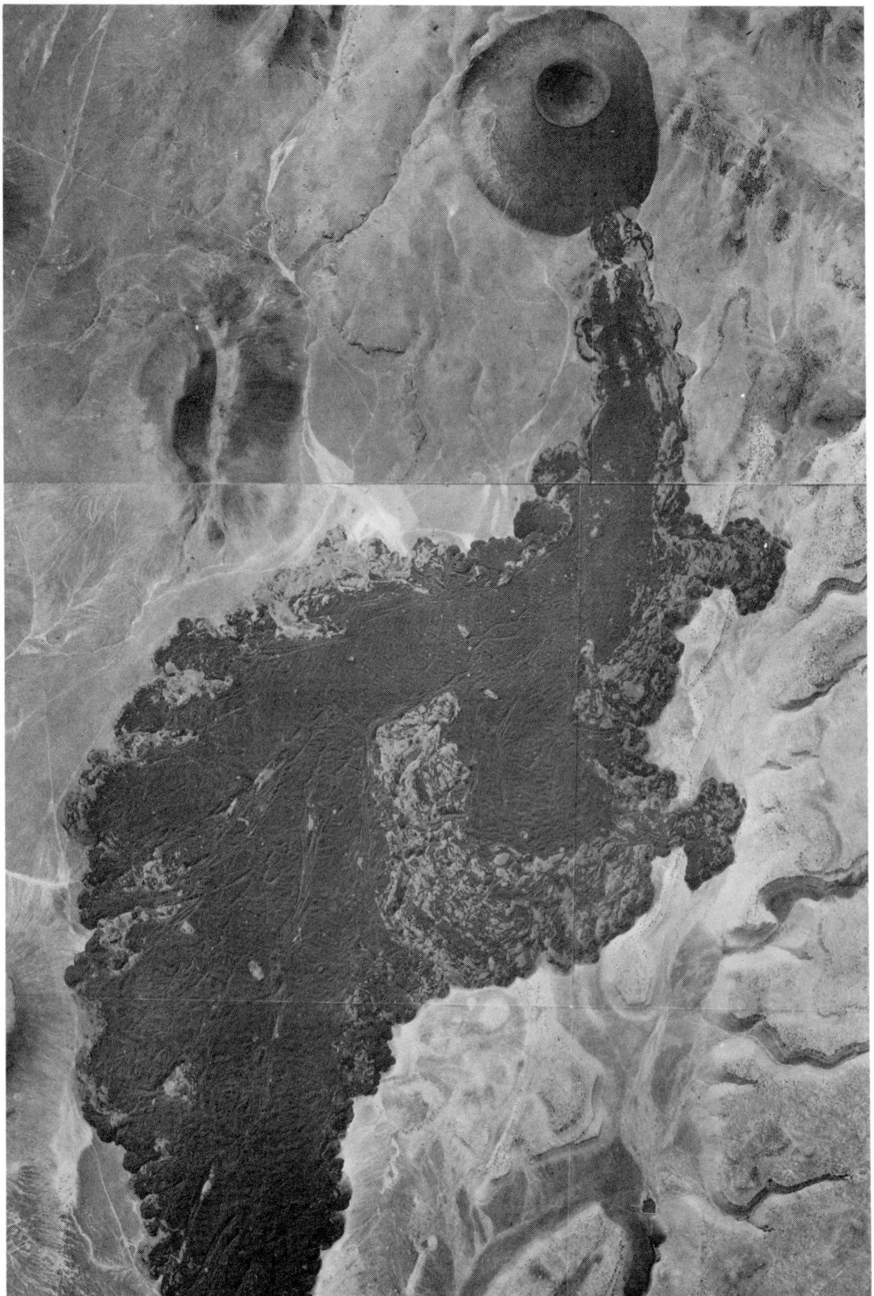

Fig. 19-12   Cinder cone and lava flow near Flagstaff, Arizona. Note the texture on the surface of the flow. *(U.S. Department of Agriculture.)*

other intrusions, as at Shiprock, New Mexico (Fig. 19-19). They resemble hogbacks of resistant sedimentary beds, but may be distinguished from hogbacks by the fact that they sometimes (1) cross one another, (2) have a radial pattern, (3) terminate abruptly, and (4) cut across sedimentary beds. Not all dikes form ridges, however. If they consist of rock less resistant than the adjacent rock, they may be etched out by differential erosion to form linear depressions.

Injections of tabular, sheetlike bodies of magma parallel to stratification of enclosing rocks form *sills*. Sills affect topography in much the same way as do resistant sedimentary beds or buried lava flows. They may form mesas, cuestas, or hogbacks, depending on their dip.

Fig. 19-13 Irregular surface of a lava flow, Craters of the Moon National Monument, Idaho. *(John Shelton.)*

A good example of a sill is the Palisades sill (Fig. 19-20) overlooking the Hudson River in New Jersey. It is up to 1,000 feet thick and over 50 miles long.

Laccoliths are concordant lenticular intrusions which have bowed up the overlying rock into a dome. When exposed by erosion, the igneous rock makes an oval dome, often surrounded by concentric hogbacks of sedimentary beds (Figs. 19-21 to 19-23).

Stocks and *batholiths* are discordant intrusions, usually of considerably greater size than any of the intrusions named above. Both are common near the cores of mountain ranges, and both tend to make topographic highs. Topography developed on batholiths is usually rather massive, lacking the distinct linear forms characteristic of sedimentary beds. The rocks in a large igneous

Fig. 19-14 Pressure ridges on a lava flow, Newberry Crater, Oregon. Note caldera in center of photograph. *(Austin Post, University of Washington.)*

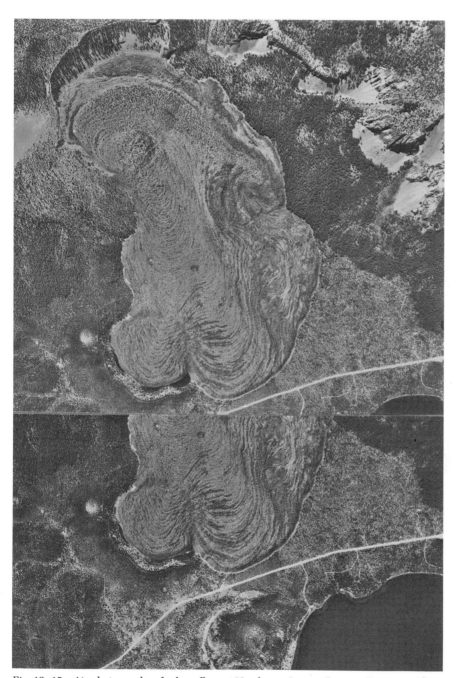

Fig. 19-15   Air photographs of a lava flow at Newberry Crater, Oregon. Pressure ridges are well shown on the surface of the flow. The upper lobe has flowed into an older crater. *(U.S. Forest Service.)*

mass are often well jointed, however, and as a result may give the topography a distinctive form (Fig. 19-24).

Columnar jointing may form in lava flows, dikes, and sills as a result of contraction upon cooling. The joint-bounded columns tend to have a hexagonal pattern. Erosion of igneous rocks having pronounced vertical columnar jointing results in vertical cliffs. Figure 19-25 shows a ridge made up of a single row of exceptionally large columns.

Fig. 19-16 Surface of a lava flow characterized by absence of surface drainage and by a rough irregular pattern, Corizozo, New Mexico. *(Wards Natural Science Establishment.)*

Fig. 19-17 Vulcan's Throne, a lava flow which cascaded into the Grand Canyon, Arizona. *(J. R. Balsley, U.S. Geological Survey.)*

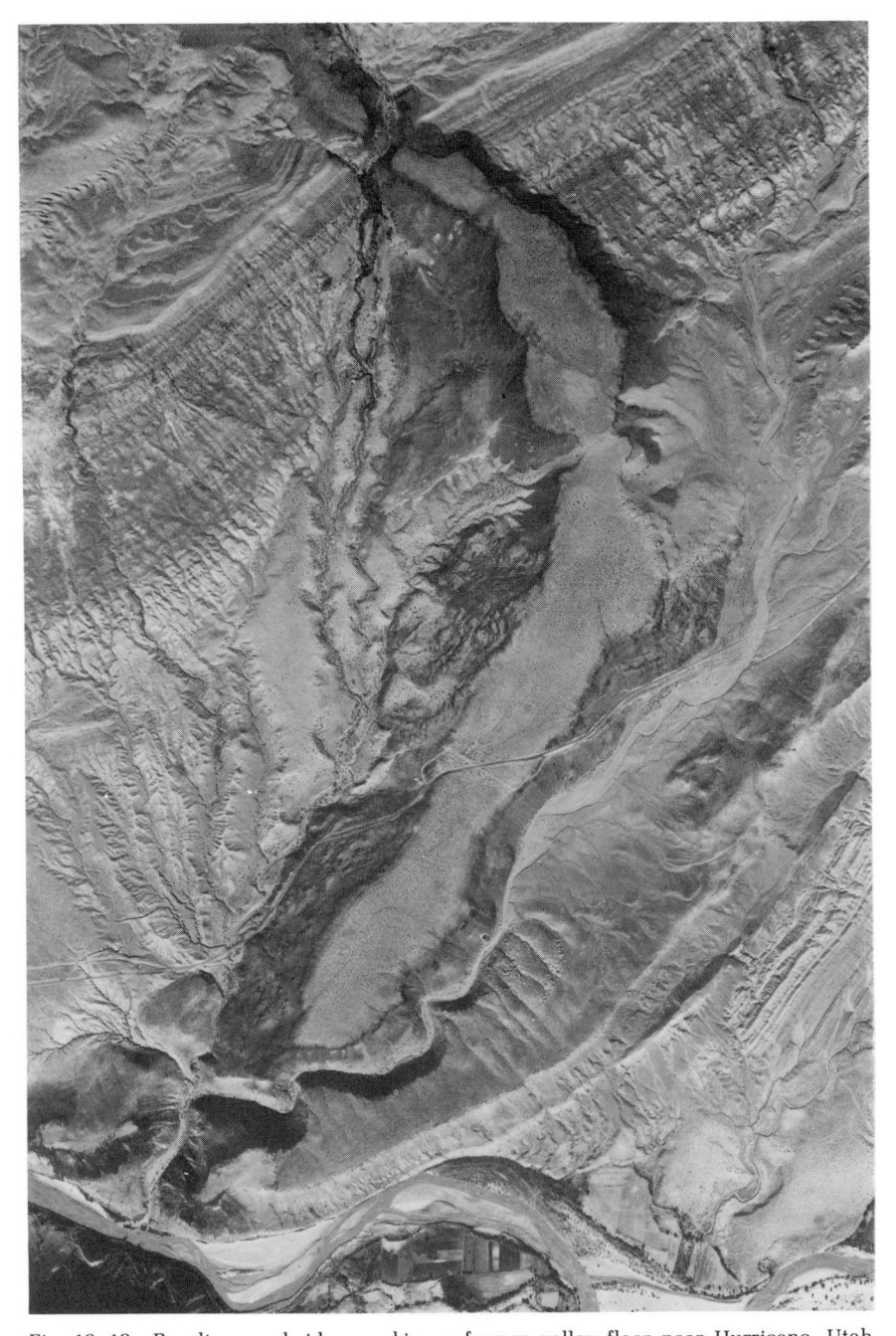

Fig. 19-18 Basalt-capped ridge marking a former valley floor near Hurricane, Utah.
An example of inversion of topography. (U.S. Department of Agriculture.)

Fig. 19-19   A volcanic neck with radiating dikes, Shiprock, New Mexico. *(Wards Natural Science Establishment.)*

Fig. 19-20 The Palisades sill along the Hudson River, New Jersey. *(Wards Natural Science Establishment.)*

Fig. 19-21   Bear Butte, South Dakota. The high central core is composed of intrusive igneous rock, surrounded by the upturned edges of sedimentary beds. *(N. H. Darton, U.S. Geological Survey.)*

Fig. 19-22 Sundance Mountain, Wyoming, a laccolith flanked by upturned sedimentary beds. *(Wards Natural Science Establishment.)*

Fig. 19-23   Green Mountain laccolith, Wyoming. *(U.S. Department of Agriculture.)*

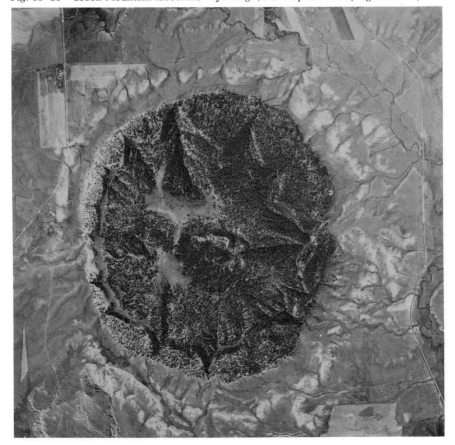

Fig. 19-24 Topography developed on jointed granite, Wyoming.

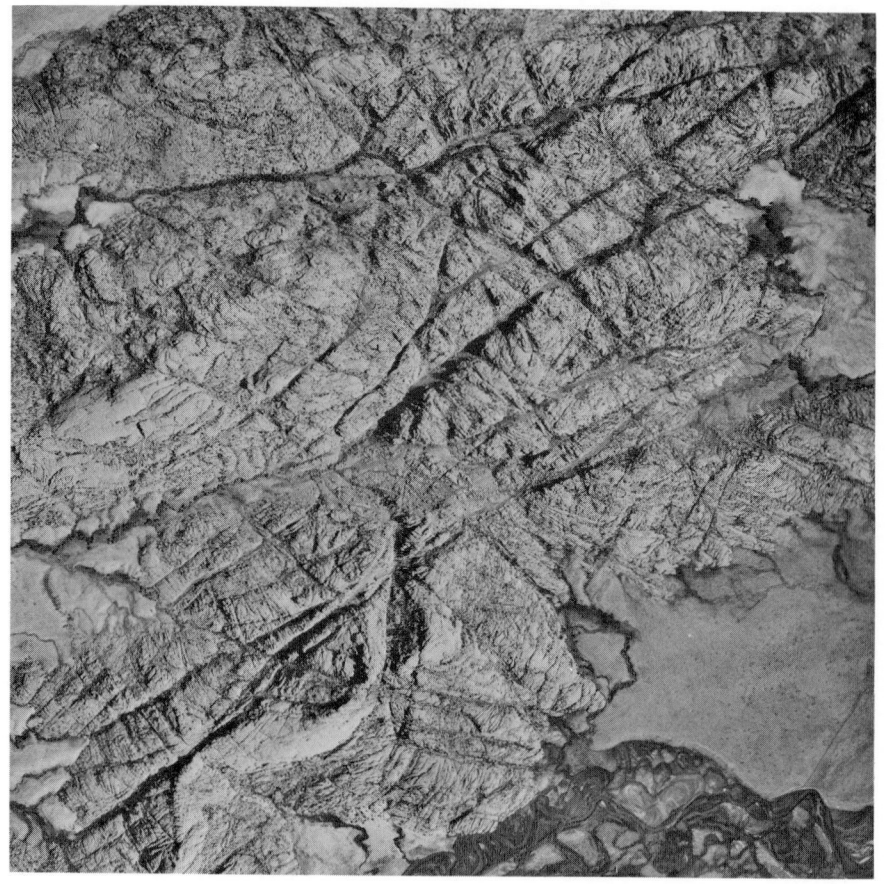

Fig. 19-25 Ridge composed of a single row of columns in a jointed basalt near Vantage, Washington.

references

Billings, M. P.: 1965. Structural Geology, Prentice-Hall, Inc., Englewood Cliffs, N.J.

Cotton, C. A.: 1952. Volcanoes as Landscape Forms, John Wiley & Sons, Inc., New York.

Diller, J. S., and H. B. Patton: 1902. The geology and petrography of Crater Lake National Park, U.S. Geol. Survey Prof. Paper 3.

Nichols, R. L.: 1939. Pressure-ridges and collapse-depressions on the McCartys basalt flow, New Mexico. Am. Geophys. Union Trans., pp. 432-433.

Putnam, W. C.: 1938. The Mono Craters, Calif., Geog. Rev., vol. 28, pp. 68-82.

Stearns, H. T.: 1924. Craters of the Moon National Monument, Geog. Rev., vol. 14, pp. 362-372.

Strahler, A. N.: 1963. The Earth Sciences, Harper & Row, Publishers, Incorporated, New York.

Thornbury, W. D.: 1954. Principles of Geomorphology, John Wiley & Sons, Inc., New York.

Williams, H.: 1932. The history and character of volcanic domes, Univ. Calif. Dept. Geol. Sci. Bull., vol. 21, pp. 51-146.

———— : 1935. Newberry Volcano of Central Oregon, Geol. Soc. America Bull., vol. 46, pp. 253-304.

———— : 1941. Crater Lake: the Story of Its Origin, University of California Press, Berkeley, Calif.

———— : 1941. Calderas and their origin, Univ. Calif. Dept. Geol. Sci. Bull., vol. 25, pp. 239-346.

———— : 1944. Volcanoes of the Three Sisters region, Oregon Cascades, Univ. Calif. Dept. Geol. Sci. Bull., vol. 27, pp. 37-83.

appendix

contours

A contour is a line which passes through points having the same elevation above sea level. If you were to start at a certain altitude on an irregular surface and walk in such a way as to go neither uphill nor downhill, you would trace out a path that corresponds to a contour line. A similar path would be made by the horizontal surface of a shoreline lapping up against an irregular coastline. Such a path will curve around hills, bend upstream in valleys, and swing outward around ridges. Theoretically, every contour must be a closed curve which returns upon itself, no matter how long it is. However, on maps of relatively small areas, contours may be either closed curves or pass off the margins of the map without closing on that particular map.

On an island in a sea, the shoreline is a contour line at zero elevation. If the island were to be raised above sea level at increments of 20 feet at a time, the successive shorelines would form a series of contour lines separated by 20-foot intervals. The vertical distance between two successive contours is the contour interval. On a map made with a contour interval of 20 feet, each contour stands 20 feet above or below the adjacent contour. Whether the

TOPOGRAPHIC MAP SYMBOLS

VARIATIONS WILL BE FOUND ON OLDER MAPS

Hard surface, heavy duty road, four or more lanes ....

Hard surface, heavy duty road, two or three lanes ....

Hard surface, medium duty road, four or more lanes ....

Hard surface, medium duty road, two or three lanes ....

Improved light duty road ....

Unimproved dirt road—Trail ....

Dual highway, dividing strip 25 feet or less ....

Dual highway, dividing strip exceeding 25 feet ....

Road under construction ....

Railroad: single track—multiple track ....

Railroads in juxtaposition ....

Narrow gage: single track—multiple track ....

Railroad in street—Carline ....

Bridge: road—railroad ....

Drawbridge: road—railroad ....

Footbridge ....

Tunnel: road—railroad ....

Overpass—Underpass ....

Important small masonry or earth dam ....

Dam with lock ....

Dam with road ....

Canal with lock ....

Buildings (dwelling, place of employment, etc.) ....

School—Church—Cemeteries ....

Buildings (barn, warehouse, etc.) ....

Power transmission line ....

Telephone line, pipeline, etc. (labeled as to type) ....

Wells other than water (labeled as to type) .... ₒ Oil .... ₒ Gas

Tanks; oil, water, etc. (labeled as to type)................................... • • ● ⊘Water

Located or landmark object—Windmill...................................... o................................ ໐

Open pit, mine, or quarry—Prospect....................................... ⚒................................ x

Shaft—Tunnel entrance.. ▫................................ Y

Horizontal and vertical control station:

    tablet, spirit level elevation.. BM △ 3899

    other recoverable mark, spirit level elevation..................... △ 3938

Horizontal control station: tablet, vertical angle elevation............ VABM △ 2914

    any recoverable mark, vertical angle or checked elevation.... △5675

Vertical control station: tablet, spirit level elevation.................. BM ✕ 945

    other recoverable mark, spirit level elevation..................... ✕ 890

Checked spot elevation... ✕5923

Unchecked spot elevation—Water elevation............................. ✕ 5657 870

Boundary: national... ▬▬ ▬ ▬ ▬▬

    state.. ▬▬ ▬ ▬ ▬▬

    county, parish, municipio.. ▬▬ ▬ ▬▬ ▬

    civil township, precinct, town, barrio................................ ▬▬ ▬▬ ▬▬ ▬▬

    incorporated city, village, town, hamlet............................ ▪-▪-▪-▪-▪-▪-

    reservation, national or state.. ▬▬ . ▬▬ .

    small park, cemetery, airport, etc..................................... -------------

    land grant.. ▬▬ .. ▬▬ ..

Township or range line, U.S. land survey................................ ▬▬▬▬▬

Township or range line, approximate location......................... ▬ ▬ ▬ ▬ ▬▬

Section line, U.S. land survey... ▬▬▬▬▬

Section line, approximate location... ▬ ▬ ▬ ▬ ▬-

Township line, not U.S. land survey...................................... ••••••••••••••••••••••

Section line, not U.S. land survey... ....

Section corner: found—indicated... +................................ +

Boundary monument: land grant—other................................ ▫................................ ▫

U.S. mineral or location monument.. ▲

Index contour..................... ▬▬▬▬ Intermediate contour......... ⌒

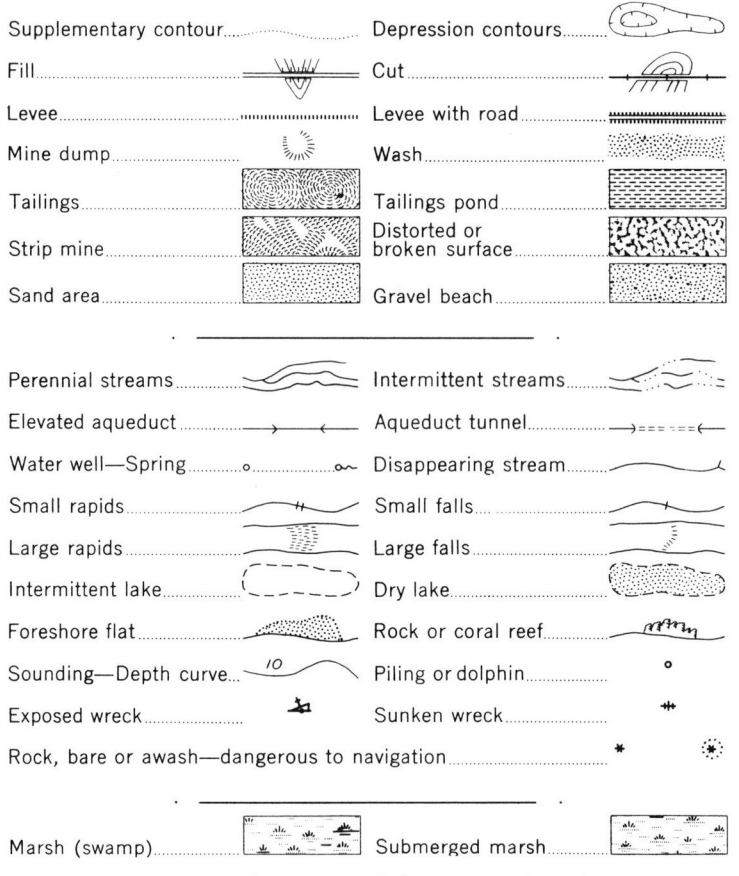

Fig. A-1 Topographic map symbols. *(U.S. Geological Survey.)*

contour interval is large or small depends primarily on the map scale and the amount of relief in an area. A small contour interval allows a fairly high amount of detail to be shown about the topography, and is generally used on maps of local areas of low relief. Maps of high relief or of small scale, covering a large area, generally have a larger contour interval, perhaps 100 feet or more.

Where the slope of the land surface is steep, contours are crowded close together; where slopes are gentle, contours are spaced farther apart. On most maps every fifth contour is somewhat wider and darker than the rest, to facilitate identification of the contours. The contours are numbered at a convenient interval, which is a specified multiple of the contour interval (Fig. A-2).

Enclosed depressions may be shown on a topographic map, but because the contours would resemble those drawn for an isolated hill, short lines, called *hachures*, are placed transverse to the contours (Fig. A-4b).

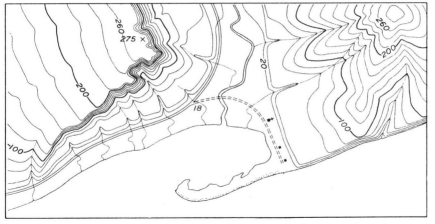

Fig. A-2   Perspective sketch and contour map of the same area. *(U.S. Geological Survey.)*

location of points on a map

latitude and longitude

Points may be located on a map by referring to a grid of intersecting lines known as *parallels* and *meridians*. Meridians are great circles passing through both poles. One particular meridian, the Prime Meridian, which passes through Greenwich, England, is used for reference. Distance east or west of this meridian is measured in degrees of longitude. Parallels are circles drawn parallel to the equator. Distance north or south of the equator is measured in degrees of latitude. Any point on the earth may be located by knowing the number of degrees of latitude north or south of the equator and the number of degrees east or west of the Prime Meridian.

township and range

Most of the United States has been subdivided into squares called *townships* having six miles on a side. Townships are laid off east and west from a principal meridian and north and south from an arbitrary base line.

A township may be divided into 36 sections, each enclosing 1 square mile (Fig. A-3).

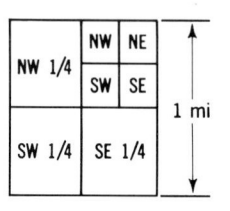

Fig. A-3 Location of an area by section, Township, and Range.

Each section may be divided into quarters, and each quarter further subdivided into quarters. The upper right corner in Fig. A-3 may be described as the NE ¼ of the NE ¼ of section 1, Township 38 north, Range 2 east.

profiles

A topographic profile is a cross section, or side view, of a portion of the earth's surface along a given line. Figure A-4 shows profiles drawn from topographic maps. The elevation along various places on the line of the profile is projected into the side view, and when these points are connected, a profile of the topography is obtained.

vertical exaggeration

Since a profile is a side view, it has both a horizontal and a vertical dimension. If both dimensions are drawn to the same scale, the profile is a true representation of the topography, but since most maps are drawn to relatively small scales, only major relief features show up. By expanding the vertical scale while leaving the horizontal scale the same, more detail may be shown on a profile. However, in expanding the vertical scale, exaggeration occurs, and a certain amount of distortion is present.

The ratio of horizontal distance to vertical distance on a profile is a measure of vertical exaggeration. Figure A-5 shows three profiles of the same terrain drawn with different vertical exaggerations. If 1 inch of horizontal distance represents 4,000 feet, while 1 inch of vertical distance represents 500 feet, the vertical exaggeration is $^{4,000}/_{500} = 8$ times. In other words, the vertical scale has been "stretched" to 8 times that of the horizontal scale. With 1 inch = 2,000 feet, the vertical exaggeration is only $^{4,000}/_{2,000} = 2$.

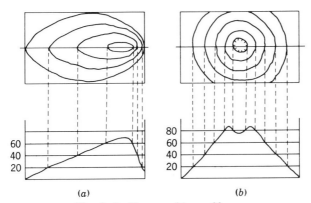

(a)   (b)

Fig. A-4   Topographic profiles.

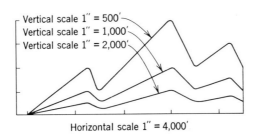

Vertical scale 1″ = 500′
Vertical scale 1″ = 1,000′
Vertical scale 1″ = 2,000′

Horizontal scale 1″ = 4,000′

Fig. A-5   Comparison of topographic profiles drawn with different vertical scales.

sources of aerial photographs in the United States

U.S. GEOLOGICAL SURVEY

For general information and photographic coverage:

Eastern United States
    Map Information Service
    U.S. Geological Survey
    Washington, D.C.

Central United States
    Central Regional Engineer
    U.S. Geological Survey
    Rolla, Mo.

Rocky Mountain States
    Rocky Mt. Regional Engineer
    U.S. Geological Survey
    Denver, Colo.

Pacific Coast States
    Pacific Regional Engineer
    U.S. Geological Survey
    Sacramento, Calif.

SOIL CONSERVATION SERVICE

Director, Cartographic Divison
Soil Conservation Service
U.S. Dept. of Agriculture
Washington, D.C.

COMMODITY STABILIZATION SERVICE

Western Laboratory, Performance Division
Commodity Stabilization Service
U.S. Dept. of Agriculture
Salt Lake City, Utah

Eastern Laboratory
Commodity Stabilization Service
U.S. Dept. of Agriculture
Washington, D.C.

U.S. FOREST SERVICE

Regional Forester
U.S. Forest Service

At the following offices:
    Missoula, Mont.
    Denver, Colo.
    Albuquerque, N.Mex.
    Ogden, Utah
    San Francisco, Calif.
    Portland, Oreg.

For eastern states:
    Chief, Forest Service
    U.S. Dept. of Agriculture
    Washington, D.C.

COAST AND GEODETIC SURVEY
U.S. Dept. of Commerce
Washington, D.C.

ABRAMS AERIAL SURVEY CORP.
Lansing, Mich.

AERO SERVICE DIVISION
Litton Industries
Los Angeles, Calif.

JACK AMMANN PHOTOGRAMMETRIC ENGINEERS, INC.
San Antonio, Tex.

TOBIN AERIAL SURVEYS
San Antonio, Tex.

glossary

Ablation. The combined processes by which a glacier wastes.

Ablation area. That part of a glacier or snowfield where ablation exceeds accumulation.

Ablation moraine. Drift deposited from a superglacial position through the melting of underlying stagnant ice.

Abrasion. The wearing away by friction.

Abstraction. Stream piracy resulting from the shifting of divides.

Accumulation area. The area of a glacier in which annual accumulation exceeds ablation.

Aggradation. The process of building up a surface by deposition.

A horizon. Zone of eluviation. The uppermost zone in the soil profile, from which soluble salts and colloids have been leached and in which organic matter has accumulated.

Alkali. Sodium carbonate or potassium carbonate or, more generally, any bitter-tasting salt found at or near the surface in arid and semiarid regions.

Alluvial fan. Low, cone-shaped deposit formed by a stream issuing from mountains into a lowland.

Alluvium. A general term for unconsolidated sediments deposited from a river, including sediments laid down in river beds, floodplains, lakes, fans at the foot of mountain slopes, and estuaries.

Alpine glacier. Glaciers occupying mountainous terrain.

Altitude. The vertical distance between a point and a datum surface or plane, such as mean sea level.

Angle of repose. The maximum slope or angle at which loose material remains stable.

Annular drainage pattern. Ringlike drainage pattern, subsequent in origin, associated with maturely dissected dome or basin structures.

Antecedent stream. Drainage established prior to deformation of beds by folding and faulting.

Anticlinal mountain. Ridge which follows on anticlinal axis, formed by a convex flexure of strata.

Anticlinal valley. Valley which follows an anticlinal axis, formed by erosional breaching of an anticline.

Anticline. Structure in which beds dip in opposite directions from the axis.

Aquiclude. A formation which, although porous and capable of absorbing water slowly, will not transmit it fast enough to furnish an appreciable supply of water.

Aquifer. A formation that is water-bearing.

Arête. A sharp-crested mountain ridge between two cirques.

Artesian water. Groundwater that is under sufficient pressure to rise above the level at which it is encountered by a well, but which does not necessarily rise to or above the surface of the ground.

Atoll. A ringlike coral island or islands encircling or nearly encircling a lagoon.

Axis of a fold. The line following the apex of an anticline or the lowest part of a syncline.

Badlands. A region nearly devoid of vegetation where erosion has cut the land into an intricate maze of narrow ravines and sharp crests and pinnacles.

Bahada. The nearly flat surface of alluvium along a mountain foot; surface of confluent alluvial fans.

Bar. An embankment of sand or gravel deposited on the floor of a stream, sea, or lake. If a spit is continued until its distal end reaches a shore, it is called a bar.

Barbed tributary. Tributary entering the mainstream in an upstream direction instead of pointing downstream, as is normal.

Barchan. A crescentic-shaped sand dune having a gentle slope on the convex side, a steeper slope on the concave, or leeward, side, and horns pointing in the direction of wind movement.

Barrier reef. A coral reef parallel to the shore but separated from it by a lagoon.

Basal till. Till carried at, or deposited from, the base of a glacier.

Base level. The level below which a land surface cannot be reduced by running water.

Basin. A depressed area.

Basin, structural. Fold in which the rocks dip toward a central point.

Batholith. A large irregular mass of igneous rock emplaced at depth, having an area of more than 40 square miles.

Bayhead beach. A beach formed at the head of a bay.

Baymouth bar. A bar extending partially or entirely across the mouth of a bay.

Beach cusp. Succession of stony or gravelly cusps with sharp points toward the water, situated on the upper part of the beach, where the waves play only at high stages of the tide.

Beach drifting. Lateral movement of beach sediments parallel to the beach.

Beach ridge. An essentially continous mound of beach material heaped up by wave action.

Bed load. Rock particles rolled or pushed along the bottom of a stream by the moving water.

Beheaded stream. The lower portion of a stream from which water has been diverted by stream piracy.

Bergschrund. The crevasse occurring at the head of a mountain glacier which separates the moving snow and ice of the glacier from the relatively immobile snow and ice adhering to the headwall of the valley.

B horizon. Zone of illuviation. The lower soil zone which is enriched by the deposition or precipitation of material from the overlying zone, or A horizon.

Blind valley. Valley enclosed at its downstream end by valley walls. Streams terminate in a tunnel or depression, at the closed, or blind, end of the valley.

Blowout. A general term for various saucer-, cup-, or trough-shaped hollows formed by wind erosion of a sand deposit.

Bolson. A basin or depression having no outlet.

Bornhardt. See Inselberg.

Boulder train. Glacial boulders derived from a single locality arranged in a line, or in several lines, streaming off in the direction in which a glacier moved.

Braided stream. A stream flowing in several dividing and reuniting channels resembling the strands of a braid.

Breached anticline. An anticline that has been more deeply eroded in the center, so that erosional scarps face inward toward the axis of the anticline.

Breaker. A wave breaking on or near the shore.

Caldera. A large circular volcanic depression the diameter of which is many times greater than that of the included volcanic vent or vents.

Caliche. Calcium carbonate precipitated in desert soils, where evaporation exceeds precipitation.

Calving. The breaking off of chunks of ice at a glacier's margin.

Carbon-14. A radioactive isotope of carbon with atomic weight 14, produced by collisions between neutrons and atmospheric nitrogen.

Carbonation. Chemical process during weathering that converts basic oxides into carbonates.

C horizon. Soil zone composed of partially decomposed parent material.

Cinder cone. A volcanic cone formed by the accumulation of volcanic cinders or ash around a vent.

Cirque. A deep, steep-walled recess in a mountain, caused by glacial erosion.

Cirque glacier. A small glacier occupying a cirque.

Clay. Particles less than $\frac{1}{16}$ millimeter in diameter; minerals that are essentially hydrous aluminum silicates, or occasionally hydrous magnesium silicates.

Climate. The sum total of the meteorological elements that characterize the average and extreme condition of the atmosphere over a long period of time at any one place or region of the earth's surface.

Coast of emergence. Shoreline of emergence. Coast made by withdrawal of the sea.

Coast of submergence. Shoreline of submergence. Coast produced by invasion of the sea.

Col. Low passes, or saddles, on the watershed between drainage systems.

Collapse sink. Caverns which become so enlarged by solution and erosion that their roofs collapse.

Colluvium. Unconsolidated deposits, usually at the foot of a slope or cliff, brought there chiefly by gravity.

Columnar jointing. Joints which form hexagonal or polygonal columns in igneous rock as a result of contraction during cooling.

Competence (of stream). The diameter of the largest particle a stream can move.

Composite cone. A volcanic cone, usually of large dimension, built of alternating layers of lava and pyroclastic material.

Composite fault scarp. A scarp whose height is due partly to differential erosion and partly to fault movement.

Cone of depression. Depression, roughly conical in shape, produced in a water table or piezometric surface by pumping of a well or by artesian flow.

Connate water. Water entrapped in the interstices of a sedimentary rock at the time the rock was deposited.

Consequent stream. A stream which follows a course that is a direct consequence of the original slope of the surface on which it developed.

Constructional surface. A surface which originates as a result of deposition of material.

Continental glacier. An ice sheet covering a large part of a continent.

Continental shelf. The shallow and gradually sloping ground from the sea margin to the 100-fathom line, beyond which the descent to abysmal depths is abrupt.

Continental slope. The declivity from the offshore border of the continental shelf at depths of approximately 100 fathoms to oceanic depths. It is characterized by a marked increase in gradient.

Contour. An imaginary line on the surface of the ground, every point of which is at the same altitude.

Contour interval. The difference in elevation between two adjacent contour lines.

Coral reef. A reef formed by the action of reef-building coral polyps, which build internal skeletons of calcium carbonate.

Corrasion. Mechanical erosion by moving glacial ice, wind, or running water.

Corrosion. Erosion accomplished by chemical solution.

Crag-and-tail. A streamlined hill or ridge consisting of a knob of resistant bedrock (the "crag") with an elongate body of glacial till (the "tail") on its lee side.

Creep. The slow downslope movement of rock fragments and soil.

Crevasse. A fissure in glacial ice formed under the influence of various strains.

Crevasse filling. Elongate kame believed to have been deposited in a crevasse.

Cuesta. A symmetrical ridge with a long gentle slope corresponding to the dip of a resistant bed and a steep slope on the cut edges of the beds.

Cuspate bar. A crescent-shaped bar uniting with shore at each end. It may be formed by a single spit growing from the shore, turning back to again meet the shore, or by two spits growing from the shore, uniting to form a bar of sharply cuspate form.

Cycle of erosion. The succession of events involved in the evolution of landforms.

Debris cone. A cone-shaped deposit of soil, sand, gravel, and boulders.

Debris flow. A moving mass of water-lubricated debris.

Debris slide. The rapid downward movement of predominantly unconsolidated and incoherent debris in which the mass does not show backward rotation but slides or rolls forward, forming an irregular hummocky deposit.

Decomposition. The breaking down of minerals through chemical weathering.

Deflation basin. Basin formed by the removal of fine material by wind.

Deglaciation. The uncovering of an area from beneath glacier ice as a result of shrinkage of a glacier.

Degradation. The lowering of the surface of the land by erosion.

Delta. An alluvial deposit, often triangular-shaped, formed where a stream enters the ocean or a lake.

Dendritic drainage pattern. Drainage pattern characterized by irregular branching of tributaries in many directions, usually joining the mainstream at acute angles.

Density current. Turbidity current. A highly turbid and relatively dense current which moves along the bottom slope of a body of standing water.

Desert pavement. When loose material containing pebbles or larger stones is exposed to wind action, the finer dust and sand may be blown away, so that the pebbles gradually accumulate on the surface, forming a veneer which protects the finer material underneath from attack.

Desert varnish. A dark surface stain of manganese or iron oxide, usually with a glistening luster, which forms on some exposed rock surfaces in the desert.

Detritus. Fragmental material, such as sand, silt, and mud, derived from older rocks by disintegration.

Diastrophism. The process or processes by which the crust of the earth is deformed.

Differential erosion. The more rapid erosion of portions of the earth's surface as a result of differences in the character of the rock or in the intensity of surface processes.

Differential weathering. When rocks are not uniform in character but are softer or more soluble in some places than in others, an uneven surface may be developed.

Dip. The angle at which a bed or other planar feature is inclined from the horizontal.

Dip-slip fault. A fault in which the movement is down the dip of the fault.

Dip slope. A slope of the land surface which conforms approximately to the dip of the underlying rocks.

Discharge. Rate of flow at a given instant in terms of volume per unit of time through a given cross-sectional area.

Dissection. The work of erosion in destroying the continuity of a relatively even surface by cutting ravines or valleys into it.

Distributary. A river branch flowing away from the mainstream.

Divide. The line of separation between drainage systems; the summit of a ridge between streams.

Doline. Rounded hollows varying from about 30 to 3,000 feet in diameter and from about 8 to 330 feet in depth; similar to sinkholes.

Dome. A roughly symmetrical upfold, the beds dipping in all directions from the crest of the structure.

Downwasting. The diminishing of glacier ice in thickness during ablation; the wearing down of a landmass by weathering and erosion.

Drainage basin. The area drained by a river system.

Drainage system. A stream and its tributaries.

Drawdown. The lowering of the water table or piezometric surface by pumping or artesian flow.

Dreikanter. Pebbles with three facets shaped by sandblasting.

Drowned valley. A valley whose mouth has been inundated by the sea and converted into an estuary by submergence of the coast.

Drumlin. A streamlined oval-shaped hill of glacial drift, with its long axis parallel to the direction of flow of a former glacier.

Dune. A mound, ridge, or hill of wind-blown sand.

Dust well. A pit in glacier ice or sea ice produced when small dark particles on the ice surface are heated by sunlight and sink down into the ice.

Earthflow. A downslope flow of unconsolidated material lubricated with water.

Effluent stream. A stream which receives groundwater.

Einkanter. Pebbles with one facet shaped by sandblasting.

Elbow of capture. The bend in the course of a captured stream where it turns from the captured portion of its valley into the valley of the capturing stream.

Eluviation. The movement of soil material from one place to another within the soil when there is an excess of rainfall over evaporation. Horizons that have lost material through eluviation are referred to as eluvial, and those that have received material, as illuvial. Eluviation may take place downward or sidewise, according to the direction of water movement.

Embayed coast. A coast with many bays formed as a result of submergence beneath the sea.

Emergence. A term which implies that part of the ocean floor has become dry land, but does not imply whether the sea receded or the land rose.

End moraine. A ridgelike accumulation of glacial drift built along the margin of a glacier.

Englacial drift. Material within glacial ice.

Entrenched meander. A meandering stream in an incised valley. Streams which have established meanders and are then rejuvenated and cut down in the old meanders.

Eolian. A term applied to the erosive action of the wind and to deposits which are due to the transporting action of the wind.

Ephemeral stream. A stream or portion of a stream which flows only in direct response to precipitation. It receives little or no water from springs and no long-continued supply from melting snow or other sources. Its channel is above the water table.

Equilibrium. A state of balance or adjustment between opposing forces.

Erratic. A transported rock fragment different from the bedrock on which it lies, generally transported by glacier ice or by floating ice.

Escarpment. A cliff or relatively steep slope separating gently sloping tracts.

Esker. A serpentine ridge of sand and gravel, deposited by a meltwater stream in a tunnel under a glacier.

Estuary. A bay at the mouth of a river, where the tide influences the river current.

Eustatic. Pertaining to simultaneous, worldwide changes in sea level.

Exfoliation. Process by which concentric sheets peel off from bare-rock surfaces.

Exfoliation dome. A large rounded domal feature produced by the process of exfoliation.

Exhumed topography. Monadnocks, mountains, or other topographic forms buried under younger rocks and exposed again by erosion.

Extended consequent stream. Extended consequents are streams of an older type which become extended across a newly emerged coastal plain.

Facet. A flat surface produced by abrasion on a rock.

Faceted spur. The end of a ridge which has been truncated by faulting or erosion.

Fan. A low cone-shaped accumulation of debris deposited by a stream descending from a ravine onto a plain, where the material spreads out in the shape of a fan.

Fault. A fracture along which there has been displacement of the two sides relative to one another.

Fault-line scarp. A scarp that is the result of differential erosion along a fault line rather than the direct result of the movement along the fault.

Fault scarp. The cliff formed by a fault.

Fault shoreline. Shoreline formed when the downthrown block of a fault is depressed to permit the waters of a sea or lake to rest against the fault scarp.

Felsenmeer. An area covered with blocks of rock.

Fetch. The continuous area of water over which the wind blows in essentially a constant direction.

Fill terrace. Terrace composed of alluvium, formed by rejuvenation of a stream in a valley fill. The surface of the terrace is constructional in origin.

Finger lake. Long narrow rock basins occupied by lakes.

Fiord. Segment of a glaciated trough occupied by the sea.

Firn. Compacted granular snow with a density usually greater than 0.4 but less than 0.82. Also called neve.

Firn limit. The highest level on a glacier to which the snow cover recedes during the ablation season.

Firn line. The line, or zone, dividing the ablation area of a glacier from the accumulation area.

Flatiron. A triangular-shaped portion of a hogback.

Floodplain. A strip of relatively smooth land on a valley floor bordering a stream, built of sediment deposited during times of flooding.

Fluting. Smooth deep furrows worn in the surface of rocks by glacial erosion.

Fluvial. Pertaining to rivers or produced by river action.

Footwall. The mass of rock beneath a fault plane; a person standing on a fault plane stands on the footwall.

Foreset beds. Inclined beds deposited by a stream on the frontal slope of a delta or channel bar.

Fringing reef. A coral reef which closely encircles the land.

Frost action. The weathering process, caused by repeated freezing and thawing.

Geomorphic cycle. Cycle of erosion. Landforms evolved with time through a series of stages from youth to maturity to old age, each of which is characterized by distinctive features.

Geomorphology. The study of physical and chemical processes that affect the origin and evolution of surface forms.

Geyser. A spring from which hot water and steam are intermittently thrown into the air.

Glacial drift. Sediment in transport or deposited directly or indirectly from a glacier or its meltwater.

Glacial striae. Usually straight, more or less regular scratches, commonly parallel, on smoothed surfaces of rocks, due to glacial abrasion.

Glacial trough. U-shaped valley shaped by glacial erosion.

Glacier. A body of ice, firn, and snow, originating on land and showing evidence of past or present flow.

Glacier table. Glacier tables occur when the general level of the ice is lowered by evaporation and melting, while the ice under a rock, insulated from the sun's rays, stays at the former level.

Glaciofluvial (fluvioglacial). Pertaining to meltwater streams flowing from glaciers or to the deposits made by such streams.

Glaciolacustrine. Pertaining to glacial lakes and sediment deposited in lakes marginal to a glacier.

Graben. A block that has been downthrown along faults relative to the rocks on either side.

Graded slope. Slopes of equilibrium developed at the least inclination at which the waste supplied by weathering can be transported.

Graded stream. A stream in which, over a period of years, slope is delicately adjusted to provide, with available discharge and with prevailing channel characteristics, just the velocity required for the transportation of the load supplied from the drainage basin. The graded stream is a system in equilibrium; its diagnostic characteristic is that any change in any of the controlling factors will cause a displacement of the equilibrium in a direction that will tend to absorb the effect of the change.

Gradient. Slope expressed as the angle of inclination from the horizontal.

Gravity flow. A type of glacier movement in which the flow of the ice is caused by the downslope component of gravity in an ice mass resting on a sloping floor.

Ground moraine. A sheet of glacial till deposited as a veneer of low relief over preexisting topography.

Groundwater. That part of the subsurface water which is in the zone of saturation.

Groundwater divide. A line on a water table on each side of which the water table slopes downward in a direction away from the line.

Hachures. A series of short parallel lines used to represent slopes of the ground, particularly depressions or embankments on contour maps.

Half-life. The time period in which half the initial number of atoms of a radioactive element disintegrate into atoms of another element.

Hanging valley. A tributary valley whose floor is higher than the floor of the trunk valley in the area of junction.

Hawaiian eruption. A type of volcanic eruption in which great quantities of extremely fluid basaltic lava are poured out, mainly issuing in lava fountains from fissures on the flanks of a volcano. Explosive phenomena are rare, but much spatter and scoria are piled into cones and mounds along the vents. Characteristic of shield volcanos.

Headland. A projection of the land into the sea, as a peninsula or promontory.

Headward erosion. Lengthening of a valley at its upper end by gullying, produced by water which flows in at its head.

Hogback. A sharp-crested ridge formed by differential erosion of a resistant bed of steeply dipping rock.

Homoclinal shifting. Migration of the divide on a homoclinal ridge as a result of more rapid erosion on the scarp face than on the dip slope.

Homocline. Tilted beds dipping in the same direction.

Hooked spit. A hook-shaped sand spit.

Horn. A high pyramidal peak with steep sides, formed by the intersecting walls of several cirques, as the Matterhorn in Switzerland.

Horst. A block of the earth's crust, generally long compared with its width, that has been uplifted along faults relative to the rocks on either side.

Hot spring. A thermal spring whose water has a higher temperature than that of the human body (above 98°F).

Hydration. The chemical combination of water with another substance.

Hydraulic. Pertaining to fluids in motion.

Hydraulic gradient. The rate of change of pressure head per unit of distance of flow at a given point. It is equal to the slope of the water surface in steady, uniform flow.

Hydrologic cycle. The cycle through which water passes, commencing as atmospheric water vapor, passing into liquid or solid form as precipitation, thence to the ground surface and to the sea by rivers, and finally again returning to atmospheric water vapor by means of evaporation and transpiration.

Hydrolysis. A chemical process of decomposition involving addition of water.

Hydrostatic pressure. The pressure exerted by water at any given point in a body of water at rest. That of groundwater is generally due to the weight of water at higher levels in the same zone of saturation.

Ice cap. A small ice sheet.

Ice-contact forms. Stratified drift bodies such as kames, kame terraces, and eskers, deposited in contact with melting glacier ice.

Ice sheet. A glacier forming a continuous cover over a land surface, moving outward in many directions. Continental glaciers, ice caps, and some highland glaciers are examples of ice sheets.

Impermeable. Rocks or sediments having a texture that does not permit perceptible movement of water under the head differences ordinarily found in subsurface water.

Incised meander. Entrenched meander. A deep, sinuous valley cut by a rejuvenated stream, the meandering course having been acquired in a former cycle.

Influent. A stream is influent with respect to groundwater if it contributes water to the zone of saturation.

Ingrown meander. Incised meander which has grown as the stream eroded its channel downward.

Inselberg. Prominent steep-sided residual hills and mountains rising abruptly above lowland surfaces of erosion.

Insequent stream. Stream system developed by random headward erosion of tributaries on horizontally stratified or homogeneous rocks.

Intercision. Drainage diversion caused by intersection of meanders of two streams.

Interglacial. Pertaining to the time between glaciations.

Interlobate moraine. A moraine built between two adjacent glacial lobes.

Intermittent stream. Streams which flow only part of the year, as after a rainstorm or during wet weather.

Joint. A fracture in a rock. A joint differs from a fault in lacking displacement on opposite sides of the fracture.

Juvenile water. Water that is derived from the interior of the earth and has not previously existed as atmospheric or surface water.

Kame. A low, steep-sided hill of stratified drift, formed in contact with glacier ice. Kames are composed chiefly of gravel or sand, whose form is the result of original deposition modified by settling during the melting of glacial ice against or upon which the sediment accumulated.

Kame terrace. A terrace of glacial sand and gravel deposited between valley ice, generally stagnant, and the valley sides.

Kame-and-kettle topography. Surface formed by a kame complex interspersed with kettles.

Karst topography. Irregular topography developed by the solution of carbonate rock by surface water and groundwater.

Karst valley. Elongated solution valley.

Karst window. Unroofed portion of a cavern, revealing part of a subterranean river.

Kettle. A depression in glacial drift, made by the melting of a detached mass of glacier ice that has been either wholly or partly buried in the drift.

Kettle moraine. A moraine whose surface is marked by many kettles.

Knickpoint. Point of abrupt change in the longitudinal profile of a stream valley.

Laccolith. A concordant lens-shaped intrusive igneous body that has domed up the overlying rocks but has a floor that is generally horizontal.

Lacustrine. Produced by, or belonging to, lakes.

Lagoon. Part of the sea nearly enclosed by a strip of land.

Landslide. The downward sliding or falling of large masses of rock.

Lateral moraine. An elongate ridge of glacial drift deposited along the sides of the glacier.

Lateral planation. Reduction of the land by the lateral swinging of a stream against its banks.

Laterite. A name derived from the Latin word for brick earth, applied to the red residual soils that have originated *in situ* from the atmospheric weathering of rocks. They are especially characteristic of the tropics, and are composed of iron and aluminum hydroxides.

Leaching. The removal in solution of the more soluble minerals by percolating waters.

Le Châtelier's principle. If conditions of a system, initially at equilibrium, are changed, the equilibrium will shift in such a direction as to tend to restore the original conditions.

Left-lateral fault. A strike-slip fault in which the movement is such that an observer facing the fault must turn left to find the other part of a displaced bed.

Levee. A bank above the general level of a floodplain confining a stream channel.

Limb. The flank of an anticline or syncline on either side of the axis.

Limestone. A bedded sedimentary deposit consisting chiefly of calcium carbonate ($CaCO_3$).

Load. The size and quantity of material transported by water or other surface processes.

Loess. Silt-size particles deposited primarily by the wind, commonly nonstratified and unconsolidated.

Longitudinal dune. Linear dune ridges, commonly more or less symmetrical in cross profile, which extend parallel to the direction of the dominant dune-building winds.

Longshore current. The inshore current moving essentially parallel to the shore, usually generated by waves breaking at an angle to the shore.

Longshore drifting. The movement of sediment parallel to the shore by longshore currents.

Marine-built terrace. A bench built be deposition of sediment seaward from marine-cut terrace.

Marine-cut terrace. Plain of marine abrasion.

Mass wasting. The downslope movement of rock debris under the influence of gravity.

Mature valley. A valley in which the width of the floodplain is approximately equal to the width of the meander belt.

Maturity. Stage in the evolution of landforms, characterized by the greatest diversity of form and maximum topographic differentiation.

Meander. Bend in the course of a stream, developed through lateral shifting of its course toward the convex side of the bend.

Meander belt. The part of a floodplain between two lines tangent to the outer bends of all the meanders.

Meander core. The central hill encircled by the meander.

Meander scar. Crescentic cut in a valley side made by lateral planation on the outer part of a meander which impinges against the valley side.

Medial moraine. An elongate body of drift on a glacier formed by the joining of adjacent lateral moraines below the juncture of two valley glaciers.

Mesa. A flat-topped mountain bounded by steep cliffs.

Meteoric water. Water which is derived from the atmosphere.

Misfit stream. Stream whose meanders are obviously out of harmony, either too small or too large, with the valley or with meander scarps preserved in the valley wall.

Monadnock. A residual hill or mountain standing above a peneplain.

Monocline. A steplike bend in otherwise horizontal or gently dipping beds.

Morainal lake. Lake that occurs behind a morainal dam which blockades drainage.

Moraine. An accumulation of glacial drift having initial constructional topography, built by the direct action of glacier ice.

Mudflow. A flow of heterogeneous debris lubricated with water.

Net slip. Total slip along a fault.

Neutral shoreline. Shoreline whose essential features do not depend on either emergence or submergence.

Neve. Snow recrystallized into granular ice.

Nickpoint. Interruption of a stream profile.

Nivation. Hollowing out of a basin by frost action and mass wasting under a snow bank.

Normal fault. Fault in which the hanging wall moves down relative to the footwall.

Nose. The apex of a plunging fold where the axis intersects the land surface.

Obsequent fault-line scarp. A scarp along a fault which faces in the opposite direction to the original fault scarp. The scarp faces the upthrown block.

Obsequent stream. A stream which flows in a direction opposite that of the consequent drainage.

Offshore bar. Accumulation of sand as a ridge built some distance from the shoreline as a result of wave action.

Old age. Stage in the evolution of fluvial landforms when the land has been reduced to low relief by erosive processes.

Oscillatory wave. Wave in which individual particles oscillate about a point, with little or no permanent change in position.

Outwash. Stratified drift deposited by meltwater streams beyond the margin of a glacier.

Outwash plain. Plain beyond the margin of a glacier, composed of stratified drift deposited from meltwater streams.

Oxbow. A crescent-shaped lake formed in an abandoned river bend by a meander cutoff.

Oxidation. Chemical process of combining with oxygen.

Paleosol. A buried soil.

Parabolic dune. A dune having a U shape, or shaped like a parabolic curve, concave toward the wind.

Parallel drainage pattern. River system in which the streams flow nearly parallel to one another.

Paternoster lakes. A chain of small glacial lakes connected by a stream.

Patterned ground. Symmetrical forms, such as polygons, nets, and stripes, characteristic of areas subject to intense frost action.

Pedalfer. Soil enriched in iron and alumina, formed in humid climates under forest cover.

Pedestal rock. A residual mass of weak rock capped with more resistant rock.

Pediment. A gently inclined erosion surface carved in bedrock, thinly veneered with fluvial gravel, developed at the foot of mountains.

Pediment passes. Narrow rock-floored passes extending from a pediment through a mountain range to a pediment on the other side of the range.

Pediplain. Widely developed plain formed by the coalescence of multiple pediments.

Pedocal. Soil enriched in lime, formed in arid or semiarid regions, where evaporation exceeds precipitation.

Peneplain. A landscape of low relief formed by long-continued erosion.

Pepino. Rounded, conical-shaped hills resulting from solution of carbonate rocks.

Perched water table. Local groundwater above the regional groundwater table and separated from it by an impervious unit.

Periglacial. Region beyond the margin of a glacier.

Permeability. Capacity of a material to transmit fluids.

Piedmont. Region at the foot of a mountain.

Piedmont glacier. A glacier formed by coalescence of valley glaciers beyond the base of a mountain range.

Piracy. Diversion of one stream by another.

Pitted outwash plain. Glacial outwash plain with many kettles.

Plastic flow. Change in shape of a solid without rupture.

Playa. Shallow lake basin in desert region intermittently filled with water which evaporates in a short period of time.

Pleistocene. The last Ice Age.

Plucking. Process of erosion whereby blocks of rock are removed from bedrock.

Plunge pool. Pothole occurring at the foot of a waterfall.

Podsol. Highly bleached soil low in iron and lime, formed under cool, moist climatic conditions.

Porosity. The ratio of the volume of pore space in a material to its total volume.

Pressure melting. Melting of ice as the result of lowering the melting point by application of pressure.

Pressure ridge. A ridge formed by horizontal pressures associated with flowage.

Prograding shoreline. A shoreline that is advancing seaward.

Push moraine. Moraine made by the plowing up of material at the front of a glacier.

Recessional moraine. End moraine formed by a stillstand of ice during recession of a glacier.

Rectangular drainage pattern. Drainage pattern characterized by right-angle bends in both the mainstream and its tributaries.

Recurved spit. Spit in which the end is curved landward.

Reef. A ridge of coral and shell debris formed in shallow, warm seawater.

Refraction. Bending of waves.

Rejuvenation. Stimulation of the erosive activity of a stream by uplift, climatic change, or change in base level.

Relief. The difference in elevation between the high and low points of a land surface.

Resequent fault-line scarp. A fault-line scarp formed as a result of differential erosion on opposite sides of a fault, the scarp facing in the same direction as on the original fault scarp.

Resequent stream. A stream which flows in the same direction as that of the consequent drainage, but which develops at a lower level than the initial slope.

Reverse fault. A fault in which the hanging wall has moved up relative to the footwall.

Ria shoreline. A shoreline formed by the submergence of a landmass dissected by numerous river valleys.

Riegel. Traverse bedrock ridge on the floor of a glaciated valley.

Rill. A very small trickle of water.

Roche moutonée. A rounded hummock of rock smoothed and striated by glacial abrasion.

Rock drumlin. Landform similar in form to a drumlin but composed of rock.

Rock fan. Landform resembling an alluvial fan but developed on bedrock by erosion.

Rock glacier. A tonguelike body of boulders, resembling a small glacier, which moves slowly downvalley under the influence of ice between the particles.

Saltation. Process by which a particle is picked up by turbulence in a stream and carried forward by a series of leaps and bounds.

Scarp. A cliff or steep slope.

Scour and fill. The process of cutting and filling of channels with variations in velocity of flow.

Sea cliff. A rock remnant isolated by wave erosion.

Seif dune. A variety of longitudinal dune, with its long axis parallel to the direction of prevailing wind.

Seismic sea wave. Tsunami. A long-period wave generated by a submarine earthquake, slide, or volcanic eruption.

Settling velocity. Velocity with which a particle sinks in a fluid.

Sheet erosion. Erosion caused by a continuous sheet of surface water.

Shield volcano. A broad, gently sloping volcanic cone, usually several tens or hundreds of square miles in extent, built by successive basalt flows.

Shutter ridge. Ridges shifted by faulting so that they block valleys on the opposite side of the fault.

Sinkhole. A depression in the surface formed by solution of limestone or other soluble material.

Sinking creek. A stream which disappears into a subterranean course.

Slip face. The steep face on the lee side of a dune.

Slope wash. Soil moved downslope by the action of gravity assisted by water.

Slump. The downward slipping of a mass of rock or unconsolidated material, usually with backward rotation.

Soil horizon. A layer of soil having observable characteristics produced through the operation of soil-building processes.

Solifluction. The process of slow downslope flowage of unconsolidated material saturated with water.

Spatter cone. A steep-sided mound or small hill built of spatter from a lava fountain.

Spheroidal weathering. Weathering which produces rounded boulders by chemical weathering along joints.

Spit. A sandbar projecting into a body of water from the shore.

Stagnant ice. A glacier in which the ice has ceased to move.

Stalactite. A cylindrical or conical deposit, usually calcite, hanging from the roof of a cavern.

Stalagmite. A cylindrical or conical deposit, usually calcite, built up from the floor of a cave.

Stratified drift. Sorted and stratified glacial drift.

Striations. Small grooves or scratches.

Strike. The compass direction of a horizontal line on a bedding plane.

Strike-slip fault. A fault in which the movement is essentially horizontal.

Stripped structural surface. A surface which owes its existence to a resistant rock layer from which weaker rock has been stripped by erosion.

Subglacial. Beneath a glacier.

Submergence. Inundation by the sea without implication as to whether the sea level rose or the land subsided.

Subsequent stream. A stream whose course is determined by selective headward erosion along weak rock belts.

Superglacial debris. Material on the surface of a glacier.

Superposed stream. A stream whose course was established on rocks at the surface, but as downcutting occurred, the stream cut through an unconformity onto older rocks having no direct relation to the original establishment of the stream system.

Surf. The wave activity between the shore and outermost limit of breakers.

Suspended load. Sediment which is transported in a current of water without contact with the bottom for considerable periods of time.

Swash. The rush of water up onto the beach following the breaking of a wave.

Swell. Wind-generated waves which have advanced into regions of calm or weak winds.

Synclinal valley. A valley which follows the axis of a syncline.

Syncline. A fold in which the beds dip inward from both sides toward the axis.

Talus. An accumulation of loose rock at the base of a cliff.

Tarn. A small mountain lake that occupies the basin of a cirque.

Tectonic basin. A basin formed by deformation of the earth's crust.

Temperate glacier. A glacier typically formed in temperate climates where recrystallization is relatively rapid and the temperature of the entire glacier is close to the melting point of ice.

Terminal moraine. A ridge of glacial till marking the farthest advance of the glacier.

Terrace. Flat, gently inclined, or horizontal surface bordered by an escarpment, composed of alluvium or bedrock.

Terra rossa. Residual red clay mantling a surface on limestone.

Thrust fault. A reverse fault having a low angle of inclination.

Till. Poorly sorted, nonstratified sediment carried or deposited by a glacier.

Tombolo. A bar connecting an island with the mainland.

Topset beds. Sediment deposited in horizontal layers on a delta.

Traction. Process of moving particles along the bed of a stream.

Translation gliding. Displacement on preferred lattice planes within a single crystal caused by compressional or tensional stresses.

Transverse dune. An asymmetrical sand ridge at right angles to the direction of prevailing winds.

Trellis drainage pattern. Drainage pattern in which tributary streams flow nearly parallel to one another and join trunk streams at right angles, similar to the pattern of a garden trellis.

Triangular facet. A truncated spur with a broad base narrowing upward to a point.

Truncated spur. End of a divide between tributary streams cut off and steepened by erosion or faulting.

Tsunami. Seismic sea wave. Giant sea wave produced by a submarine earth quake, slide, or volcanic eruption.

Turbidity current. A current due to difference in density between sediment-laden water and clear water.

Turbulent flow. Type of flow in which streamlines are thoroughly distorted by mixing of flow.

Underfit stream. A stream which appears too small to have eroded the valley in which it flows.

Unpaired terraces. Terraces on opposite sides of a valley which are unmatched in elevation.

Uvala. Large sinkhole formed by the coalescence of several sinkholes.

Vadose water. Subsurface water in the zone of aeration above the zone of saturation.

Valley train. A long narrow body of glacial outwash deposited downstream from a glacier.

Varve. A sedimentary bed deposited within one year's time, usually consisting of a coarse, light-colored layer deposited in the summer and a fine-grained, dark-colored layer deposited in the winter.

Ventifact. A pebble or boulder shaped by the abrasive action of wind-blown sand.

Viscosity. Resistance to flow in a fluid caused by internal friction.

Volcanic neck. Solidified magma filling the vent of an extinct volcano.

Water gap. A pass in a ridge through which a stream flows.

Water table. The upper surface of the zone of saturation.

Wave base. The depth at which wave action no longer moves the sediment on the bottom.

Wave-built bench. Gently sloping bench built by wave and current deposition of sediment.

Wave-cut bench. Beveled bedrock surface produced by wave erosion.

Wave height. The vertical distance between a wave crest and an adjacent trough.

Wave length. The horizontal distance between two successive wave crests or troughs.

Wave period. The time for two successive waves to pass a given point.

Wave refraction. Bending of waves.

Weathering. Disintegration and decomposition of rocks by surface processes.

Wind gap. A notch in a ridge made by a former stream.

Yazoo river. A tributary stream which flows for some distance parallel to the main channel because levees prevent it from entering the mainstream.

Youthful topography. Stage in the evolution of topography characterized by narrow V-shaped valleys, waterfalls, rapids, and broad divides.

Zone of aeration. Subsurface zone above the zone of saturation in which pore spaces are not filled with water.

Zone of flowage. Zone in a glacier where stresses are accommodated by flowage of material.

Zone of fracture. Zone in a glacier where stresses are accommodated by fracturing.

Zone of saturation. Subsurface zone in which all openings are filled with water.

indexes

name index

Agassiz, L., 75, 76
Alden, W. C., 86, 225, 234
Anderson, J. G., 234
Ault, W. V., 334

Bagnold, R. A., 141, 290-295, 303
Balsley, J. R., 127, 150, 169, 171, 172, 347, 375, 401
Bascom, W., 307, 308, 333
Baulig, H., 195, 240, 245
Beadnell, H. J. L., 303
Beckinsale, R. P., 14
Berkey, C. P., 303
Bernhardi, A., 75
Billings, M. P., 378, 410
Birman, J. H., 72
Blackwelder, E., 69, 71, 72, 214, 234, 273, 285, 378
Borland, W. M., 142
Bretz, J. H., 176, 177, 264, 265
Brown, W. H., 223, 234
Brush, L. M., 143
Bryan, K., 245, 273, 276, 285
Buckman, H. O., 214
Bull, W. B., 141

Capps, S. R., 221, 234
Carr, D. R., 108
Chamberlain, R. T., 52
Chamberlain, T. C., 14, 29
Chapman, R. W., 214
Charlesworth, J. K., 108
Charpentier, J., 75, 76
Chorley, R. J., 14
Clayton, L., 86, 88, 89, 108
Cloud, P. E., 334
Cook, J. H., 72
Cooper, W. S., 303
Cotton, C. A., 72, 195, 239, 245, 378, 410
Cox, A. V., 96, 97, 108, 221, 235
Crandell, D. R., 71, 72, 230, 233, 234
Curray, J. R., 102, 108

Dalrymple, G. B., 96, 97, 108
Daly, R. A., 331-334

Dana, J. D., 334
Darton, N. H., 405
Darwin, C., 329-331, 333, 334
da Vinci, L., 145
Davis, W. M., 4, 11, 14, 141, 162, 165, 174-177, 195, 241, 242, 245, 262, 265, 285, 334, 362, 378
Demorest, M., 52
Denny, C. S., 141, 234
Dicken, S. N., 265
Diller, J. S., 386, 387, 410
Dobrin, M. B., 334
Doell, R. R., 96, 97, 108
Donn, W. L., 107, 109
Dunn, A. J., 14
Dury, G. H., 133, 141
Dutton, C. E., 166, 177

Easterbrook, D. J., 109, 251, 265
Eaton, J. P., 334
Embody, D. R., 235
Emery, K. O., 334
Esmark, J., 75
Ewing, W. M., 107, 109

Faul, H., 109
Finkel, H. J., 303
Fisher, J. E., 72
Fisk, H. N., 303
Flint, R. F., 33, 45, 52, 72, 83, 102, 103, 106, 108, 109, 234
Foth, H. D., 214
Frey, D. G., 110
Friedkin, J. F., 125, 126, 128, 141
Fuller, R. E., 378

Geikie, A., 14
Gilbert, G. K., 14, 43, 46, 65, 103, 104, 109, 117, 142, 160, 166, 177, 206, 228, 239, 245, 273, 285, 295, 334
Gilluly, J., 273, 285
Goldich, S. S., 214
Grady, N. C., 214
Gravenor, C. P., 84, 86, 88, 109
Greenfield, M. A., 214

447

Guilcher, A., 334
Gutenburg, B., 102, 103, 109

Hack, J. T., 176, 177, 303
Hadley, J. B., 234
Hadley, R. F., 285
Hansen, H. P., 92, 109
Hardman, G., 104, 110
Hayden, E. W., 224, 235
Hjulstrom, F., 121, 142
Holmes, C. D., 72, 109, 142, 176, 177,
    235, 245, 285
Horberg, L., 265
Horton, R. E., 13, 14, 146, 162, 243,
    245
Howard, A. D., 285
Howe, E., 235
Hubbert, M. K., 265
Hunt, C. B., 345
Hutton, J., 3

Ives, R. L., 235

Jefferson, M., 142
Jenny, H., 214
Johnson, D. W., 14, 29, 52, 72, 195,
    241, 242, 245, 273, 285, 286, 319,
    321, 334, 378
Johnson, W. D., 271
Jones, F. O., 229, 230, 235, 392

Kamb, B., 52
Keefer, W. R., 225, 235
Kesseli, J. E., 235
King, C. A. M., 334
Kuenen, P. H., 334
Kulp, J. L., 108
Kupsch, W. O., 86, 109

Ladd, H. S., 334
Lane, E. W., 142
Leighton, M. M., 303
Leliavsky, S., 142
Lemke, R. W., 84, 109

Leopold, L. B., 13, 14, 114-116, 119,
    125, 133, 142, 162
Lewis, W. V., 57, 72, 84, 109
Libby, W. F., 109
Lobeck, A. K., 359
Long, J. T., 303
Love, J. D., 225, 235
Lovering, T. S., 346
Lyell, C., 14
Lyon, T. L., 214

McCann, F., 285
McGee, W. J., 273, 286
Mackin, J. H., 13, 14, 29, 125-128,
    130, 131, 142, 160, 162, 177,
    192, 196, 275, 276, 285, 378
Maddock, T., 115, 116, 119, 133, 142
Matthes, F. E., 199, 201-203, 214,
    370
Matthes, G. H., 122, 142
Meier, M. F., 52, 84, 109
Meinzer, O. E., 265
Melton, M. A., 243, 246
Meneley, W. A., 84, 109
Meyerhoff, H. A., 242, 246
Milankovitch, M., 107, 109
Millar, C. E., 214
Miller, C. F., 30
Miller, D. J., 235, 310-312, 335
Miller, J. P., 114, 115, 142, 162
Miller, V. C., 30, 286
Moffit, F. H., 45, 52
Morisawa, M., 143
Morris, F. K., 303
Mullineaux, D. R., 71

Nelson, R. L., 72
Nichols, R. L., 410
Nye, J. F., 52

Paige, S., 286
Patton, H. B., 410
Penck, W., 176, 177, 241-243, 246
Perkins, B., 334
Peterson, W. L., 235
Playfair, J., 14, 163
Post, A. S., 3, 9, 33, 35, 36, 55, 56, 58,
    62, 68, 186, 224, 398

Powell, J. W., 15, 146, 163, 166, 167, 177
Putnam, W. C., 104, 110, 410

Quraishy, M. S., 143

Rapp, A., 243, 246
Ray, R. G., 30
Reiche, P., 214
Rich, J. L., 196, 286
Richmond, G. M., 71, 73
Richter, D. H., 334
Rubey, W. W., 120, 125, 143
Russell, I. C., 83, 110

Schumm, S. A., 122, 143, 235, 241, 243, 245, 246, 286
Sharp, R. P., 38, 40, 45, 52, 73, 286, 303
Sharpe, G. F. S., 235
Shelton, J. S., 63, 78, 81, 92, 93, 123, 168, 181, 258, 272, 278, 325, 330, 339, 342, 343, 352, 353, 381, 387, 390, 393, 394, 397
Shepard, F. P., 321, 335
Shumskii, P. A., 34, 41, 52
Smith, K. G., 286
Snavely, B. L., 334
Snyder, C. J., 104, 110
Stacy, J. R., 218
Stearns, H. T., 410
Stose, G. W., 220
Strahler, A. N., 13, 15, 67, 80, 243, 246, 327, 329, 359, 410

Swinnerton, A. C., 265

Tabor, S., 214
Tator, B. A., 286
Terzaghi, K., 223, 235
Thomson, J., 143
Thornbury, W. D., 15, 256, 257, 262, 266, 335, 359, 410
Tracey, J. I., 334
Tuan, Y. F., 286
Turk, L. M., 214

Varnes, D. J., 235
Von Engeln, O. D., 246, 359

Wahrhaftig, C., 221, 235
Waldron, H. H., 230, 233, 234
Waters, A. C., 378
Werner, P. W., 143
Wiens, H. J., 335
Williams, H., 387, 390, 410
Willis, B., 379
Willman, H. B., 303
Witkind, I. J., 217, 218, 236, 363, 366, 379
Wolman, M. G., 13, 14, 114, 115, 125, 142, 143, 162
Woodford, A. O., 143
Wright, H. E., 110

Zdenek, F. F., 104, 110

subject index

A horizon of soils, 213
Aberdeen, South Dakota, 78
Ablation, 41-45
    zone of, 43-45
Ablation till, 48, 49, 79
Abrasion, 46, 47, 132, 134, 315
Absaroka Range, Wyoming, 131
Absteigende Entwicklung, 176
Abstraction, 160
Accumulation, zone of: on glaciers,
    41, 43-45
    in soils, 213
Adel, Oregon, 374
Adiabatic cooling, 270
Adiabatic heating, 270
Adriatic Coast, 256
Aeration, zone of, 249
Aerial photographs:
    criteria for interpretation, 24, 25
    distortion in, 21, 22
    mosaics, 23, 24
        controlled, 24
        uncontrolled, 24
    scale, 19, 21, 22
    sources of, 418, 419
    status of, 24
    stereoscopic pairs, 22, 23
Aftonian interglaciation, 99
Aggradation, 132
    zone of, 273
Air photos (see Aerial photographs)
Alaska Range, Alaska, 221
Aleutian Islands, 382
Alfalfa Center, Nebraska, 124
Alluvial fans, 135-141, 276, 277
Alluvial scarps, 363
Alpine glaciation, 55-73
Altiplanation, 174
Andes Mountains, 382
Angle of repose, 226, 294
Angular drainage pattern, 151, 153
Annular drainage patterns, 152, 155
Antarctica, 10, 11, 34, 36-38, 74,
    100, 102
Antecedent stream, 420
Antelope Peak, Arizona, 274, 275
Anticlinal ridges, 340, 341, 356, 357,
    359

Anticlinal valleys, 342, 359
Anticlines, 344, 345, 348-350, 355,
    357-359
Appalachian Mountains, 4, 148, 343,
    345, 349, 358, 359
Aquiclude, 253, 254
Aquifer, 253, 254
    confined, 253, 254
Arches National Monument, Utah, 211
Arctic Ocean, 107
Arete, 57
Argon-40, 95, 96
Arid erosion cycle, 278-285
Aroostock County, Maine, 172
Artesian well, 253
Arvada, Wyoming, 119
Asgard Range, Antarctica, 10
Ash, volcanic, 71, 96, 382
Athabaska Glacier, Alberta, 61
Atolls, 330, 333
Aufsteigende Entwicklung, 176

B horizon of soils, 212, 213
Backwasting:
    of glaciers, 44, 45
    of slopes, 241-245
Badland National Monument, South
    Dakota, 153, 158, 243
Badland topography, 153, 157, 158,
    241
Bahada, 276, 282
Banff National Park, Alberta, 137, 344
Bank erosion:
    in braided streams, 125, 126
    in meandering streams, 125, 126
Bar:
    baymouth, 317, 326-328
    offshore, 319, 322, 323, 327
Barchan, 294-299
Barnard Glacier, Alaska, 87
Barren River, Kentucky, 27
Barrier beach, 319
Barrier reef, 329, 330, 333
Basalt, 382, 391, 402
Base level, 166, 167
    local, 167
    sea level, 167

Basin and Range province, 104
Basins:
    deflation (see Blowouts)
    structural, 347
Bathhurst Inlet, Canada, 85
Batholith, 398
Bauxite, 213
Baymouth bar, 317, 326 - 328
Beach, 317
    barrier, 319
    bayhead, 319, 325, 327
    storm, 319
Beach drifting, 317
Bear Butte, South Dakota, 405
Beaumont, California, 171
Bed load, 120 - 122
Bedrock terraces, 186, 188, 193
Bells Canyon, Utah, 373
Benson Creek, Kentucky, 161, 162
Bergschrund, 55 - 57
Berkeley Hills, California, 228
Beta particle, 94
Big Horn Basin, Wyoming, 128, 160,
    275
Big Horn River, Wyoming, 160, 162
Bikini Atoll, 333
Birdfoot delta, 135
Black Hills, South Dakota, 348
Black Rapids Glacier, Alaska, 45
Black Swamp Branch, Kentucky, 27
Blind valley, 261, 263
Blowouts, 295, 296
Bonneville, Lake, 103 - 105
Book Cliffs, Utah, 347
Boothbay Harbor, Maine, 326
Bottomless Lakes, New Mexico, 231
Bottomset beds, 135, 139
Bowling Green, Kentucky, 258
Braided stream, 64, 124 - 126
Breaker, 310
    plunging, 312
    spilling, 312
Breton Sound map, 136
British Columbia, Canada, 97
Brunswick, Maine, 325
Buried erosion surfaces, 176
Buried soils, 71, 95
Buttes (mesas), 343

C horizon of soils, 212, 213
Calcite, 204, 206, 208, 256, 265
Calcium-40, 95
Calderas, 383 - 390
Caliche, 214
Calving, 43
Campti quadrangle, Louisiana, 165
Capulin Mountain, New Mexico, 386
Carbon-14, 93 - 95
Carbonation, 207, 208
Carlsbad Caverns, New Mexico, 261,
    264
Casa Grande quadrangle, Arizona, 281
Cascade Mountains, 97, 99, 382
Caves, 261, 262, 264, 315
Cayuga Lake, New York, 76
Cedar Spring Valley, Kentucky, 261
Centripetal drainage patterns, 153,
    156
Channel characteristics, 122 - 127
Channel deposits, 170
Channels:
    braided, 122 - 126
    meandering, 122, 123, 125, 126
Chelan Falls, Washington, 186
Chemical weathering (see Weath-
    ering)
Cincinnati arch, 348
Cinder cone, 382, 384 - 387, 396
Circle Ridge Dome, Wyoming, 341
Cirque, 54 - 57, 66
Cirque glaciers, 34
Clayton Valley, Nevada, 156
Cliffed headland, 327
Climatic changes, 180
Coast Range, California, 371, 373
Coastal morphology, 307 - 335
Coconino Plateau, Arizona, 175, 181
Col, 57
Colluvium, 245
Colorado Plateau, 166
Colorado River, 133, 135, 181
Columbia Plateau, Washington, 299,
    340, 391, 392
Columnar jointing, 400, 409
Comb Ridge, Utah, 352
Commonwealth Glacier, Antarctica,
    10

Compaction, 34, 48
Composite scarp, 362
Composite volcanic cone, 382, 389
Compound shoreline, 320, 321
Cone of depression, 253
Connate water, 249
Consequent streams, 144 - 148
Continental glaciation, 74 - 110
    criteria for recognizing glacial
        stages, 92 - 98
        lithology, 97, 98
        paleomagnetic correlation, 96, 97
        pollen analysis, 92, 93
        potassium-argon dating, 95, 96
        radiocarbon dating, 93 - 95
Contour interval, 411, 414
Contours, 411 - 415
Convict Lake, California, 63
Copper River, Alaska, 221
Coral reefs, 327 - 333
    atolls, 330, 333
    barrier, 329, 330, 333
    fringing, 329, 333
    origin of, 330 - 333
Corizozo, New Mexico, 400
Correlation:
    of glacial events, 67 - 72, 92 - 98
    of stream terraces, 188 - 192
Coteau des Prairies, South Dakota, 78
Cottonwood Creek, Wyoming, 185
Crag-and-tail, 83
Crater Lake, Oregon, 383, 386, 387,
    390
Craters of the Moon, Idaho, 380,
    387, 397
Creep, 219 - 221, 227, 239, 240,
    241
Crevasse fillings, 67, 86 - 88
Crevasses, 40, 67, 86 - 88
Critical tractive force, 120
Cross-sectional area in stream
    channels, 114, 115, 131
Cub Run quadrangle, Kentucky, 194
Cuesta, 343, 347, 352
Cuny Table East, South Dakota, 159
Curie temperature, 96, 97
Cut-in-fill terraces, 187, 188
Cut-off meander, 8, 123, 126

Cycle of erosion, 165, 166
Cyclic erosion surface, 182, 184 - 196
Cyclopian stairs, 59, 60

Dakota hogback, 346, 348
Dakota sandstone, 346, 348
Darcy's law, 252
Death Valley, California, 139, 278
Debris flow, 233
Deccan, India, 391
Degradation, zone of, 273
Delta, Utah, 288
Deltas, 63, 65, 80, 134 - 139
Dendritic drainage, 144, 148 - 150
Desert landforms, 268 - 303
    depositional, 276
    erosional, 270 - 276
    evolution of, 278 - 285
Desert pavement, 292
Deserts, 268 - 286
    origin of, 269, 270
Devils Lake, North Dakota, 93
Diastrophism, 4
Differential erosion, 5, 6
Dike, 23, 395, 403
Dip slope, 342
Dipolar magnetic field, 96
Discharge, stream, 114, 115, 119, 127,
    130 - 132
Disintegration ridges, 86 - 89
    circular, 87 - 89
Disintegration trenches, 86, 88, 90
Dissolved load, 118
Divides, stream, 155
Domes, 340, 347, 353
Donau glaciation, 99
Down-dip method, 378
Drag, 369, 372
Drainage density, 153, 155, 157, 159
Drainage patterns, 144, 148 - 156
Drainage texture, 155, 157, 159
Drayton Harbor, Washington, 16
Dreikanters, 291
Drift, glacial, 48 - 51
Drifting:
    beach, 317
    longshore, 317

Drowned valleys, 323, 327
Drumlins, 80 - 83, 98
Dry Falls, Washington, 204
Dunes, sand, 294 - 298
Dynamic equilibrium (see Equilibrium)

Earthflow, 233
Effingham quadrangle, Illinois, 149
Elbow of capture, 160
Elkhorn Creek, Kentucky, 162
Emblem Bench, Wyoming, 189, 192,
    193
Emergent shorelines, 319, 327, 329
Emery County, Utah, 152
Emmons Glacier, Washington, 229
Empirical approach, 13
End moraines (see Moraines)
Englacial debris, 48
Eniwetok Atoll, 333
Entrenched meanders, 192
Eolian landforms, 288 - 303
Equilibrium, 12, 61, 77, 128, 130 - 132,
    170, 171, 176, 243, 278, 283, 285,
    293, 315, 316
Erratics, 98
Esker, 64, 65, 67, 80, 84 - 86
Estrella quadrangle, Arizona, 280
Eustatic sea-level change, 179
Everett quadrangle, Pennsylvania,
    355
Evolution:
    of coral reefs, 330-333
    of desert landforms, 278 - 283
    of fault-line scarps, 364
    of fluvial landforms, 164 - 177
    of shorelines, 324 - 327
    of slopes, 239 - 246
    of topography on folded beds, 339
Exaggeration of profiles, 416, 417
Exfoliation, 200 - 202
    arch, 198
    dome, 201, 203
Exhumed surfaces, 176

Facets:
    glacial, 47
    triangular, 368, 371

Fairweather Fault, Alaska, 310
Falmouth, Massachusetts, 92
Fan-head trenches, 138, 140, 282
Fans, 63, 65
Farmdalian substage, 99
Farmington, New Mexico, 181
Fault-line scarp, 361, 362, 366,
    369
Fault scarp, 361 - 367, 371
Faulted shorelines, 320
Faulting, topography associated with,
    26, 360 - 379
Felsenmeer, 200, 201
Fetch, 307
Fill terraces, 186, 188
Finger Lakes, New York, 76
Finger lakes, 76
Firn, 34
Firn line, 43, 45
Fissure eruption, 391
Flagstaff, Arizona, 396
Flood basalt, 391
Flood plains, 170, 172
Floods, 132 - 133
Fluted topography, 83 - 84
Fluvial morphology, 112 - 196
Folded sedimentary rock, 339 - 359
Folds:
    differential erosion of, 340 - 350
    initial forms, 339 - 340
    plunging, 344 - 346, 348 - 350, 355,
        357, 359
Footwall, 361, 362
Foreset beds, 135, 139
Fort Pierce, Florida, 322
Fort Valley, West Virginia, 359
Frank, Alberta, 216 - 225
Frankfort, Kentucky, 161, 162
Freezing and thawing, 200
Fremont County, Wyoming, 341
Friction, 114 - 117
Fringing reef, 329, 333
Frost heaving, 220, 221

Gardner Creek, Kentucky, 261
Geologic structure, interpretation
    of, 350 -359
Geomorphic process, 5

Geysers, 254 - 256
Gila Butte quadrangle, Arizona, 281
Glacial abrasion, 46, 47
Glacial advances, reconstruction
    of, 98, 99
Glacial control theory, 332
Glacial drift, 48
Glacial erosion, 46, 76
Glacial grooves, 39
Glacial polish, 46, 47
Glacial quarrying, 46, 48, 57
Glacial striations, 47
Glacial till, 48 - 50
Glacial trough, 58 - 60, 66
Glacier Bay, Alaska, 43
Glacier National Park, Montana, 226,
    377
Glaciers:
    ablation, zone of, 43 - 45
    accumulation, zone of, 41, 43 - 45
    economy, 41 - 46
    movement of, 37 - 41
        basal slip, 38, 39
        intergranular shifting, 40
        intragranular shifting, 40
        marginal slip, 38, 39
        plastic flow, 40, 41
        shearing, 39
    origin of, 33
    regimen of, 43
    surging, 45
    types of, 34
        alpine, 34, 35
        continental, 36, 38
        piedmont, 36
Glacioeolian deposits, 51
Glaciofluvial deposits, 50, 62 - 67
    88 - 92
Glaciolacustrine deposits, 50, 51
Glaciomoraine deposits, 49, 50
Gleichformige Entwicklung, 176
Goldfield map, Nevada, 156
Graben, 370, 373, 374
Graded stream, 126 - 132, 170, 171
Gradient, stream, 116, 117, 127,
    128, 130
Grand Canyon, Arizona, 5, 133, 169,
    180, 182, 183, 401
Grand Coulee Dam, Washington, 229

Grand Island, Nebraska, 123
Grand Tetons, Wyoming, 2
Granite Mountain, Utah, 271
Granite weathering ratio, 71
Grass Creek Basin quadrangle,
    Wyoming, 185
Gravity, 200
Gravity slope, 242, 243
Great Barrier Reef, Australia, 330
Great Bear Lake, Northwest Territo-
    ries, 332
Great Salt Lake, Utah, 103
Great Terrace, Columbia River, Wash-
    ington, 186
Green Bay, Wisconsin, 78
Green Mountain, Wyoming, 407
Green River, Kentucky, 27
Green River, Utah, 150
Greenland, 75, 100, 102
Greenland Gap quadrangle, West
    Virginia, 356
Greenwich, England, 415
Greenwood quadrangle, Mississippi,
    129
Greybull River, Wyoming, 131, 160,
    162, 189 - 192
Groin, 319
Gros Ventre slide, 223 - 225
Ground moraine (see Moraines)
Groundwater, 248 - 266
    connate water, 249
    juvenile, 249
    movement of, 252
    origin of, 249
    primary, 249
Gunnison, Colorado, 57
Gunz glaciation, 99
Gypsum Valley anticline, Colorado,
    343

Hachures, 415
Half Dome, Yosemite National Park,
    California, 201, 203
Half-life, 94, 95
Hanging valleys, 60, 66, 369
Hanging wall, 361, 362
Hayes Glacier, Alaska, 33
Headland, 313, 314, 327

Headward erosion, 7, 146, 160
  random, 146 - 148
  selective, 146, 147, 151
Hebgen Lake, Wyoming, 363, 366
Helical flow, 122, 123, 125, 126
Hematite, 205, 208
Henry Mountains, Utah, 166, 273, 345
Hidden Glacier, Alaska, 65
High Peak, West Virginia, 359
Hilo, Hawaii, 309
Hogbacks, 343, 346, 350
Holy Cross, Colorado, 66
Homoclinal, 342 - 359
Homoclinal ridges, 342 - 345, 351,
  354, 359
Homoclinal valleys, 342
Horn, 57, 58
Horse latitudes, 270
Horst, 370, 373
Hot springs, 254 - 256
Hudson Bay, Canada, 103
Hudson River, 398 - 404
Humboldt Current, 270
Hurricane, Utah, 402
Hyannis, Massachusetts, 91
Hydration, 207, 208
Hydraulic gradient, 252, 253
Hydraulic lift, 120
Hydraulic plucking, 134
Hydraulic pressure, 314

Ice:
  deformation of, 40, 41
  elastic limit, 41
Ice Age, 76
Ice-contact deposits, 63 - 67, 84 - 90
Ice-marginal lake, 65, 67, 80
Ice sheets, 36
Ice-walled lake plain, 85, 86, 89
Icebergs, 43, 44
Idaho Falls South quadrangle, Idaho,
  302
Igneous intrusions, 394, 395, 397,
  398, 400
Igneous landforms, 380 - 410
Illinoian glaciation, 99
Illinois River, 128
Impact, 200

Incised meanders, 178, 189, 194, 195
Inclosed meanders, 195
Inertia, 122, 125, 126
Ingrown meanders, 193
Inselbergs, 275, 281, 282
Insequent streams, 146, 147
Intercision, 162
Interlobate moraines (see Moraines)
Intertill deposits, 71
Inversion of topography, 393, 394, 402
Inward Point, Massachusetts, 321
Inyo Craters, California, 394
Irontown, Missouri, 238
Isostatic rebound, 102
Isotope, 93 - 96

Jamestown, North Dakota, 90
Jasper National Park, Alberta, 351
Jellico quadrangle, California, 367
Jenny Lake, Wyoming, 2
Joints, 25, 151, 152, 225, 408, 409
Jordan, Narrows, Utah, 105
Juvenile water (see Groundwater)

K-Ar, 95 - 96
Kaibab limestone, 175, 181
Kaibab Plateau, 175, 181
Kame and kettle topography, 84
Kame delta, 67, 80
Kame terraces, 67, 84, 85
Kames, 65, 67, 84, 85
Kansan glaciation, 99
Kaolin, 207
Karst district, Yugoslavia, 256
Karst topography, 248, 256 - 265
Karst windows, 261
Kentucky River, 161, 162
Kettle Moraine, Wisconsin, 78
Kettles, 64, 65, 67, 79, 80, 84, 85,
  89, 90
Klamath Falls, Oregon, 368
Klippe, 376

Laccolith, 398, 405, 407
Lac de Gras, Canada, 85
Lagoon, 317, 319, 327, 330

Lahar, 232
Lake Bonneville, Utah, 103 - 105
Lake Geneva, 135
Lake Mead, Nevada, 135
Lake Michigan, 77, 78
Lakes:
    cirque, 55, 66
    due to faulting, 365, 369
    glacial, 55, 56, 61, 66, 67, 76, 77, 80,
        87 - 89, 103, 104
    landslide, 218, 224
    morainal, 61, 66
    oxbow, 8, 123, 126, 127, 172
    playa, 277 - 279
    sinkhole, 260
Laminar flow, 118
Lander, Wyoming, 353
Landslide, 223 - 225, 231, 234
Lateral moraine (see Moraines)
Lateral planation, 174 - 175, 272, 273,
    275, 276
Laterization, 213
Latitude, 415
Latosols, 213
Lava:
    dating of, 96
    flows, 385, 389, 391 - 402
    tunnels, 393
Leaching, zone of, 213
Lee Vining Canyon, California, 68
Leestown, Kentucky, 162
Levees, natural, 172
Lewis thrust fault, 377
Limestone, 4, 256, 257, 261, 262, 340
Limestone Plateau, South Dakota, 348
Little Crease Mountain, West Virginia,
    359
Little Muddy quadrangle, Kentucky,
    28
Little Sinking Creek, Kentucky, 261
Little Wabash River, Illinois, 149
Lituya Bay, Alaska, 310 - 312
Load in streams, 127
    and the graded profile, 130 - 132
    transportation of, 118 - 122
Locate, Montana, 116
Lodgement till, 48, 79
Loess, 51, 71, 72, 299 - 301
Longitude, 415

Longitudinal dunes, 296 - 298, 301,
    302
Longshore currents, 317
Longshore drifting, 317
Loveland, Colorado, 6

McLean County, North Dakota, 7
McMurdo Sound, Antarctica, 11
Madison River, 217, 218
Madison slide, 217, 218
Magnetism, 96, 97
Malisping Glacier, 36
Mammoth Cave, Kentucky, 248, 259,
    261
Mammoth Hot Springs, Wyoming, 256,
    267
Maps:
    contour, 411 - 415
    scale, 18 - 20
    symbols, 412 - 414
    topographic, 17, 411 - 415
Marine terraces, 102, 306, 316,
    327, 330, 331
Martha's Vineyard, Massachusetts,
    328
Mass movement, 217 - 236
Matterhorn, 58
Mature topography, 164, 166, 170, 171,
    173
Maui, Hawaii, 388
Meander belt, 170
Meander core, 162
Meandering, 112, 122 - 126, 128, 129
Meanders:
    cutoff, 8, 123, 126
    entrenched, 192
    incised, 178, 189, 194, 195
    inclosed, 195
    ingrown, 193
Mechanical weathering (see Weath-
    ering)
Medial moraines (see Moraines)
Menan Buttes, Idaho, 384 - 385
Mendel glaciation, 99
Meridians, 415
Mesas, 343
Misfit streams, 133
Mississippi Delta, 136

Mississippi River, 118
Missouri River, 131
Moenkopi shale, 345
Mojave River, California, 272
Monadnock, 172, 173
Mono Craters, California, 393
Mono Lake, California, 69, 71, 393
Monocline, 340, 342, 346, 352
Monomoy Point, Massachusetts, 320,
    321
Morainal lakes, 61, 66
Moraines:
    end, 2, 60-62, 66, 68-70, 77,
        78-80, 91, 98, 101
    ground, 61, 62, 79-81
    interlobate, 78, 80
    lateral, 60, 61, 63, 68, 69
    medial, 32, 60
    recessional, 61, 78, 80
    terminal, 60, 61, 63, 66, 80
Moses Lake, Washington, 297, 298
Mosquito Valley, 345
Mount Cleveland, Montana, 58
Mount Dome quadrangle, California,
    365
Mount Erabus, 11
Mount Mazama, Oregon, 390
Mount Rainier, Washington, 71, 229,
    230, 233
Mount Rushmore, South Dakota, 159
Mount Shasta, California, 389
Mount Shukson, Washington, 54
Muddy Creek, Kentucky, 27
Mudflow, 228-233
Multiple glaciations, recognition of,
        67-72, 92-99
    lithologic criteria, 97, 98
    paleomagnetic correlation, 96, 97
    pollen, 92, 93
    potassium-argon dating, 95, 96
    radiocarbon dating, 93-95
    stratigraphic criteria, 71, 72
    topographic criteria, 67-70
    weathering criteria, 70, 71
Murkin Bench, California, 367

Nashville dome, 348
Natural bridge, 211, 261, 315, 393

Natural levees, 172
Nebraskan glaciation, 99
Neck, volcanic, 23, 395, 403
Netarts Bay, Oregon, 318
Neutral shorelines, 320
Neve, 34
Newberry Crater, Oregon, 398, 399
Nile Delta, 135
Nippenose Valley, Pennsylvania, 345
Nitrogen, 93, 94
Nivation, 55
Nolin River, Kentucky, 194
Noncyclic erosion surfaces, 180-184
Nonpaired stream terraces, 182, 184
Normal fault, 361, 362
Normal magnetic field, 96, 97
North White Deer Ridge, Pennsyl-
    vania, 349
Nova Scotia, 83

Obsequent fault-line scarp, 362, 364,
    366
Obsequent stream, 147, 148
Obsidian flow, 394
Offshore bar, 319, 322, 323, 327
Old age topography, 172, 173
Old Faithful geyser, 255, 256
Orbital motion, 308
Organic activity, 202, 203, 205
Oro Grande, California, 272
Osceola mudflow, Washington, 233
Outwash channels, 90, 93
Outwash deposits, 62, 63, 88-92
Outwash plain, 80, 88-92
    pitted, 90-92
Overbank deposits, 170
Owyhee River, Oregon, 168
Oxbow lakes, 8, 123, 126, 127, 172
Oxidation, 205, 206, 208

Paleomagnetism, 96, 97
Palisades sill, New Jersey, 397, 404
Palmyra, New York, 82
Palos Verdes Hills, California, 330, 331
Palouse Falls, Washington, 392
Panamint Range, California, 140

Parabolic dunes, 295, 296, 298 - 300
Parallel drainage pattern, 153, 157
Parallel slope retreat, 241 - 245
Parallels, 415
Park City, Kentucky, 260
Passage Creek, West Virginia, 359
Paternoster lakes, 60
Pedalfers, 213
Pediment pass, 281, 282
Pediments, 268, 271 - 285
Pediplain, 283
Pedocals, 214
Peneplain, 165, 166, 173, 175
Perched water table, 252, 253
Perihelion, 107
Permeability, 153, 154, 250 - 252
Peyto Lake, Alberta, 137
Piezometric surface, 253, 254
Pitted outwash plain, 90 - 92
Platte River, Nebraska, 123, 124, 131
Playas, 277 - 279
Playfair, 145
Pleistocene glaciations, cause of,
    104 -108
Pleistocene lake fluctuations, 103, 104
Plucking, 48, 56, 57, 314
Plug dome, 391, 393, 394
Podsols, 213
Point bar deposits, 112, 122, 123, 126,
    129
Point Fermin, California, 232
Polar wandering, 108
Polarity, 96, 97
Pollen, 92, 93
Porosity, 250, 251
Potassium-argon dating, 95 - 97
Potassium-40, 95, 96
Powder River, 116, 119
Pressure ridges, 392, 398, 399
Price Glacier, Washington, 54
Primary water, 249
Prime meridian, 415
Profile:
    of equilibrium, 128, 130 - 132
    graded, 128, 130 - 132
    soil, 213
    stream, 128, 130 - 132, 184, 188,
        189, 192
    topographic, 416, 417

Provo shoreline, Utah, 103
Puget Lowland, Washington, 97, 230
Pumice, 383
Pyroclastic ejecta, 382

Qualitative geomorphology, 12
Quantitative geomorphology, 12
Quarrying, 48, 56, 57, 314

Radial drainage pattern, 151 - 154
Radiocarbon dating, 93 - 95
Range, 416
Rangell Mountains, Alaska, 221
Rational approach, 13
Recessional moraines (see Moraines)
Rectangular drainage pattern, 150 - 152
Red Rock Pass, Utah, 103
Red Valley, South Dakota, 348
Reed terrace, Washington, 229, 230
Reefs (see Coral reefs)
Refraction, 306, 310, 312 - 314
Regimen of glaciers, 43 - 46
Rejuvenation, 179 - 196
    causes of, 179, 180
        climatic, 180
        eustatic, 179
        tectonic, 179, 180
Remnant magnetism, 96, 97
Resequent fault-line scarp, 362, 364
Resequent stream, 147, 148
Resurrected surfaces, 176
Reverse fault, 361, 362
Reversed magnetic field, 96, 97
Rhone River, 135
Ria coasts, 323
Riegel, 60
Right-lateral strike-slip fault, 375
Rills, 146, 272, 273
Ripples, 292, 293
Riss glaciation, 99
Robert Scott Glacier, Antarctica, 38
Roche moutonée, 439
Rochester, New York, 81
Rock-defended terraces, 182 - 184
Rock drumlins, 83
Rock fall, 226, 227
Rock fans, 272
Rock glaciers, 221 - 223
Rock tables, 42

Rocky Mountains, 71, 99, 148
Rule of V's, 353, 354

Sacaton Mountains, Arizona, 268, 281
Sahara Desert, 270
Salt Lake City, Utah, 371, 373
Saltation, 120, 291, 292
Salton Sea, 145, 373
San Andreas fault, California, 373,
    375, 376
Sand dunes, 294 - 298
Sand storm, 290, 291
San Francisco earthquake, 375
San Francisco Plateau, Arizona, 206
Sangamon interglaciation, 99
San Juan County, Utah, 178, 301
San Juan Island, Washington, 39
San Juan Mountains, Colorado, 223
San Juan River, Utah, 181, 195, 352
San Luis, California, 316
San Raphael Swell, Utah, 342
Santa Barbara County, California, 313
Santa Cruz, California, 306
Saturation, zone of, 249, 250
Scandinavia, 99, 102, 103
Scarp face, 342
Scarps:
    composite, 362
    fault, 361 - 367, 371
    fault-line, 361, 362, 366, 369
Scientific method, 26
Scour-and-fill, 130, 132
Sea arches, 315
Sea caves, 315
Sea cliffs, 315, 316
Sea level changes, 100 - 102
Sea stacks, 315, 316
Seif dunes, 295
Seismic sea waves, 309, 310
Selective transportation, 132
Seneca Lake, New York, 76
Settling velocity, 118, 119, 289, 290
Shackleton Glacier, Antarctica, 74
Sheep Mountain anticline, Wyoming,
    338
Sheet wash, 272
Sherman County, Oregon, 295
Shield volcanoes, 382, 388

Shinarump conglomerate, 345
Shiprock, New Mexico, 23, 181, 397,
    403
Shorelines:
    classification of, 319 - 324
    compound, 320
    emergent, 319, 327, 329
    evolution of, 324 - 327, 329
    neutral, 320
    prograding, 319
    submergent, 319, 325, 326, 328
Shoshone River, Wyoming, 128
Shutter ridges, 375
Sierra Nevada Range, California, 69,
    99, 104, 200, 201, 370
Signal Knob, West Virginia, 359
Sill, 397, 398, 404
Sinkhole, 258 - 263
    compound, 260
    ponds, 260
Sinking Creek, Kentucky, 261
Sinking creeks, 261, 263
Sixth-power law, 120
Slave River, Canada, 138
Slip face, 294
Slope (see Gradient)
Slope profiles, 239 - 245
Slope wash, 184
Slump, 227, 228
Smith Grove quadrangle, Kentucky,
    263
Snake River, Idaho, 103
Snake River, Wyoming, 187
Snow, 33, 34
Soil, 71, 211 - 214
    horizons, 212, 213
Solifluction, 221
Solution, 134, 203, 204
Solution valleys, 261
South White Deer Ridge, Pennsyl-
    vania, 349
Spheroidal weathering, 201, 204
Spits, 17, 317, 318, 320, 321, 327
Springs, 253, 254
Squeeze-up, 392
Stage of development, 10, 165, 166
Stages of evolution of drainage sys-
    tems, 168
    criticisms of concept, 174 - 177

Stages of evolution of drainage
  systems:
  maturity, 170, 171, 173
  old age, 172, 173
  youth, 169, 170, 173
Stagnant ice, 63 - 67, 84 - 90
Stalactite, 265
Stalagmite, 264, 265
Stansbury shoreline, Utah, 103
Steens Mountain, Oregon, 278
Stock, 398
Stokes' law, 118, 119
Strasburg quadrangle, West Virginia,
  358
Stream:
  capture, 155, 158, 160 - 162
  channels, 114 - 116
  discharge, 114, 115, 119, 127,
    130 - 132
  divides, 155
  erosional processes, 134
  gradient, 116, 117, 127, 128, 130
  load, 118 - 122
  ordering, 146
  origins, 145 - 148
  patterns, 144, 148 - 156
  processes, 112 - 143
  terraces, 181 - 196
Strike-slip fault, 362, 371
Stripped structural surfaces, 175,
  180 - 183
Structure, geologic, 4, 338 - 410
Subsequent streams, 146, 148
Sundance Mountain, Wyoming, 406
Sunspot cycle, 106
Sunwapta Pass, Alberta, 351
Superglacial debris, 48, 49
Superglacial lakes, 85 - 88
Superglacial streams, 88
Superposed stream, 440
Suspended load, 118, 119
Swells, 309
Synclinal ridge, 346, 351, 359
Synclinal valley, 340, 359
Synclines, 344, 346, 348, 349, 355,
  357 - 359

Table Mountain, Wyoming, 189, 192
Taku Glacier, Alaska, 45, 46

Tallahatchie River, Mississippi, 8, 112.
Talus, 226, 227
Tapeats sandstone, 181, 182
Tarn, 55, 66
Tectonic uplift, 179, 180
Terminal moraines (see Moraines)
Terra rosa, 261
Terraces:
  marine, 102, 106, 316, 327, 330, 331
  stream, 181 - 196
    bedrock, 186, 188
    correlation of, 188 - 192
    cut-in-fill, 187, 188
    fill, 186, 188
    nonpaired, 182, 184
    profiles of, 189, 192
    rock-defended, 182 - 184
    stripped structural, 180 - 183
Thermal expansion and contraction,
  202
Thorn Hill, Kentucky, 162
Thrust faults, 362, 376, 377
Till:
  ablation, 48, 49, 79
  fabric, 48, 49
  lodgement, 48, 79
  plain, 81
Tisbury Great Pond quadrangle,
  Massachusetts, 328
Tombolo, 317, 326, 327
Tonto Platform, Arizona, 181, 183
Topographic maps, 17, 411 - 415
Topset beds, 135, 139
Township, 416
Traction, 120
Transverse dunes, 298
Trellis drainage pattern, 148, 151
Triangular facets, 368, 371
Trochoid, 308
Truncated spurs, 59
Tsunamis, 309, 310
Turbulent flow, 118
Turtle Mountain slide, Alberta, 216,
  225
Tuscorora sandstone, 345
Two Creeks interval, 99

U-shaped notches, 356, 357
U-shaped valleys, 59

Unconformities, 176
Underfit streams, 160

Vadose water, 249
Valders glacial advance, 99
Valley train, 62
Vantage, Washington, 409
Varves, 51
Velocity:
    stream, 113 - 117, 122, 127, 128, 130, 131
    wave, 307
Venetz, 75, 76
Ventifacts, 291
Virgin anticline, Utah, 345, 350
Volcanic ash, 71, 96, 382
Volcanic cones, 154, 382 - 390
Volcanic landforms, 380 - 394
Volcanic neck, 23, 395, 403
Vulcan's Throne, Arizona, 401

Walker Lake, California, 69
Warren County, Kentucky, 262
Warrior Ridge, Pennsylvania, 357
Wasatch Range, Utah, 371
Wash slope, 242, 243
Water table, 249, 250, 252
    perched, 252, 253
Wave base, 315
Wave-cut bench, 315, 316
Wave erosion, 314 - 316
Wave height, 309, 310
Wave length, 309, 310
Wave motion, 307 - 314
Wave period, 309, 310
Wave refraction, 306, 310, 312 - 314
Weathering:
    chemical, 203 - 207
    differential, 210
    effects of, 210 - 214
    mechanical, 199 - 203
    rate and character of, 207 - 210
Weiser Pass quadrangle, Wyoming, 354
Wellington quadrangle, Utah, 283, 284

West Bay fault, Northwest Territories, 361
Wetted perimeter, 114, 115, 131
White River, Arkansas, 127
White River Valley, Washington, 230
White Sands National Monument, 299
Whitwell quadrangle, Tennessee, 244
Width-depth ratio, 126
Williamsport quadrangle, Pennsylvania, 349
Wind River Mountains, Wyoming, 56, 70
Wind transport, 289 - 293
Wind velocity, 289 - 293
Windgap, 160
Wisconsin glaciation, 99 - 101
Wolf Glacier, 45
Wood River, Wyoming, 131
Woodfordian glacial stage, 99
Woolsey Hollow, Kentucky, 261
Wurm glaciation, 99

Yalobusha River, 129
Yarmouth interglaciation, 99
Yazoo River, Mississippi, 129, 172
Yazoo streams, 172
Yellowknife, Northwest Territories, 360
Yellowstone National Park, 217, 218, 223, 254 - 257
Yosemite National Park, 46, 59, 198, 201 - 203
Youthful topography, 166, 168 - 170, 173
YU Bench, Wyoming, 189 - 192

Zone:
    of ablation, 43 - 45
    of accumulation, 41, 43 - 45, 213
    of aeration, 249
    of aggradation, 273
    of degradation, 273
    of leaching, 213
    of saturation, 249, 250

This book was set in Uranus by University Graphics, Inc., printed on permanent paper by Halliday Lithograph Corporation, and bound by The Book Press, Inc. The designer was Marsha Cohen; the drawings were done by John Cordes. The editors were Bradford Bayne and Eva Marie Strock. Robert R. Laffler supervised the production.

ID646417

Interrupted by God

Interrupted by God

glimpses from the Edge

Photographs and Essays by

Tracey Lind

THE
PILGRIM
PRESS
Cleveland

— to Emily

The Pilgrim Press, 700 Prospect Avenue, Cleveland, Ohio 44115-1100

pilgrimpress.com

Copyright © 2004 by Tracey Lind

Scripture quotations, unless otherwise noted, are from the New Revised Standard Version of the Bible,
© 1989 by the Division of Christian Education of the National Council of Churches of Christ in the
United States of America and are used by permission. Changes have been made for inclusivity.

Copyright material taken from *A New Zealand Prayer Book—He Karakia Mihinare o Aotearoa* is used with permission.
Quotes from *The Book of Common Prayer* (1979) of the Episcopal Church, U.S.A., are used with permission.

Printed in the United States of America on acid-free paper

09   08   07   06   05   04   5   4   3   2   1

Library of Congress Cataloging-in-Publication Data

Lind, Tracey, 1954-
Interrupted by God : glimpses from the edge / photographs and essays by Tracey Lind.
    p. cm.
    Includes bibliographical references.
    ISBN 0-8298-1622-4 (cloth : alk. paper)
    1. Christian life—Anecdotes.   2. Christian life—Meditations.   I. Title.

BV4501.3.L55 2004
283'.092—dc22

2004049415

Contents

Foreword   vii

Preface   xi

Introduction: The Question   1

One   It was born on a winter day . . .   17

    The Light of Darkness   19

    Just Another Homeless Family   29

    A Baptism to Remember   37

    A Child Shall Lead Us   45

Two   It lived and died . . .   59

    Why I Do Ashes   61

    The Preacher's Prayer on the Eve of War   69

    The Garbage Tree   77

    Stolen Shoes   87

Three   It rose again . . .   97

    The Unfinished Story   99

    Welcome the Risen Christ   107

    Life Happens in the Interruptions   111

Four	It continues to rise . . .	117
	Mountain Musings	119
	Crossing the Great Water	127
	Keep on Singing	139
Five	It nourishes . . .	149
	Bread of Life	151
	Sally's Feast	159
	Beckoned to the Banquet	167
Six	It challenges . . .	175
	The Demolition Contractor	177
	Love for the Fallen Flower	191
	The Trinity of Love	199
	Confessions of an Evangelical Universalist	207
Seven	And it calls out . . .	219
	All-ee All-ee in Free	221
About the Photographs		223
Notes		229

Foreword

In his book, *Ministry and Imagination*, the late theologian Urban Holmes suggested that the spiritual imagination is an act of pilgrimage whereby the hungry soul goes "outside the city" to find God's presence. Leaving the city in order to see God is an ancient Christian practice, one that can be traced back to the fourth century. As the faith became increasingly conflated with imperial Roman values, faithful believers who could still imagine a life of intentional spirituality made their way to the countryside and desert where they hoped to better envision and practice a way of life in God.

Today, "the city" embodies the system of what is, the religious patterns and institutions with which contemporary Americans live—the accepted way of doing things, approved faith narratives, and proven programs of piety. Although such ways of being may be popular or pragmatic, much of what currently passes for Christianity has lost the power of imagination. It cannot connect people to God's transcendent beauty, embody Christ's love and justice, or open the heart to peace and wholeness. For the vast numbers of Americans, the Christian way of life differs little from a middle-class, white, suburban, and politically conservative way of life. Giving into the power of that "city" has flattened Christian identity, theology, and spirituality into narrow—and often materialistic or exclusionary—categories. Yet, amid this loss of imagination—or perhaps because of it—a deep spiritual longing pulls the hearts of many seekers and believers who dream that there exists a clearer vision of God and a healing way of life. It may well be that, in these days, Christians again need the ancient wisdom of imagination and pilgrimage. Only by traveling beyond the contemporary city to the edges of society can pilgrims finally *see* the deeper meanings of life experienced and promised in the good news of Jesus Christ.

Tracey Lind's book is a series of suggestive and episodic meditations on life at edge of the contemporary city, visual images captured in her own photographs, from which she unpacks spiritual and theological meaning. Unlike the ancient Christians who fled their cities to find faith, Tracey Lind stays in the geographical city and discovers God in the homeless, prostitutes, immigrants, and even in graffiti, run-down buildings, and urban garbage. Through the camera lens, interwoven with the insights of spiritual imagination, she moves beyond what St. Augustine called the "City of Man" to the often-elusive City of God. In this grace-filled pilgrimage, she invites others on a journey to see God's city with her—opening a compassionate and compelling vision of Christian faith that is transcendent and welcoming to all.

The good news of *Interrupted by God* is that seekers and believers need not imitate their ancient ancestors and flee the geographical city in order to see God and practice faith. And unlike her liberal Protestant forebears, Lind does not envision a God who is indistinguishable from the world and the secular social order. Thus, the spiritually hungry need neither completely forsake nor fully embrace the city. In these pages, God emerges as a God who is both in the world but not of it. Spiritual wholeness—with all its creativity, passion, and imagination—is a pilgrimage of seeing and living into this truth.

In a final flourish of paradox, Lind confesses to being an "evangelical universalist" as she powerfully reclaims a distinctive and loving way of life in and through Jesus Christ. In doing so, she joins her voice with the voices of others who are beginning to proclaim that a new way of being Christian is arising, a way that finds God's truth in the shadowed edges of light beyond the borders of the city. Lind testifies that God, indeed, is alive—a being of infinite personal, transforming, and challenging love who can be found everywhere and may be known through the intentional exercise of spiritual imagination.

Sometimes the most remarkable pilgrimages are simply walking outside the door of one's own house and seeing the world through the eyes of God's spirit.

Diana Butler Bass
Alexandria, Virginia
Author of *Strength for the Journey: A Pilgrimage of Faith in Community* and
Broken We Kneel: Reflections on Faith and Citizenship

Preface

Come to the edge.
    We might fall.
Come to the edge.
    It's too high!
Come to the edge!

And they came
And he pushed
And they flew.[1]
—Christopher Logue

The English poet Christopher Logue aptly describes what happened to me one cold January day in a McDonald's restaurant on Forty-second Street in New York City. I was interrupted by an unfamiliar voice that called out to me from within the depths of my soul. It probed and prodded, provoked and persuaded, pulled and pushed at all that I was and all that I hoped I would become. "Come to the edge," the voice said. "No, I am afraid." "Come to edge," it insisted. So I went to the edge. The voice pushed, I flew, and I've been soaring ever since.

As an Episcopal priest and dean of a major urban cathedral, I live at the center of an established church and privileged society. Yet in my very being, as a child of an interfaith marriage, a lesbian, and one who has spent a great deal of time with the homeless, I belong to the edge, to the fringe, to the people who are never certain if, when, or where they fit into the great scheme of things. Staying close to the edge, I see all kinds of things that I couldn't see if I only lived in the center of safety and privilege.

A journalist once asked me what it is like to live with "double vision." As a person who lives in the center but is drawn to the borderlands and boundary waters of the margins, I see the truth of life in various shades of grey. There is no black and white. Nothing is absolute, and there is always opportunity for something new to emerge in both the darkness and the light. As St. Paul, a spiritual ancestor whose life and vocation was also interrupted from the edge, once said, "Now I know only in part; then I will know fully, even as I have been fully known" (1 Cor. 13:12).

Over the years, I have learned to claim the edgy religious, sexual, social, economic, and political paradoxes of my existence and of those whom I have encountered in my daily life and ministry. As a person of faith, I have searched for the good news and truth within those paradoxes. This book is an attempt to speak honestly about the paradoxes that call me from the center to the edge and back again.

The stories and photos I've included here illuminate gospel truths and divine revelations from my perspective on the edge of exclusion and embrace. In the introductory chapter, I explore the question of passing or dying for my faith, of claiming or denying the essence of myself. This question, asked of me over three decades ago, was my first interruption from God, and the search for its answer has afforded me frequent glimpses of God from the edge.

In the chapters that follow, I introduce you to Lisa, the homeless Christmas angel; Siah, the infant hope of her ancestors; Mike, the child who understands the wisdom of the wind; Sally who feasts on communion remnants; the Garbage Tree on Ellison Street; Yvonne, the Good Friday interruption; Bacardi, the drunken Christ; and the sisters of mercy who beckon us to the banquet—all of them interruptions of the holy. I explore my desire to light up the world, why I do ashes, how I pray for peace, and the power of a five-dollar bill. I invite you to embrace the darkness, encounter the demolition contractor, remember the power of baptism, climb mountains, cross the great water, love the fallen flower, and keep on singing. I unabashedly describe myself as an Evangelical Universalist who claims

Jesus as my way, finds life in the Trinity of love, but believes that the journey with Christ is not the only saving truth. In the closing chapter, I describe what I understand to be the essence of the Gospel in the children's game of hide-and-go-seek.

In an article for *The Witness Magazine*, newspaper columnist Ina Hughes wrote, "It is not by accident that all great teachers of every religion used stories to get their message across. You can preach me a sermon, show me a doctrine, recite me a creed—and I might be impressed. But tell me a story, and I will remember."[2] I have been blessed with the gift of many stories. A few of my stories are out of the imagination of my heart and head, but most of them are true stories from real life, usually from intrusions into my daily routine.

A number of years ago, I picked up a camera and went looking for God in my neighborhood. Through my camera lens, I discovered a spiritual discipline for myself. Photography is not my profession, nor is it my hobby. Rather, the making of photographs is for me a form of prayer and meditation. As a raging extrovert, I am usually actively engaged in the world around me. Photography is a way that I distance myself enough to see what is happening. When I examine life through the eye of a camera, I am forced to step back, slow down, concentrate, and become deeply attentive to the situation. Looking through my viewfinder, I can't allow myself to focus on simply what lies in the middle of the frame; I must explore the circumference, the corners, and the edges in order to really see the entire situation.

As one who moves quickly, the art of photography does not come naturally to me. I must work at it. But I have discerned that photography is the work of my soul, not my ego, and I have learned the hard way that whenever my ego gets involved, it distorts the picture. Photography keeps me humble and often frustrated, but is an important metaphor for the rest of my life. Photography has taught me to slow down, wait for the moment, and open myself up to the Spirit so that I may experience God in unexpected people and places.

Over the years, I have discovered that I can look for and find glimpses of the Holy through a camera lens, but I must be very respectful of what is found and seen. I don't "take shots" and I don't "capture the moment." Rather, I enter into sacred relationship with the other, and with respect and permission I make a portrait of what I see, and then I pray that my photograph will honor the essence of what my eye saw.

I have been amazed and awestruck by what I've seen and heard, and I have tried to be faithful in the rendering and interpretation. In working on this book, I have become convinced that if we allow them, the poor can be our wisest teachers, the wounded can be our most powerful healers, and the oppressed can be our strongest liberators. I now know for certain that the last can be first, the despised can be loved, the outcast can be welcomed, the dead can be raised, and that which the world deems to be garbage can be made holy if we are open to the unexpected grace of God.

Interrupted by God has long been in the making and would not have been possible without the encouragement, love, and support of a great many people. I want to say thanks, and I ask you to accept my apologies for this long list of names and to feel free to skip over these pages and get on to the introduction and the stories. But if you do, please know that I am nothing without these individuals.

I offer special thanks to those who took the time to read the very first draft of this book: Dan Schoonmaker, Angela Ifill, Robin Hitchcock, Richard Gildenmeister, and especially Jon Wakelyn, who told me to let it go and send it to the publisher. I am grateful to John Dominic Crossan, who looked at the photographs, read the first few pages, and whispered in my ear, "It's a sacrament. Go for it!" I am most appreciative to Pamela Johnson for her editing, support, and inspiration. I thank Janice Brown and Robyn Henderson for photo editing, production, and design. To Michael Lawrence, Timothy Staveteig, and the rest of the team at The Pilgrim Press, thank you for having faith in my project and making this book a reality.

I am especially indebted to my teachers over the years. I will never forget Janet Raleigh Smith, who in high school shared with me the love of books, music, and art, and encouraged me to ask hard questions and not accept easy answers. I offer special thanks to Phyllis Trible, Dorothee Soelle, Carter Heyward, Walter Brueggeman, James Forbes, Marcus Borg, James Cone, Raymond Brown, Ardith Hayes, Paul Henry, and Marge Lotspeich—scholars who before, during, and after my years in graduate school and seminary encouraged me to think critically and out-of-the-box.

I am ever thankful for some wonderful bishops I have known, admired, and loved in the Episcopal Church. To Paul Moore, who told me on the morning of my ordination, "All you have to do is love them." To Jack Spong, who challenges the church, encourages its scholarship, expects excellence from its clergy, and said to this young priest, "If you wait for a sabbatical to write, it will never happen. You have to get up every morning and do it." To Jane Holmes Dixon, who has shown me the power of passionate truth telling in the name of gospel justice. To Clark Grew, Arthur Williams, and David Bowman, with whom I have shared ministry in the Diocese of Ohio. And to Richard Shimpfky, who taught me how to be a priest and still loves me unconditionally.

I am grateful to some remarkable ecumenical and interfaith colleagues. Doug Fromm advised me to "dance at their weddings and cry at their funerals." Joan Brown Campbell invited me to the summer pulpit of the Chautauqua Institution, thus opening new doors and avenues for preaching. Arthur Waskow helped me understand the sacred significance of the fringe, and Howard Ruben welcomed me home.

I have been blessed with many friends and companions along the journey. I offer special thanks to the Women's Sewing Circle for friendship, laughter, and support; and to the Gang of Four for wit, wisdom, and wisecracks. I will always be indebted to Diana Beach for patient listening, deep digging, and wise counsel over the course of two decades. The photographs in this book would not have been possible without Clelia Belgrado, Doug Beasley, Meg Meyers, and Jennifer Jones, who taught me how to see through the lens of a camera. I am grateful to Nancy and Bill Dailey for schooling me in the art of radical hospitality, to Joe Russell for the gift of storytelling, and to Jane Russell for teaching me about long-term loving. Special thanks go to my cathedral colleagues: Greg, Tina, Kurt, Joyce, Dan, Mike, Marcia, Tricia, Rebecca, Rufus, James, Roderick, KG, Melonie, Ed, Rosemary, Twanna, Agnus, and Joanne for being a great staff team, ever patient with my foibles. I also count my blessings for Lucinda, my sermon and shopping buddy; Kathleen, Annamarie, Jamie, and Eric, who helped me say "yes" to God; Fran, who accompanied me on the way; Bob, who taught me to hug trees; Michael, who sang a new church with me; Wiley, who introduced me to Village Cleveland; Kate, who showed me her Chautauqua; and Karen and Tracey for being good pals. To the rest my friends and companions along the way—you know who you are—thanks for teaching me about life and love.

And who are we without family? Thanks to Kathleen for long walks in the cul-de-sacs; Tom and Jon for granting me the family memory; Jesse, Gillian, Jake, and Phillip—the next generation; Kathi for loving my brother; Missy and Bob for believing in me since I was a child; and Paul, Jean, Melanie, Margaret, Steve, and Anne for welcoming me into the Ingalls family.

And finally, this book and its author simply would not have been without my parents, Stanley and Winne Lind, who gave me the best they had to give; my "second mother" Marge Christie, who shares my love of church and beach; Emily Ingalls, my partner on the journey come hell or high water; and the people of Christ Church, Ridgewood, St. Paul's Church, Paterson, and Trinity Cathedral, Cleveland, who over the past twenty years have shared their lives, loves, and losses with me.

This book is offered in love to those who are willing to receive it. I hope that this marriage of word and photograph will help you see glimpses of God with your own eyes and validate your own interruptions from the edge.

Introduction
the Question

Thirty-five years ago I sat in a classroom with fifty other adolescents watching the movie *Let My People Go*. It was the first documentary of the Holocaust, and I'll always be haunted by the memory of emaciated corpses being pushed down a slide in the Warsaw Ghetto. At the end of the film, a young rabbi tried to elicit responses from a stunned and silent class of usually loud and obnoxious ninth graders. I'll never forget the moment when he looked at me, the only kid with a non-Jewish parent, and said: "Tracey, you don't look Jewish. You could have passed. What would you have done? Would you have died for your faith or denied it?" I didn't have an answer. I didn't know what it meant to pass. I didn't know what it meant to die for one's faith. I didn't really know what my faith was. I only knew that I was angry, embarrassed, confused, and alone. So I just stared back at him and finally said, "I don't know."

That accusatory statement, "You could have passed," followed by the probing question, "What would you have done?" has haunted me all the days of my life. It has permeated my dreams; it has kept me awake; it has stood with me in the pulpit; and it has influenced every major life decision I have made. And just when I think I have put the accusation to rest and answered the question, it reemerges as a beast from the deep recesses of the ocean called my unconscious. This question, "will I pass or will I claim who I am and what I believe regardless of its cost, even to death?" is the angel with whom I wrestle, causing me to walk with a limp. It is the burning bush in front of which I stand barefooted, the slow burning flame that keeps alive my passion but does not consume my spirit. Whenever I travel into the

1

wilderness of my soul, I am tempted to avoid this question's pain and confusion. It remains the blinding flash of light that forces me to my knees when I try to run away.

The question of passing or claiming one's faith, even to the point of death, is for me *The Question*. It is what Buddhism calls my koan. Phyllis Trible, one of my favorite seminary professors, used to say that we all need a myth to guide our lives; and that if we don't claim a myth, one will claim us. I think everyone needs both a myth and a question. In fact, I believe that everyone has a myth and a question; we just don't always choose to acknowledge and explore these truths in our conscious self.

To pass has many implications. Passing can be as simple and as seemingly innocent as allowing a racist, sexist, anti-Semitic, or homophobic remark to go unnoticed. It can mean worshipping a homeless man on Sunday and walking by without seeing dozens of homeless men, women, and children during the rest of the week. Passing can cause one to hide in all kinds of closets for all kinds of reasons. Passing can simply mean taking the easy way out, even at the cost of one's soul.

To live and die for one's faith—now that is another matter. That's true discipleship. That's the stuff that Dietrich Bonhoeffer, Martin Luther King, Oscar Romero, Mother Teresa, Dorothy Day, and the other great prophets and disciples of the ages are made of. Dying for one's faith is having the courage to really live a life worth living.

For two years, I slept in Dietrich Bonhoeffer's old room at Union Theological Seminary in New York City. I used to lie awake at night wondering what Bonhoeffer would have been thinking about when he couldn't sleep—whether to return to his native land and confront the evil of Nazi Germany or stay and teach in the safety of the academy. Bonhoeffer chose the former, and it cost him his life. I still don't know if I have that kind of courage; nor do I know if that sort of action is my calling. What I do know is that Bonhoeffer's witness has deeply influenced my vocation, and has on more than one occasion called me to account for my decisions and actions in this world.

There is a famous rabbinical saying: "Consider three things and you will not fall into the grip of sin. Know where you come from, to where you are going, and before whom you are destined to be judged and called to account."[1] Knowing one's history, one's story, and one's God is incredibly important. Without this knowledge, we are lost—both individually and collectively. Our ancestors in faith knew their God and their story; the Bible is their recorded history.

Each of us has a sacred story, much of which is inherited through ancestry and birthright. It is a by-product of our race, ethnicity, religion, gender, and class. It is planted and cultivated in the landscape and geography of our birth and upbringing. It is engraved in our soul and frequently lost in our memory. Sometimes, to my family's chagrin and my church's discomfort, I have set about the task of searching out, retrieving, claiming, speaking, and discerning the life-giving truth of my story.

Much of life-giving or gospel truth is about transforming contradiction into paradox—like water into wine, brokenness into wholeness, scarcity into abundance, the last into the first, and death into life. Paradox makes sense of things that don't make sense and holds divergent things in tension. Turning contradiction into paradox is choosing both/and rather than either/or, seeing shades of gray instead of black and white. My life is one big contradiction seeking to become a paradox. In many ways, I think the American story itself—all sorts and conditions of people coming to this land on many ships but being in the same boat now—is a contradiction trying to become a paradox. It is both our blessing and our curse. And I know it well, since I personify the American story.

My mother's family came from England. To the best of our knowledge, her ancestors arrived in Jamestown during the seventeenth century, settled in the upcountry of Virginia, and eventually moved west to southern Ohio in the late eighteenth century. My mother was born in a place called Greasy Ridge,

Ohio—a hamlet located twenty miles north of the Ohio River. She was the youngest of seven children born to Noah and Minnie Heffner, two of the hardest working coal miners and farmers you'd ever want to meet. My mother was raised in the Methodist tradition; her grandfather and great-grandfather served as lay leaders of a little backwoods church called Locust Grove. My mother hated those mountains; she hated the poverty, ignorance, superstition, and old-time religion of rural Appalachia. At the age of fifteen, having graduated from a one-room school, determined to make a better life, my mother left the mountains for good and moved north to the city.

My father's ancestors were Austrian-German Jews who came to this country with the great wave of nineteenth-century immigration. One of my great grandfathers was a cigar roller and reputed Torah scholar who lived out his years in America on the Lower East Side of Manhattan. When I toured Ellis Island, I could see his face in the photos of Jacob Riis. My other great grandfather peddled his way to Zanesville, Ohio, where he settled and raised up six sons to become successful merchants in a booming town through which the Old National Road passed, only decades later to be bypassed by the interstate.

Wealth and poverty became a paradoxical symbol in my family. At the dinner table, my father used to recall how his mother had to let go some of the household help during the Great Depression. My mother would sit silent and never speak of the real poverty she knew during those painful years when her father couldn't find work in the mines. To this day, when I view the Farm Service Administration photos of that era, I can't help but look for the faces of my mountain ancestors.

My parents met shortly after World War II. He was a young businessman, and she was a nurse. They fell in love on their first date and married six weeks later. Unfortunately, since he was Jewish and she was Christian, no clergy would perform the wedding. So they eloped and were married by a judge, and for their honeymoon, my mother accompanied her new husband on a business trip to Cleveland. Upon hearing of the marriage, my mother's family, who had never known any Jews, was upset and unable to welcome my father. My father's family received the young couple, so my mother agreed to raise her children in the Jewish community, without completely letting go of her basic Christian teachings and customs.

The way my parents raised my younger brother and me religiously was really very simple. They taught us to say our prayers at night: "Now I lay me down to sleep, I pray to God my soul to keep; if I should die before I wake, I pray to God my soul to take." They taught us the Great Commandment, otherwise known as the Golden Rule: "Love God with all your heart, soul, and might, and love your neighbor as yourself." They sent us to Sunday school at the Reform Temple where we would learn about God without all the superstition and myth about Jesus. We attended synagogue services on Rosh Hashanah and Yom Kippur, and we would celebrate an amalgamation of other religious holidays. To this day, I joke about lighting Hanukkah candles under the Christmas tree and finding Easter eggs at the Passover Seder. In sixth grade, I was sent to a private girls' school where we went to daily chapel, a simplified version of Morning Prayer, Episcopal-Presbyterian style.

I think of myself as half Jewish and half Christian, and I consider my rich heritage a mixed blessing. As a child, I wanted to be a preacher—I just wasn't sure whether I should be a rabbi or a minister. Since my mother wasn't Jewish, I wasn't considered a real Jew, so I didn't think I could be a rabbi—and anyway, I assumed I could never learn Hebrew. When I imagined becoming a minister, I couldn't figure out how to do that either—you see, I wasn't baptized, so I wasn't a real Christian, and I didn't want to be baptized because I had learned somewhere (an untruth, I now believe) that the Nazis baptized Jewish babies and then sent them to the gas chambers. Anyhow, I was a girl, and back then girls couldn't be ordained either as rabbis or ministers. But I loved being in the house of God, and I loved playing "Saturday-go-to-temple" and "Sunday-go-to-church." I still remember setting up the chairs in our family room, putting my stuffed toys and dolls in straight rows, and preaching to the silent, appreciative, and complacent congregation of inanimate worshippers.

By the time I was in ninth grade, I was sitting through daily chapel at school, confirmation classes at the Temple, and playing my guitar for the weekly folk mass at the local Episcopal Church. No wonder I couldn't answer *The Question* asked by the rabbi that day. I really didn't know who I was. On Sundays while everyone was kneeling for communion at the altar rail, I was singing Gordon Lightfoot's words: "I'm standing at the doorway, my hat held in my hand, not knowing where to sit, not knowing where to stand . . ."[2] It took a long, long time for me to figure out where to sit and where to stand. And there are still days when I feel like a rabbi in a clerical collar.

Jesus, in the Book of Thomas the Contender, says: "whoever has known himself [or herself] has simultaneously come to know the depth of all things."[3] In a lifelong effort to answer *The Question*, I have spent a lot of time, energy, and money getting to know the depths of myself and making connections to the world around me.

I never made the connection between my birthday and history until I walked into the Martin Luther King Museum in Memphis, Tennessee. And there before my eyes, as big as life itself, was the front page of the *New York Times*, dated May 17, 1954, with a headline that boldly proclaimed: "BROWN VS. THE BOARD OF EDUCATION." I have always felt that the complexity of racial justice was implanted in my soul and grafted into my unconscious, and in that moment I knew why. I came into this world on a decisive day in the life of the civil rights movement, and I have spent my entire life in the midst of that struggle.

Like lots of suburbanites, I never knew many people of color. In fact, I knew very few people of any ethnicity other than WASP or Jew. What I did know was that the maids and gardeners who were black did not have last names we remembered (or even knew), and that it was best to stick to your own kind (even if your own kind was not purebred). I also knew the anger and rage in my father's face when the race riots found their way to our city, and his inner city furniture warehouse suffered smoke damage from a nearby fire. When I was in fifth grade, the public schools in Columbus, Ohio, were integrated, and I was sent to private school. My parents insist this was a coincidence, but I believe there is no such thing as coincidence; it is just God (or the devil) at work unbeknownst to us. So off I went to receive an excellent education in an even more pristine, privileged, and guarded community of affluence and homogeneity. As I entered the hallowed halls of the Columbus School for Girls, *The Question*—that haunting question of passing or claiming myself—was lurking in the wings.

In the summer of 1970, something happened that changed the vocational direction of my life. While my friends were off at summer camp, Europe, and Outward Bound, my parents insisted that I get a job—volunteer or paid, it didn't matter. So I went to work with a big, angry attitude as a volunteer teacher's aid in a new program called Head Start. I was assigned to the Ohio Avenue School, where I spent two months accompanying a wonderful group of "underprivileged" preschoolers through their daily routine. I would prepare their snack, ready them for naptime, read to them, and play with them. And sometimes, I would make home visits with their teacher to see how and where my young friends really lived. On those field trips, I saw firsthand the pain of the poor, and I learned about the struggles of raising children in urban poverty.

It was also the summer I got my driver's license and a car. With wheels came the newfound freedom to explore beyond the boundaries of my neighborhood and to go where I wanted. I drove through communities that my parents said were not safe, trying to see and experience the danger for myself. I drove to the Ohio State University campus, getting involved in the antiwar and student rights movement. I drove through the foothills of the Appalachians, touching roots I still had not yet discovered. That summer changed my life. My eyes were opened; my sense of exploration and adventure was awakened; my awareness of poverty, racism, and oppression was provoked; and my passion to work for justice was born.

As the passion and anger of the sixties brewed and boiled over, so did my own passion and anger. How could there be such extremes of wealth and poverty in our nation? How could there be such hatred and fear among blacks and whites? How could women be told to stay in the kitchen? How could young men be sent to fight in a war that shouldn't be? How could I go back to school and act like nothing had happened?

The decade of the seventies was an endless, exhausting, but exhilarating marathon of running away from all that I had known to a world beyond myself. I did a lot of hard growing up in a very short time. At eighteen, my father was diagnosed with cancer, and I responded to the news by jumping into an ill-advised marriage. Less than two years later, the marriage was dissolved, and I returned home, a prodigal daughter, to be with my father, who died a year later. By the age of twenty-one, I felt older than my years, and I was angry with God for letting "bad things happen to good people."[4]

As a young adult, perhaps to compensate for my rebellious nature, I had a strong need to achieve, excel, and prove myself worthy of *The Question*. By the time I graduated from college, I had organized and directed a landlord/tenant agency and coordinated the development of a neighborhood revitalization program. Following college, I went to graduate school to study community planning, and for a few years I worked in a variety of community institutions, ranging from the United Way to the Girl Scouts. All the time I still was angry with God, but God was quietly watching over me—guiding me through every mess I got myself into without intruding into my fierce independence and self-determination.

During my early twenties, I also began coming to terms with my sexuality, realizing that I was a lesbian. It's hard to understand the coming out process if you're not gay. I liken the journey of coming out to a second adolescence with particular rites of passage that often include falling in and out of love, and sometimes looking for love in the wrong places. Fortunately, with support and acceptance from friends and family, I sorted out my sexual identity issues, and those confusing years actually have made me a more responsive and compassionate pastor to young adults and their families. By the time I rounded the quarter century mark, things were finally beginning to make sense, but *The Question* was becoming more complicated as the issue of passing took on new significance and meaning.

At the age of twenty-five, I found my way to Boston, where I lived *The Question* of faith and started getting some answers. Along the way, I decided it was safe to be baptized—I was finally convinced I wouldn't be sent to the gas chamber, and I wanted to belong. Moreover, I wanted to be an ordained minister. After years of struggle, I felt pushed in the direction of seminary, and every door seemed to open without a key.

My first semester at Union Theological Seminary in New York City was a wrestling match with God. Exhausted from taking on someone bigger and stronger than me, I found myself walking down Forty-second Street one day in January asking God to let me go. And then it happened. Suddenly, a voice called out to me from within me saying, "I'm not going to let go of you." "What do you want with me?" I asked. "I want your life," the voice answered. "Why me?" I responded. "Why not?" the voice replied. At this point, I realized that something was happening and I needed to stop and pay attention to this voice. I went into a nearby McDonald's restaurant, ordered my usual cheeseburger, fries, and coke and began frantically scribbling down a conversation with this voice from within. The voice called me by name, identified itself as God, confronted me with my own issues and private wounds, contradicted my theology, answered lots of questions, called me to the ordained priesthood, and reassured me when I protested. The voice said, "I brought you to New York for a reason, to look beyond yourself and those like you. . . . I want you to celebrate my Eucharist. . . . You must feed my people. . . . You will guide people to come to me through this and other acts. . . . You will help people to love each other and me. . . . You've changed; why can't others. . . . It's a loving revolution, so be my hands and my mouth, not your own."

In the course of the conversation, I questioned why the voice was talking with me, and it responded, "Because you've been asking for it." It was true. I had been asking, begging, even challenging God to be clear with me, to help me answer *The Question*. And here I was—on a cold January afternoon, sitting in a McDonald's restaurant on Forty-second Street in Manhattan, having this private conversation with a voice. At the end of our time together, I asked, "If you're inside of me, then how can you be God?" The voice replied in words I'll never forget, "What's so special about me is that I'm inside of anyone and everyone who wants to know me. And, if the world would hear me and follow me, my kingdom would come." With that comment, the conversation ended. I got up and walked home in quiet amazement, wondering if I had really spoken with almighty God. Like Mary, I kept silent and treasured these words, pondering them in my heart.

A few days later, one of my professors, Dorothee Soelle, told our class that faith is a two-way street: it is both a gift from God and our decision to accept the gift. I didn't know if I had talked with God, but in a letter to a friend I wrote, "If I don't accept the voice of God on faith now, I don't think I'll ever get a more direct message." *The Question* was finally beginning to be actively addressed.

From the moment I left the McDonald's, I was determined to follow the voice wherever it would lead me. It led me to a summer internship at Trinity Church on Wall Street. It took me to Bronx Youth Ministry and St. Margaret's Church in the South Bronx where I was sponsored for ordination. Following graduation from seminary, it invited me home to the suburbs to begin my ordained ministry at Christ Church in Ridgewood, New Jersey. It then called me to become Rector of St. Paul's Church in Paterson, New Jersey.

For over a decade, I had the privilege of serving a community of "all sorts and conditions" of people who gather for worship, service, and witness in a glorious historic landmark church located in a poor neighborhood in one of our nation's old industrial cities. During my twelve-year tenure, the congregation became incredibly diverse. We were black, white, Hispanic, and Asian. We were native born and newcomers from many lands. We were young, old, and in-between. We were gay and straight, married and single, bisexual, and even transgendered. Some of us dressed up for church; others came in blue jeans. Some members of our church had beautiful homes in lovely neighborhoods; others lived on the streets. A number in our company had large investment portfolios; but most of us got by from paycheck to paycheck or welfare check to welfare check. Many of us took freedom and citizenship for granted; but some knew all too well its precious price. There were those among us who worked in the judicial system; and there were others who had done time in the system. We had students and their teachers, employers and their employees, doctors and their patients all sitting in the same pews. St. Paul's became a living testament that "the things which divide us from each other may be overcome in the oneness of God."[5]

At St. Paul's Church, God frequently interrupted me with glimpses of the holy from the edge. Over the course of my time as pastor of that old church, I met the Risen Christ over and over again. S/he wandered in and out of our sanctuary, shelter, food pantry, and even my office day and night, sometimes in disguises that I found disturbing and threatening, and on other occasions, humorous and inviting. The work was hard but rewarding; the pain of the community was great, but so was its joy.

People ask me why I stayed at St. Paul's so long. My answer is simple. I hadn't been called to leave, and how could I leave when Christ was lurking in the shadows bidding me to remain? As the larger church and society became more polarized over issues of race, gender, class, sexuality, and creed, St. Paul's Church remained a community of faith where we had the opportunity to make a difference, to demonstrate that we could live together in all of our diversity because God had bid each of us welcome and called us to build a better world.

When I was ordained, my bishop and hero Paul Moore said to our deaconate class: "All you have to do is love them, really love them. That's all you have to do." It was the best and most challenging advice I ever received. Love in the public context of the church is never easy, especially if you're gay or lesbian. How does one obey Jesus' command to love those who hate you, especially when those who hate you say, "We love you and that's why we want to save you from your sin."

I learned the lesson of "love the sinner but hate the sin" the hard way in the fall of 1995 when once again I was confronted with *The Question* of claiming myself or passing. I am a lesbian who is not called to a vocation of celibacy but has been called to the vocation of ordained ministry. Remarkable as it might seem, somehow, some way, my sexuality about which I've been relatively open (everything is relative) since I was in my early twenties, did not get in the way of my ordination process or parochial ministry. I always handled it with discretion, but I was always truthful and honest about it, answering questions when asked and volunteering information when it seemed appropriate. I guess I was lucky, or perhaps God simply had another idea in mind.

My friend and fellow Union alumnus Barry Stopfel wasn't so fortunate. His ordination became a subject of debate within the Episcopal Church, and his ordaining bishop Walter Righter became the subject of the second heresy trial in our denomination. I knew that this trial was not just about Barry or Walter. Rather, it was about all of us who were gay and lesbian, and all those who stand in solidarity with us. In the words of singer-songwriter Holly Near, "It could have been me, but instead it was you."[6]

For many years, I had said that if on a given Sunday, everyone who was gay or lesbian could turn purple in church, the issue would be over. When the threat of a heresy trial became a reality, I realized this vision of purple was not going to be the case. God was not going to do our labor of liberation for us. No, God was calling us to do our own work. In my humble opinion, it was time for all of us gay and lesbian clergy who were in positions of power and relative security in the church to come and make public witness about being gay, being Christian, and being called by God to be full participants in the church.

The Gospel of Jesus Christ commands us to pick up our crosses and carry them. The cross of being denied full inclusion and open participation in church and society because one is gay or lesbian is, was, and will be for some time to come, I presume, a cross to bear, and I could not allow anyone else to carry it alone. Because St. Paul's is a church that welcomes and affirms the gay-lesbian-bisexual-transgendered community, the Vestry (the elected governing body of the congregation) determined it was also the parish's cross to bear. Together, we decided that we would be as public as we needed to be to stand with Barry, Walter, and the Diocese of Newark during the course of the heresy trial.

Shortly after making this decision, as fate would have it, I read an essay by lawyer and Episcopal layman William Stringfellow. The essay, entitled, "Living Humanly in the Midst of Death," is about why people resisted the Nazis. It captures what I believe to be the essence of resistance to oppression: "To exist, under Nazism, in silence, conformity, fear, acquiescence [and] collaboration—to covet 'safety' or 'security' on the conditions prescribed by the state—caused moral insanity, meant suicide, was fatally dehumanizing, [and] constituted a form of death. Resistance was the only stance worthy of a human being, as much in responsibility to oneself as to all other humans, as the famous commandment mentions."[7] Stringfellow argued that while resisting oppression ensured risk and peril, nonresistance or acquiescence "involved the certitude of death—of moral death, of the death to one's humanity, of the death to sanity and conscience, of the death that possesses humans profoundly ungrateful for their own lives and for the lives of others . . ."[8]

For the first time, I had words to express what I knew in my heart. To exist in a homophobic society in silence, conformity, fear, acquiescence, and collaboration; to hide in our closets for fear of being caught, rejected, fired, abused, disowned, disinherited, ridiculed, and despised; to covet "safety" or "security" on the conditions prescribed by the state or the church causes moral insanity and the death of one's soul. To come out, to state honestly and clearly who one is and who one loves is *not* to flaunt one's sexuality, but rather, to be faithful to one's integrity, to choose freedom over oppression, and to claim life in the midst of death.

On October 15, 1995, I broke a pact I had made with God, the world, and myself when I came out in the pulpit. The unspoken bargain I had made went something like this: if I had to be gay in this society, then I would be the very best gay person I could be, and I would never do anything to embarrass anybody or make anyone feel uncomfortable about my sexuality. The heresy trial caused me to realize that this was a pact with the devil, not with God. So in a crowded sanctuary, in front of newspaper reporters and television cameras, I spoke aloud from our ten-foot-high pulpit the truth of my life. I turned myself purple on that Sunday morning, finally answering *The Question*. I'll never forget the closing words of my sermon: "And now, to answer your question, God: No, I will not pass! Yes, I am ready and willing to claim who I am and to live and die for my faith!" And the people responded with thunderous applause and a loud *Amen!*

As Dean of Trinity Cathedral in Cleveland, Ohio, I live in the center of the institutional church. As a priest and pastor, I carry the keys to unlock and open the church doors so that the stranger passing by may enter. I preach the word of God so that those who listen may know the good news of God's justice, love, and mercy for all creation. I stand at God's table and make Christ known in the breaking of the bread and the pouring of the wine so that God's hungry people may be fed. And I pronounce God's blessing upon those who seek it so that they may experience the gift of God's creative love.

It would be very easy to exclude people: to make some feel welcome and others not, to feed some and turn others away, to bless some and curse others. Like any human being charged with such a daunting task and awesome responsibility, I run that risk each and every day.

Whenever I am tempted to lock up God's house, to gate God's table, or to refuse God's blessing, I am confronted with *The Question*. The memory of my own exclusion, separation, and alienation, and that of my ancestors in flesh, faith, and spirit jolts me. These memories, painful as they may be, remind me of Jesus' mission in this world: to bring the love of God to those who seek it; to show the way to God to those who want to follow; and to extend the covenant of promise and salvation to all God's people.

A long time ago, I was asked a question I could not answer: would you have died for your faith or denied it by passing? I have struggled with this question ever since. It has shaped my life and directed my ministry. It grounds my theology and informs my ethics, provoking me to listen to the voices from the edge and pay attention to the fringe. It is at the heart of this book. Thirty-five years later, *The Question* still holds me accountable. I hope it always will. In the life, death, and resurrection that surrounds me everyday, I am beginning to glimpse my answer.

One

It was born on a winter day...

the Light
of Darkness

I am a fool for holiday lights. I love the candles of Advent, Hanukkah, winter solstice, Christmas, and Kwanza. I enjoy seeing holiday lights as I drive through various communities, noting the diversity as I move from one neighborhood to another. I also like festive downtown office buildings and department store windows. And I even appreciate what some people call "tacky" Christmas displays—the bigger, the better, I say, setting aside for the season my concern over energy conservation.

I am particularly fond of a suburban home in Mahwah, New Jersey, with a decorated pond and a singing Elvis on the roof. But my all-time favorite was an unassuming cottage located across the street from a Fraternal Order of Police Hall in Cuyahoga Heights, Ohio. This cottage was so well lit that it could be seen for miles, even from the freeway. As we neared the house, there were literally dozens of people crossing the street, walking up the driveway, and paying two dollars apiece to ooh and ahh at the array of thousands of sparkling lights in all shapes, sizes, and colors. Among the various displays were an American flag, a jack-in-the-box, and a gingerbread house. There were numerous Christmas trees, choirs, snowmen, and Care Bears. There was a crèche complete with the holy family, attending shepherds, barn animals, and angels by the dozen. And of course, there was Santa Claus and his playful elves and reindeer. The entire display was constructed of twinkling, multicolored holiday lights. Christmas carols were blasting out of stereo speakers, and volunteers collected money for local charities.

As we walked away, I asked myself, what makes folks go to all of this effort and expense? Moreover,

I wondered, what makes people like me travel a distance in the cold of the night to witness such extravagant displays of holiday cheer? The answer is quite simple. We need light. During the bleak midwinter, we human beings develop a craving for light. When the sun retires, we light candles and turn on artificial lights; when the trees are a leafless brown, we bring fresh evergreens inside; when the cold wind blows, we drown out its howling with music; when the harvest is over and the fields are bare, we feast; when the days are short, we party long into the night. No wonder we overindulge at the holidays; we're trying to compensate for the dark and barren days of the winter season.

I love decorating my own home for the holidays. I look forward to putting candles in the windows and lighting the Advent wreath. I like the ritual of picking out and cutting down the "perfect" tree that is never perfect when we get it home. I enjoy the challenge of stringing lights on the tree, only to realize that at least half of them don't light up when they are plugged into the socket. I love hanging the ornaments, especially the shiny red bulbs with gold crochet made by my grandmother. And I really relish the moment when we turn off the other lights and turn on the tree lights. If left to my own devices, I will play Christmas music on the stereo and sit and look at the tree for hours upon hours as it twinkles in the surrounding darkness.

One of the best things about my job as a parish priest is that I get to help decorate a really big house—God's house—and then periodically sneak in for a private glimpse of several trees lighting up the darkness. I usually get the inspiration for my Christmas sermon during these private moments.

One year, on the day before Christmas Eve, I wandered into the church to turn on the lights and stare at the trees. As fate would have it, when I plugged in the lights, I found that one of the trees had fallen over. Unsuccessful in my attempt to upright the tree by myself, I went into the men's shelter and recruited a helper. My plan to fix one Christmas tree turned into a few hours of readjusting all the trees, moving some sanctuary benches, and having a lengthy conversation about the real meaning of Christmas. Unfortunately, I came home perplexed about what I would say in my Christmas Eve sermon.

After dinner that evening, still in search of a Christmas sermon, I decided that I needed to buy additional lights to hang on the bushes in front of our house. I jumped in my car and with Christmas music blasting on the radio drove to the drug store. I ran into the store, bought a half a dozen boxes of lights, practically threw my money at the sales clerk, leaped back into my car, and drove home. After hanging the new lights, I plugged them in, and they didn't work. I had purchased several boxes of defective lights. I tore the lights off the bushes, threw them in a bag, and drove back to the store, only to find a young man locking the door. "We are closed," he said. "Come back tomorrow." "I can't come back tomorrow. Tomorrow is Christmas Eve. I've got to work and I just bought these defective lights, and I simply want to exchange them. Please let me in. It will take me just a minute." As the aggravated clerk shook his head and I was about to burst into tears, the store manager walked by, recognized my panic-stricken face, looked at the lights in my hands, and opened the door. "Come on in," he said with a tired smile. "Let's get you some working lights." This kind man actually took the time to open the boxes and test the lights. As they twinkled, my face lit up like a Christmas tree, and I started to weep like a child. Embarrassed by my unexplainable behavior, I thanked the generous store manager and apologized to the disbelieving store clerk. I went home, hung the lights and climbed into bed—still without a Christmas sermon.

The next morning I got up, turned on the Christmas tree, sat down in front of it, and opened my Bible to the passage that is read in the dark every Christmas Eve just around midnight. "The light shines in the darkness, and the darkness did not overcome it" (John 1:5). As I read those words from John's gospel, I realized what was going on inside of me. I was trying to overcome my own darkness. I was doing my best to dispel the dark shadows of night from my own life.

I don't like the dark. In fact, I've always been a little afraid of the dark, fearful of the bad and scary things that go bump in the night. I love going to the movies, but I don't like sitting in a dark theater waiting for the film to begin, and I can't stand watching the credits roll on a dark screen at the end of the movie. I enjoy dining by candlelight at home, but I don't especially like dark bars or dimly lit restaurants. I don't like sleeping in pitch darkness, driving down dark streets, or walking in dark woods. I didn't like trick-or-treating as a child because you had to walk around the neighborhood in the dark; and as an adult, I am a passionate photographer who can't stand working in the darkroom. I dread the short days and long nights of winter, and I get anxious when the sky darkens before a storm. If the truth were told, I think I suffer from light deprivation, and my deepest fear is being imprisoned in a dark and dreary cell or getting trapped in an underground tunnel.

That Christmas Eve day as I sat in my study still struggling with my sermon, I got honest with myself and admitted that it had been a difficult year, and I was stuck in the infamous "dark night of the soul."[1] St. John of the Cross, a sixteenth-century mystic who coined this phrase, was plunged into darkness and despair when he was imprisoned for supporting a reform movement within his Carmelite Order. For nine months, he was beaten, starved, and confined to a monastery cell "with no other light than that which came in through the diminutive opening high up in the wall of the tiny cell."[2] During his imprisonment, John of the Cross encountered the complete and total absence of God to the point that he could no longer pray.

Though I was not imprisoned in a dungeon or being tortured for my religious convictions, I was wrestling with the question of passing or claiming my faith. In the midst of the Episcopal Church heresy trial over the issue of gay/lesbian ordination, I was struggling with the institutional church and its tendency toward exclusivity, tokenism, scapegoating, and conflict avoidance at the cost of justice. I was engaging the deeper and more systemic issues of urban poverty and violence and found myself rethinking the role of the church in the city and my own ministry as an urban priest. Ten years out of seminary, I was running on empty. I was so exhausted that I came down with pneumonia and was confined to bed for much of Advent.

Darkness had intruded upon my life as an uninvited and unwelcome guest. I had journeyed to that place of emptiness, loneliness, and gloom where "the night [had stripped] away the surface of my world."[3] It had been a long season of patiently waiting, watching, and hoping for God to light up my darkness. And when Christmas was upon me, with no end to the darkness in sight, I had to do something to overcome it. I had to confront the darkness head-on without divine intervention. I had to light up my own world.

Over the years, I've looked back on that crazy pre–Christmas Eve with a modicum of laughter and embarrassment. What a fool I made of myself running into the drugstore at closing, insisting like a mad woman that I had to exchange my defective Christmas lights when those tired employees were trying to lock up and go home for the night. You would have thought that I needed a prescription from the pharmacy to save my life. Maybe I did. Maybe those lights were antibiotics to ward off the evil spirits of darkness that had invaded my soul and interrupted my life.

Darkness is not an evil spirit. Rather, darkness is a primal element. It existed before light. Darkness is the background, the underpinning, and the fabric for the quilt of creation. The Book of Genesis tells us that in the beginning, "darkness covered the face of the deep" (Gen. 1:2). Creation began in the dark of night. It was out of darkness that God gave birth to the rest of the created order, including light. And it is in the darkness of the womb that life is conceived.

According to Edith Hamilton, the ancient Greeks believed that "Long before the gods appeared, in the dim past, uncounted ages ago, there was only the formless confusion of Chaos brooded over by unbroken darkness. Night was the child of Chaos and so was Erebus, which is the unfathomable depth where death dwells. In the whole universe there was nothing else: all was black, empty, silent, endless. . . . And then a marvel of marvels came to pass. In some mysterious way, from this horror of black boundless vacancy the best of all things came into being. . . . From darkness and from death, Love was born, and with its birth, order and beauty began to banish blind confusion."[4] Was love trying to be born anew in me that long, dark Advent?

In most cultures, primal darkness is considered chaotic. The North Australian aborigines say that "In the beginning, all was darkness forever. Night covered the earth in a great tangle."[5] The poet John Milton spoke of the primal Chaos as "the vast immeasurable abyss, outrageous as a sea, dark, wasteful, wild."[6] And yet, the Bible tells us that God created both light and darkness (Gen. 1:5, Isa. 45:7). Was God creating something new in the chaos of my darkness?

The Fourth Gospel tells us, "in the beginning was the Word, and the Word was with God, and the Word was God. . . in [the Word] was life, and the life was the light of all people" (John 1:1, 4). The King James Version reads, "The light shineth in the darkness, and the darkness comprehended it not" (John 1:5).[7] How about that? Both darkness and light were there in the beginning of creation, and yet the darkness did not understand the light. Was God simply doing a new thing in my life that the dark shadows of my unconscious did not yet comprehend?

This light, that enlightens everyone who receives it, was said to be the Word of God. It was the Word that was with the Eternal One when the world was created. It was the Word spoken by the Creator to human beings since Adam and Eve lived in the Garden of Eden. It was the very Word that called

Abraham and Sarah to birth a chosen people. It was the Word that rescued their descendants from slavery and led them through the wilderness to a promised land. It was the Word that became Torah, a new way of life. It was the same Word that, through the prophets and priests of old, disciplined God's people when they went astray and called them to renewal and right relationship over and over again. But for reasons as varied as our humanity, the Word was not always heard and followed. And when the Word fell on deaf ears, the Light that accompanied it became dim and the world grew dark. Was God renewing the Word in my life and my darkness could not understand it?

According to Christian tradition, in the fullness of time, God decided to do something radically new: to send the Word, the Light, into the world as a human being. So on a dark and cold winter night over 2,000 years ago, a baby was born, and the Divine Word, the Eternal Light, came among us and became one of us. Jesus shared the Light and spread the Word wherever he went. He shed the Light on the poor, the sick, the outcast, the oppressed, and the marginalized. He shared the Word with both the powerful and the powerless, those living in the center and on the edge. He showed the way to all who would follow.

The scriptures tell us that he was received by some and rejected by others, so that eventually the wood of the cradle became the wood of the cross. At his death, the Word was silenced and the world once again became dark. But the essence of the Divine Light remained. God's Word among us could not and would not be entombed by death and evil forever. Christ rose from the grave, and with him the Light ascended in the morning sky and the Word was heard again. Was God resurrecting the Light of the Word in me, and I could not hear or see it yet?

Throughout human history, the Word of God has been with us, and the Light of Christ has never been extinguished. It has dimmed in places of war and times of terror, but whenever we act in faith against oppression, hatred, and poverty, we echo the Word and rekindle the Light. Whenever we lead another to God's love, we become a beacon in the night and a flashlight illuminating the way. Whenever we gather together to proclaim the love of God for the world, we become a bonfire of joy and a chorus of angels. Was God helping me to understand the contemplative, creative energy of the dark so that I could appreciate more deeply the Light of the Word?

That Christmas Eve I went to church for our traditional midnight mass. At the end of the service, the lights were turned off, and by the flicker of a single candle I read aloud from the prologue of John's Gospel. As I said the phrase, "The light shines in the darkness, and the darkness did not comprehend it," I thought to myself, maybe the light did not comprehend the darkness either. Maybe it was a mutual misunderstanding between God's two original beloved creatures. Maybe that's where the original power struggle began. And then I settled into the darkness. When we sang "Silent Night" by candlelight in a darkened church that year, for one brief moment time stopped and the world felt safe. The darkness felt safe.

In the prayer book of the Anglican Church in the Province of New Zealand, there is a prayer to be said before retiring for the night.

> *Lord,*
> *It is night.*
>
> *The night is for stillness.*
> > *Let us be still in the presence of God.*
>
> *It is night after a long day.*
> > *What has been done has been done;*
> > *what has not been done has not been done;*
> > *let it be.*
>
> *The night is dark.*
> > *Let our fears of the darkness of the world and of our own lives*
> > *rest in you.*

The night is quiet.
   Let the quietness of your peace enfold us,
   all dear to us,
   and all who have not peace.

The night heralds the dawn.
   Let us look expectantly to a new day,
      new joys,
         new possibilities.

In your name we pray.
Amen[8]

That Christmas Eve I realized that the darkness was not so bad. In fact, in the darkness my fears could rest in God, the quietness of God's peace could enfold me, and I could wait for the dawn of new life and love to be born.

Each year, Trinity Cathedral hosts the Boar's Head and Yule Log Festival, a Christmas tradition dating back to the fourteenth century at Queen's College in Oxford, England. Following the great procession and adoration of the Christ Child, at the end of the festival, the Dean and a young sprite skip out of the darkened cathedral carrying a candle into the night. Some think this is cute, others believe it's irreverent, and the most cynical say it's downright hokey. Personally, as the one appointed to carry out this annual task, I take it very seriously and almost literally. It is not only my responsibility but also my privilege to bound joyously into the world with a child in hand bearing the light of Christ. What a perfect role for one who, as much as is humanly possible, wants to light up the midwinter night and sing out to remind the world that the Word of God is very alive. Sometimes as I skip down the aisle of the cathedral into the darkened night, I feel like singing, "This little light of mine, I'm going let it shine, let it shine, let it shine, let it shine."[9] But instead I give thanks for the interruption of the dark because I now know that the two cannot be separated. Without the darkness, the light cannot shine.

Just Another Homeless Family | PATERSON, NEW JERSEY, 1998

Just Another
Homeless Family

It was early winter, not too cold but cold enough. Becky and Bill had come from Detroit. They had heard there was work in Paterson, and Becky had some family that might be able to help. Anyway, there was no reason to stay in Detroit.

Bill was an unemployed autoworker. He had been laid off and looking for work for over two years. He'd had a few odd jobs: packing cartons in a warehouse, night clerking at a convenience store, washing dishes in a diner. His unemployment had long since run out. He was hoping for an extension, but now it looked hopeless.

Becky once had a steady job in an office, and then at a department store. However, with the local economy suffering from so many plant closures, she couldn't even find a full-time waitress job. Besides, she was now nine months pregnant. Recently, nobody had been willing to hire her.

Life had been hard for the past few years. First, the auto plant had closed. Then the house had sold with just enough profit to replace the transmission on their late model used car, pay some overdue bills, and make a deposit on an apartment—the deposit that the landlord kept after they got evicted for back rent. To make matters worse, their health insurance had expired months ago, and they didn't know how they were going to pay for the baby about to be born. So they packed up the few belongings that hadn't been sold or repossessed and headed for New Jersey.

They had been in Paterson for a few days, staying in a motel on Broadway, the one with the broken coke machine and the sign advertising weekly, daily, and hourly rates, the one across from the big church with the men's shelter. They had received help from its food pantry, but without a kitchen, it had been hard to cook, and eating out had taken every last dime.

Now their money had run out, and they were running on empty. The hotel manager said that they had to go. "NO CREDIT—PAYMENT IN ADVANCE" read the sign behind the Plexiglas barrier at the registration desk. The manager's wife whispered something to her husband about Becky's state of pregnancy, but all she got was a scowl and a muttered, "I told you that we can't keep doing this."

Bill and Becky went down to the local welfare office. Having patiently waited in line for over two hours, they were met by a haggard and hurried social worker trying to get away for the holiday weekend. Politely but curtly, as if to avoid eye contact, she said without even looking up from her desk, "It's going to take a few days for your paperwork to be processed. Come back early next week."

"Where do we stay for now?" they asked. The caseworker opened her jam-packed file cabinet and pulled out a list of family shelters as well as the men's shelter at the big church on Broadway. "What about a place for my wife to stay if we have to be separated?" asked Bill with a worried wrinkle on his brow. Still avoiding eye contact, the social worker shrugged her weary shoulders and replied, "Unfortunately, there are no shelters for women in this county. Our only one was closed a few months ago. Perhaps you can find something in the next county over."

Becky and Bill looked at each other in the determined way couples do, knowing the unspoken thoughts of the other. No, they wouldn't be separated—no matter what. Resolved that somehow things would work out, they left the welfare office and spent all day looking for a shelter that would take them both. They had no luck; the shelters were full.

They were sitting in a restaurant, drinking cups of coffee, staying warm, and wondering what to do next. They needed to find a place to go before dark. They walked out into the parking lot and saw a woman in her mid-twenties standing alone, leaning on a building across the street and staring at the passing traffic. Every now and then, she would run up to an approaching car, yell, "Hi there," and lean into the window to talk with the driver. She seemed like she knew her way around, and she had a friendly smile.

Becky and Bill looked at each other and thought, "What do we have to lose?" They walked up to her and asked: "Do you know where we might stay for the night? We're not from around here; we've run out of money; we're waiting for welfare; the shelters are full; we're expecting a baby; and we've got to find a place—just for tonight." With expectant eyes, they pleaded, "Can you please help us?"

The woman looked at them for a few seconds, and then with a grin she said, "Sure, follow me." Off they went, following this stranger, not knowing exactly where they were going, but knowing that they really had no options left. As they walked, the woman introduced herself, saying her name was Lisa. She barraged Becky and Bill with a constant stream of friendly questions, but interrupted their answers with a running commentary on the neighborhood.

Eventually, Lisa led them to a big, dilapidated house on Van Houten Street. Paint peeled from the broken shingles, garbage filled the overgrown yard, and several abandoned mailboxes hung on the front porch. The house was boarded up, but a piece of plywood had been preyed loose from one of the windows. Lisa and Bill helped Becky through the window and then climbed in behind her.

Once inside, as their eyes adjusted to the darkness of a building without windows, they could see many rooms and lots of stuff in various states of age, dirt, and decay: clothing, newspapers, mattresses, blankets, dishes, pots and pans, beer and wine bottles, along with some discarded syringes and empty crack vials. There was a hose running through the wall from a spigot outside the house next door. Somebody had even hot-wired electricity, thus allowing a few single light bulbs hanging from old ceiling fixtures and wall sconces to light up the interior maze of rooms.

As they looked around, Becky and Bill realized that other people were living in this supposedly abandoned house. Lisa introduced her new friends to the others and explained their situation. She showed them to a soiled mattress surrounded by clothes, pillows, blankets, and bags. "This is my space. You can stay here. Nobody will bother you. I'll be back in a while." And then she crawled out the window they had just crawled in.

Becky and Bill cautiously sat down on the mattress. They were exhausted, too tired to speak and lost in their own thoughts. Bill was reminiscing about days past, better days, and wondering if he would ever see them again. Why did the plant have to close? Why did they travel to New Jersey? Where was their family when they were most needed? And, why, Lord, did Becky have to get pregnant? Leaning his sore back against the dirty wall, Bill recalled the discussion, actually the argument they had about abortion so many months ago. Without the utterance of words, he wondered: "Did we make the right choice?" "Too late now," he concluded.

Angry, frustrated, and scared, Bill's thoughts turned to money. They didn't have enough money for a hotel room, much less a hospital bed. "How will we handle this one?" he asked himself. Evading his own question, he thought, "At least we have a few more days before the baby is due." Bill felt more alone than he had ever felt in his whole life. He just looked at Becky asleep on the mattress and sighed.

Meanwhile, Becky lay quietly on the stained mattress. She couldn't sleep; she was too tired and too scared. Thinking to herself, lots of questions raced through her mind. "Where are we? Who are all these people in this house? Are we safe? Were we foolish to follow Lisa here?" She too remembered the argument in the early days of her pregnancy. "Were we stupid to have this baby?" Glancing over at Bill, she was thankful they were off the streets and relatively warm. "Fortunately, the baby isn't due for a few days. We'll figure things out." Becky fell asleep.

Suddenly, in the middle of the night, Becky awoke to the breaking of her water—all over Lisa's mattress. The contractions began coming fast and furious. Becky was frightened, and Bill didn't know what to do. People in the house began to stir. Someone turned on a broken lamp and brought it over to their corner.

Shouldn't they go to the hospital? This was the question on everybody's mind. But nobody had enough money for a cab, and it was too late to walk. And if they called for an ambulance, they would risk losing their safe haven. "Could she have the baby here?" somebody asked.

Lisa had returned and was frantically running around trying to decide what to do. After all, they were her responsibility now. A middle-aged woman staying in the room upstairs came down. Her name was Pearl. Standing next to her was a sleepy young boy, about the age of five. Pearl looked at Becky and Bill and then at Lisa and declared with the wisdom and authority of age, "When I was growing up, babies were born at home. I guess she'll have to do it here." Taking charge, Pearl instructed Lisa to get some hot water and some towels. She told Bill to calm down and hold Becky's hand.

The contractions started coming harder and faster. Becky was screaming and crying. Bill was shaking. A small group of people staying in the house began to form a circle around them. Lisa waved them away, back into the shadows.

After an hour and a half, Becky pushed hard, and a baby was born. Pearl took the baby, cut the umbilical cord with a kitchen knife, and placed the baby on Becky's breast whispering, "Here's your angel child. He's a boy."

As the group stood quietly around the mattress, each with his or her own thoughts, Pearl's child crept up to Becky and her infant. He leaned over them, kissed the baby on the cheek, and whispered in his ear, "I hope you find a place to live."

Becky gazed at Bill smiling with tears in his eyes. She then looked up at Pearl and her child, Lisa, and all the people standing in the sacred circle. Quietly she asked, "What shall we name this baby?" Lisa smiled and said, "How about Jesu? And on that cold, winter night, in an abandoned house, in a poor city neighborhood, a child was born, a son was given, and his name was Jesús.

Becky and Bill are fictional characters. And yet, I meet them almost every day in my ministry. They come to our churches for food, shelter, clothing, and money and sometimes for prayer and counsel.

Lisa, on the other hand, was a real person. In December 1991, Lisa and her companion, Ivan, were homeless. Actually, they were living in an abandoned storage trailer in the parking lot of a factory across the street from my church. They both had been on the streets for some time, in and out of the shelter and jail systems, and they had become my friends. I had been trying to convince them to get off the streets and into permanent housing. I feared they wouldn't survive the winter months. Each time we talked about it, they laughed and told me not to worry.

On Christmas Eve, I asked Lisa if she would like to be the angel in my Christmas message. In her excited manner, she was delighted. In appreciation, I gave Lisa and Ivan money to have a shower, a meal, a new set of clothes, and a bed for the night in the hotel across the street from the church. I then invited them to attend Christmas Eve services and hear the story.

As I rose to the pulpit that night, I saw Lisa and Ivan sitting in a pew in the middle of the nave. Both of them were freshly showered and wearing relatively clean clothes. When I first mentioned Lisa's name in the sermon, her eyes lit up, and by the end of the story, she was grinning from ear to ear.

The next day, Lisa and Ivan were arrested for trespassing. Because of bench warrants, they were locked up in the county jail and had a warm place to sleep for the next several months. Maybe God was watching out for them. Both Lisa and Ivan have since died and are now real angels in heaven. I know they still love each other.

A Baptism
to Remember

Before performing a baptism, the minister approached the young father and said solemnly, "Baptism is a serious step. Are you prepared for it?" "I think so," the man replied. "My wife has made appetizers, and we have a caterer coming to provide plenty of cookies and cakes for all of our guests." "I don't mean that," the minister responded. "I mean, are you spiritually prepared?" "Oh, sure," came the reply. "I've got a keg of beer and a case of wine."

This stupid pulpit joke makes a serious point. Many of us get our priorities confused when it comes to baptism. We often get caught up in the christening parties and the outfits, and we lose the real meaning of the sacrament into which we're entering. We fail to remember the power of this liminal and formative experience.

How many of us remember our baptism? Do we recollect when or where we were baptized? Can we recount who witnessed this sacred event? Do we remember the priest or the minister who performed the baptism? Do we even recall the water being poured over our head and the invisible sign of the cross being engraved upon our forehead? Most of us probably answer "no" to these questions. We don't remember our baptisms. Many of us were too young to remember since we were baptized as infants or toddlers, long before our conscious memories took shape.

Isn't it a shame that we don't remember this rite of passage, that we can't recollect the promises made or the love offered on that very special day. In fact, most of us can't recall whether we laughed, cried, screamed, smiled, or slept through one of the most important events of our entire life.

Our baptisms are worth remembering, even by reconstruction. Although the act of baptism lasts only a few minutes, and the baptismal party is over in a few hours, the incredible, divine love made visible in our baptism lasts forever. The love that welcomes, bathes, cleanses, redeems, and saves us is also a love that will sustain us all the days of our lives, even unto our deaths. Even though we might not remember the initiation of this love, it forever protects, envelops, and enfolds us. This love that gives meaning to our lives and the world around us *is* worth remembering.

Martin Luther, the great sixteenth-century reformer, believed that the sacrament or sign of baptism was quickly over, but the "spiritual baptism" lasts as long as we live and is completed only in our death.[1] Therefore, Luther intentionally remembered his baptism every morning. When he washed his face he would say, "I am baptized." It was a way of reminding himself that living out his baptism was a daily event.

Once a year, on the Sunday after the Epiphany, the church intentionally remembers Jesus' baptism. On this feast day, we celebrate with baptisms and the renewal of our baptismal covenant. And it gives the preacher a good excuse for talking about the meaning of baptism.

One snowy January morning, I baptized Jamelle Siah Phillips, whose middle name means "firstborn girl." Siah, as she came to be called, was the firstborn and American-born infant daughter of two Liberian immigrants in my Paterson congregation who were seeking political asylum in the United States for the duration of the Liberian civil war. Siah's parents were bright young adults, descendants of slaves returned to West Africa after the American Civil War. They were the great grandchildren of the founders of Liberia. They had come to study in the United States with every intention of returning home to live, work, and raise their family. But circumstances changed, and while they were here, they got caught in the crossfire of war in Liberia and were unable to return home.

The night before Siah's baptism, I saw the powerful, painful, and hopeful film, *Amistad*. The movie is based on a true story about a freedom mutiny on an African slave ship, and the ensuing trial that challenged the very foundation of our legal system by calling into question the basic right of freedom.

At one point in the drama, the leader of the slave revolt, Cinque, was preparing for trial with former President John Quincy Adams, who came out of retirement to fight for the Africans' cause in the United States Supreme Court. They had an extraordinary conversation. John Adams said to Cinque: "We're about to go into battle with a lion that's threatening to rip this country apart. And all we have on our side is a rock. . . . The test ahead of us is an exceptionally difficult one."

Through a translator, Cinque replied: "We won't be going in there alone. My ancestors will be with us. I will call into the past . . . far back to the beginning of time. And I will beg them to come and help me. At the judgement, I will reach back and draw them to me. And they must come. For at this moment, I am the whole reason they have existed at all."[2]

When I heard that line—*I will call upon my ancestors, and they must come. For at this moment, I am the whole reason they have existed at all*—I received a new insight, a new understanding about baptism. Baptism is a sacred moment in time when God, the saints on earth, our ancestors, and all the company of heaven come together as a great cloud of cosmic and earthbound witnesses to welcome a new member into the community of faith, to claim kinship, to declare belonging.

Baptism may be compared to an hourglass—when the past and the future meet in the present; when all the cosmic energy of the universe focuses upon one individual and proclaims: you are a child of

God; you are the culmination of the past, the power of the present, and the hope for the future. In that moment, Siah was the reason we had come together; on that Sunday morning, she held the past in her hand and the future in her heart.

In Jesus' baptism by John, the heavens opened, and the Holy Spirit descended upon him like a dove. And a voice came from heaven proclaiming, "You are my Son; the Beloved; with you I am well pleased" (Mark 1:11, Luke 3:22). When Jesus was baptized, all the cosmic energy, all the God-stuff of the universe came together and descended upon him: firm and clear, yet gentle like a dove. When Jesus was baptized, I imagine that his ancestors in faith—Abraham and Sarah, Isaac and Rebekah, Jacob, Rachel and Leah, Moses and Miriam, Joshua, Naomi and Ruth, Jesse, Eli, Hannah, Samuel, David, Solomon, Job, Isaiah, Jeremiah, Ezekiel, Daniel, Amos, Micah, Jonah, Ezra, Esther, Zachariah, and all those who came before in the name of God—all stood and witnessed this event by River Jordan. For in that moment, *Jesus was the whole reason they had existed at all.*

Throughout the centuries, as individuals have been presented to Christ at the font of baptism, this family lineage we call the apostolic succession—the hand that touched the hand that touched the hand—has grown and expanded across oceans and deserts, through fields and forests, on slave ships, in prisons, and in courtrooms. In the waters of baptism, generations have been drowned and raised to new life, marked with the invisible, yet permanent, brand of the cross. In this sacrament of new birth, generations have affirmed an ancient and yet living covenant to be one with God and one with neighbor, saying "yes" to righteousness and "no" to injustice. In this act of initiation, generations have passed from the chains of bondage to the mantle of freedom. And whenever and wherever baptism takes place, the significance of this action, the power of this sacrament, is the same. God says, "I will have you as my own; you will be my beloved; and I will send Jesus, my anointed one, to be your guide and teacher."

And we respond saying, "I will have you as my own; you will be my God; and I will follow Jesus, your anointed one, as my guide and teacher, my Lord and Savior."

As I held Siah and looked into her eyes, and those of her parents, I saw the future. Preparing to wash a small and helpless infant with the waters of baptism and to anoint her with the oil of chrism (an act which she would not even remember), I exclaimed that perhaps I was baptizing a future president of Liberia. Every Liberian immigrant and every other immigrant in a church filled with immigrants from all over the world smiled, applauded, and proclaimed, "Amen."

On that January morning, during an ordinary Epiphany service, in an Episcopal Church in Paterson, New Jersey, one congregation and its rector realized the promise of the incarnation. Through the baptism of Jamelle Siah Phillips, we came to understand that, "in every child who is born under no matter what circumstances and of no matter what parents, the [potential] of the human race is born again."[3]

When we gathered to baptize this infant, we were not alone. Her ancestors were with us. In fact, the whole company of heaven was present. They had come. For in that moment, as she began her life in Christ, as she started her journey of faith, she was the whole reason they had existed. In her, the good news of the Gospel was born again.

I was baptized as a young adult. I was so embarrassed and self-conscious that I only vaguely recall one of the most important events of my life. So every once in a while I intentionally remember my baptism. I sit down on the floor of my study with a bowl of cool water. I read the baptism service from the Prayer Book. I rehearse the baptismal covenant. I say the prayers for the candidates. I bless the water. And then, I close my eyes and breathe deeply asking God to be with me. I think about Jesus coming to the River Jordan. I imagine his face as his cousin John baptized him with the muddy water. And then, with my eyes still closed I put my hands in the bowl and gently splash water on my face, repeating to myself God's words: "You are my beloved child with whom I am well pleased." As the water wets my face, I remember that God loves me more than I will ever know. Sitting alone on the floor of my study, I remember that no matter what is going on in my life at present, God's love surrounds me, and my ancestors in faith are there to guide and watch over me through life's journey.

The waters of baptism nourish us each and every day whether or not we remember. The mark of baptism is there forever whether or not we remember. We are God's beloved children, whether or not we remember. And that good news is worth remembering.

A Child Shall Lead Us | PERU, 2001

A Child
Shall Lead Us

Martin Bell tells a story, "What the Wind Said to Thajir."[1] Thajir was a little boy who liked the sunshine, the grass, and the trees, and most especially the wind. Whenever there was wind, Thajir would go for a walk. One day the wind spoke to him and told him to remember three basic truths about life. First, "everything that is, is good, and at the center of things, life belongs to life." Second, "decisions that you make today . . . will validate or invalidate everything that has gone before, and make possible or impossible everything that is to come." And the most basic truth, according to the wind: "It is in dying that one lives. It is in losing life that you find it . . . you will be united with the origin and aim of life precisely as you expend yourself on behalf of the entire world."

Thajir took these words to his heart and promised not to forget them. That night Thajir's mother read him a story about elephants. When she finished, she asked him what he recalled about it. Thajir told his mother: "It was most important to remember that whatever hurt elephants hurt him, and that whatever helped elephants helped him. He added that at the center of things, life belonged to life, and this meant that he and the elephants shared in the same experience and were somehow united with one another. He went on to say that it was good to be an elephant, and it was good to be Thajir."

His mother did not question what he had said. But when he went to sleep, she stared at him and said, "I'm afraid for him! Oh God, I'm afraid for him. What will become of my Thajir?

This story reminds me of what Mary must have thought when she found Jesus at the age of twelve sitting in the Temple among the teachers, listening to them and asking questions. Perhaps it is what Elizabeth pondered when she heard Zechariah's prophesy about their son John the Baptist. I can imagine Ann saying the same thing when she learned of her daughter Mary's pregnancy. It is what David's mother probably said to herself when her young son volunteered to slay Goliath. I'm sure Rachel worried as Joseph began to dream. And I can hear Hagar's silent prayer in the wilderness, "Oh God, I'm afraid for him. What will become of my Ishmael?"

Mothers and fathers fear for their children, and for good reason. We do not want to see our children get hurt, but we know it will happen. It is a fact of life. We get hurt. We also do not know what our children will become when they grow up, especially in these uncertain and changing times. And so as we celebrate their lives, we also find ourselves fearing for our children and wondering what will become of them.

But God does amazing things. Edmund McDonald, a Presbyterian minister, once wrote:

> When God wants an important thing done in this world or a wrong righted, [God] goes about it in a very singular way. [God] doesn't release thunderbolts or stir up earthquakes. God simply has a tiny baby born, perhaps of a very humble home, perhaps of a very humble mother. And God puts the idea or purpose into the mother's heart. And she puts it in the baby's mind, and then— God waits. The great events are babies, for each child comes with the message that God is not yet discouraged with humanity, but is still expecting goodwill to become incarnate in each human life.[2]

That's what the wind was saying to Thajir. With each birth, the human potential is born again. With each new day, our own potential is born again. Every day, we are given a chance to start over, to make decisions and take actions that will validate or invalidate everything that has gone before, and make possible or impossible everything that is to come. It really is the wisdom of the twelve-step movement. "Today is the first day of the rest of your life." Today is the first day of the rest of creation. We can start anew. The wind understood this wisdom of grace, and so did Thajir.

Both children and nonhuman creatures have special wisdom to offer human adults about the creation in which we reside. Just look at a group of children lying on their backs, simultaneously soaking up the energy of both earth and sky. Witness a boy throwing pebbles in the water, fascinated with its expanding circles of dynamism. See a girl climb or hug a tree, literally touching its strength and healing power. Stand with a classroom of preschoolers and watch the expressions on their faces as they discover that the plant on the windowsill or the gerbil in the cage has died; they grieve in ways far beyond our understanding. Watch kids mimic animals in the zoo, chase squirrels in the park, or talk back to ducks in the pond. I am convinced that children intuitively understand the first secret of the wind that everything created is good, and it's all interconnected.

Creation itself has much to teach us. In the first place, it teaches that we are made of a common substance, that in the words of the wind, "at the center of things, life belongs to life." We humans come from the earth and to earth we shall return. The dirt (the humus of the earth), the sun, and the water freely expend themselves, so that other creatures might live. They know the wind's most important teaching—the meaning and purpose of sacrificial love—that in losing one's life, one finds it, and that in giving one's life away, one is united with the origin of life.

The creatures of the earth also have much to teach us if we are willing to observe and learn. The tides of the ocean, with their natural clock, show us the structure of time. From the trees, we learn balance. The whales, our close relatives in the sea, show us much about our own immune system. The bats teach us about metabolism, night vision, and living in close quarters. The deer demonstrates the art of adaptation. The eagle shows us how to soar, the loon instructs in the way of tranquility, the rainforest is a dance teacher, the owl demonstrates watchfulness in the night, the wolf gives us permission to howl, and the birds of the sky teach us how to sing. We humans would be richer and wiser if we could learn from the wisdom of the rest of creation.

Unfortunately, many children are growing up without any exposure to or experience in the world of nature and are thus not learning nature's wisdom. Far too many young people never see the ocean or even a lake. They've not walked in the woods or hiked up a mountain. They've not planted a garden or set up a bird feeder. Because far too many children around the globe are growing up alienated from the natural world, their natural intuition is thwarted and their natural learning is inhibited. How will our children know about the importance of protecting natural resources, the environment, and the global commons if they've never experienced it?

And for those children who are exposed to nature as they grow up, they all too often forget what they have learned. Have you ever tried to pull a fourteen-year-old away from the television, the computer, or the phone to go for a walk in the woods? The invasive commercialization of our shared airwaves, sightlines, and public lands are creating havoc and turmoil for those who rear children. I can't count the number of times I've said that I won't take the church youth group to the amusement park. I want to take them to a real park. But there is still hope for our children; it's not too late.

There was a young boy in my congregation who suddenly developed Toxic Shock Syndrome while on a camping vacation with his family. When I first visited Mike in the hospital's intensive care unit, I found frightened and anxious parents and a tired but restless child who wanted to have something to eat and go home. We talked for a few minutes; he asked me if I had any pets. I told him that I had an old cat; he said he wanted a dog and his parents promised him a dog when he came home from hospital. I looked at the stuffed dog lying by his side, and I suggested that maybe this could be his dog while he was in the hospital. He smiled and nodded. As it was time for me to leave, I offered a prayer that the angels watch over him as he slept and that he would have good dreams through the night.

The next morning I visited Mike again. I inquired about his night, and he said he had dreamed about a dog. It was running with him. I suggested that this dog might be trying to help him to get well. He liked that thought. I also told him that lots of people were praying for him to get well. As I spoke these words, I wondered: was the rest of creation praying for him? Were the dogs, the cats, the trees, the sky, and even the wind sending their healing energy to this hospital room?

I visited Mike the next day. This time he was sleeping. When I spoke his name, he opened his eyes slightly. I gave him a rock in the shape of a heart and told him to hold onto this piece of the earth for it had the power to heal. He smiled and went back to sleep. His smile told me that in his young spirit, he understood; and in that moment, I was sure that all of creation was praying for him in whatever way creation prays.

Mike comes from a rich ethnic mix of Native American, European, and African. When I asked him what kind of dog he wanted, he told me that he would like to have a mixed breed like himself because they were smart. I laughed and thought it's probably true. Mixed breeds live on the edge, on the fringes, and thus have a unique perspective on life. Mike was saying something about himself through his choice of a dog. He intuitively was making a connection with the rest of creation, and in doing so, he taught me something about myself. Mike understands the teaching of the wind, that "at the center of things, life belongs to life," and that he and other mixed breeds share in the same experience and are somehow united with one another.

Mike also understands that he has a responsibility to care for the earth and the other creatures of life. Though he might know it in his spirit, he has to learn how to do it from his elders. And therein lies our responsibility. As the book of Proverbs teaches: "Train up children in the right way, and when old, they will not stray" (Prov. 22:6).

Children need to experience firsthand the wonders and joys of creation for if they don't walk in the woods, stand at the water's edge, or talk to the animals, they won't learn creation's wisdom. We have the responsibility to provide those opportunities, especially to urban youth. Children need to learn how to care for the rest of creation, how to be responsible residents of the global commons, and to not be afraid to speak out when the earth is being abused and/or overused. Children need to learn that the decisions they make and the actions they take can and will change the world. We have the responsibility to teach environmental stewardship at home, in school, and in our faith communities. Perhaps most importantly, our children need know that they hold the future in their hand, for with the birth of every child, the human potential is born again. It is our responsibility to remind our children how valuable, how precious, how important they are in our eyes, in the mind of God, and in the rest of creation.

Throughout my ministry, I have taken seriously the presence of children in my congregations. They have been my teachers, my friends, my playmates, and sometimes my soul mates. They have always had a special place in the gathered community around God's table.

At Trinity Cathedral, we have a custom of children con-celebrating with the priest at the nine o'clock service. When the table is set, young children are invited to join the clergy at the altar and the older kids and adults form an ever-growing circle around God's table. Sometimes as many as a dozen toddlers and young children stand around the altar extending their little arms over the plate of bread and the cup of wine. During the offertory anthem, children joyously dance in circles like Miriam after crossing the Red Sea. When the priest lifts her or his arms to invite the people into prayer, the children lift their arms. During the singing of the Sanctus, a few children skip and sway around the altar, unconsciously mimicking the seraphim and cherubim flying about the holy throne. At the Words of Institution, the children all hold their hands out to bless the bread and wine. At the great "Amen," the children clap and stomp their feet in wild and abandoned applause for the goodness of God. During the Lord's Prayer, the children bow their heads and try to recall the words. And sometimes a child will follow the priest as communion is distributed.

One Easter we had about 250 people at the nine o'clock service. There were far too many of us to stand in a circle around the altar. At the announcements, I instructed everyone to stay in their place until it was time to receive communion. But I could see the disappointed looks on some young faces, and so spontaneously, I invited the children to come up and help us bless the bread and wine. "We can't do it without you," I exclaimed. And up came some two or three dozen young children, including a few toddlers and infants in their parents' arms. With the presence of our young assistants, we proceeded to celebrate the Easter feast.

We had a visitor at that nine o'clock service that did not like what he experienced and wrote me an e-mail letting me know what he thought. He told me that he had visited many contemporary worship services in the Episcopal Church, but he had never seen one where the priest actually invited the children to "con-celebrate" the Eucharist. He went on to ask me since when did an Episcopal priest announce that she couldn't bless the bread and the wine without the assistance of children. I wrote back politely explaining that this was the custom of the nine o'clock service and perhaps he should try our more formal and traditional worship at eleven o'clock. To my knowledge, he has never visited Trinity Cathedral again.

Now, the truth is that of course I could bless the bread and wine and celebrate communion without the assistance of children, but I did not want to preside at the altar without our children. It would not have been the nine o'clock service had I not invited the children to join us at God's holy table. Let me tell you why.

A few years ago, when I was still new to the cathedral, two families from another city in the diocese came to our then relatively new nine o'clock service. One family was Roman Catholic, and they were trying to find a home in the Episcopal Church. I welcomed them to the Cathedral and asked what brought them such a distance. They explained that their daughter had been saying that she wanted to be a priest when she grew up, but she had never seen a woman priest. So they came to see the woman priest and dean of the cathedral in Cleveland. I smiled and introduced myself to this shy girl standing by her parents.

When it came time for communion, I asked my young visitor if she would like to stand with me at the altar. Without hesitation, she responded in the affirmative. And up she came. Then something remarkable happened. As I started to lift my arms to pray, she lifted her arms. When I raised my hands over the bread and wine, she raised her hands. As I celebrated communion, she intensely mimicked every move I made. Another little girl joined her; this one danced around the altar. At the end of the service, both sets of parents both told me what an extraordinary gift I had given their children.

And thus it began. After that, I always invited the children to come up to the altar and help at the nine o'clock service. Then my cathedral clergy colleagues started inviting the children to help them. Within a few months, children at the altar were a custom at Trinity Cathedral and some kids didn't even need to be invited. They just came on up. The nine o'clock service was growing, and many people were coming because of the children. Congregants and visitors alike would comment, "This is remarkable. I feel the intense power of the spirit in the presence of the children at the altar." How could I dare to change that on Easter morning?

Over the past few years, I believe our children literally have transformed our nine o'clock worship.

Liturgical dancers, complete with blinking shoes, have blessed us. We've had dolls sitting on the altar with their arms and hands held out to bless the bread and wine—what better place for Barbie in her ridiculous fashion or GI Joe in his combat uniform to sit. One young con-celebrant often wears a pirate hat to church. Older siblings lift younger ones up to reach the altar. We've had plenty of fingerprints on the altar linens and smudges on the silver but never a complaint from the Altar Guild; frankly, I think that's a miracle in an Episcopal cathedral. There's usually a little bit of chaos, but always a lot of joy, laughter, tears, and grace at God's table.

If the truth were told, I am no longer sure that I can celebrate communion at nine o'clock without the assistance of my younger companions. Why? Our children teach us something essential about the gospel. In their very presence at God's holy table, they embody Jesus' words: "Whoever welcomes one such child in my name welcomes me, and whoever welcomes me welcomes not me but the one who sent me" (Mark 9:36–37).

For centuries scholars have pondered what on earth did Jesus mean when he invoked the child? I think Jesus choose a child as his teaching tool for two reasons: to stand in solidarity with the least and to hold up the openness and imagination of children as a model of faith.

Children, in both Jesus' society and ours, signify the least of our brothers and sisters. In the Greco-Roman world, children were the lowest of the low, servants of the servants, and the very last members of society. They had no rights, and for most children, there was no such thing as an easy childhood. Unfortunately, that's true for so many children around the world today. Even here, in the United States—the wealthiest country in the world—approximately one hundred thousand children are homeless; over ten million children have no health insurance; and almost fifteen million children live in poverty.[3] As child advocate Marian Wright Edelman reminds us, we who live in the richest nation of the world make our children "the poorest group of citizens."

And yet children—those who are the most vulnerable, the most dependent on others—can be the most powerful teachers of us all. At the altar on Sunday mornings, our children are ministers of God's sacraments. Their hands, eyes, arms, and breath bless the bread and wine and invoke the very presence of Christ in our midst.

As one who does not have biological children, I need children around me, especially the children in the congregations I serve. I depend on them to enlarge my heart, to make me unselfish, to remind me to laugh and smile, to keep me in touch with the childlike grace and openness of Jesus, and to teach me the truth of the gospel.

During the 1995 heresy trial in the Episcopal Church, when I made my public witness about being an openly gay priest, it was the children who disarmed the media. At St. Paul's Church in Paterson, the children went to Sunday school for the first part of the worship service and then joined us for communion. On that particular Sunday, my sermon was lengthy and the children appeared just as it was ending. At its close, the congregation rose to its feet with a strong and sustained standing ovation. I was overwhelmed with emotion and started to cry. By the time I walked down the steps of that grand old pulpit, I was weeping. One of our children exclaimed, "Look, Rev. Tracey is crying." The next thing I knew, my young friends surrounded me. They were hugging my legs so hard that I dropped to the ground and there we sat. That was the photograph above the fold on the front page of the next morning's newspaper whose headline read, "Congregation Embraces Gay Priest."

Children don't care whether or not I am gay or straight. I'm just "Rev. Tracey" to them. I make faces at them through the door during Sunday school and wink at them during a long scripture reading or a boring hymn. I invite them to hold the water pitcher as we baptize their younger siblings. I explain to them that they might see their parents cry during a funeral, and then ask them to put dirt on the grave of the departed grandparent. I stoop down to listen as they tell me an important story during coffee hour, and I tell them stories during the sermon.

In the words of Frederick Buechner, "Children live with their hands more open than their fists clenched . . .they are so relatively unburdened by preconceptions that if somebody says there's a pot of gold at the end of the rainbow, they are perfectly willing to go take a look for themselves . . ."[4] And so, on that memorable Sunday morning, the children of St. Paul's saw their priest crying and they went to take a look themselves and give her a hug.

We think that our children don't pay attention to what is done and said in church. But I know for a fact that they do listen and they incorporate what is proclaimed in both word and deed. One time a father called me during the week to tell me what his kids were doing. They had created a new game. It involved going to the kitchen with their wagon and filling it up with canned food. Then they went to the hall closet and put gloves and hats in the wagon. Their parents could not figure out what they were doing. Eventually they asked and were told, "We're taking food and clothes to the homeless, just like Rev. Tracey told us to do in her sermon."

On more than one occasion, I've had parents report to me that their children were playing communion at home with their dolls and stuffed toys. They would set a little table with cookies and juice and then share it with their imaginary congregation. I'm willing to bet that if encouraged, a few of them will enter the ordained ministry as adults.

My favorite kid story of all time is about the little boy who stood on his head during communion one Sunday. We were just about to bless the bread and the wine when a young child did a headstand right in the middle of the sacred circle. I don't know why he did it, but it was funny. We all laughed, continued with our worship, and thought nothing of it. But it has stayed with me for a very long time and it has helped me to understand an important dream I had during seminary. I was in a very big church with lots of people. During the offertory anthem, the ministers of communion processed to the altar walking upside down on their hands with their vestments blowing in the breeze, and hundreds of ducks came up to receive communion. I could never comprehend the meaning of that dream. But some twenty years later, with the unconscious assistance of a child, I now understand its sacred message. The ministry of Jesus is about turning the world upside down, living the gospel with imagination, treating all of creation with radical hospitality, and receiving the wisdom of the wind.

The prophet Isaiah proclaimed the hope of God that "the wolf shall live with the lamb, the leopard shall lie down with the kid, the calf and the lion and the fatling together, and a little child shall lead them" (Isa. 11:6). I depend on our children to embody this hope: to make us laugh when we're taking ourselves too seriously; to say "Amen" like we mean it; and to cry when we hurt, to frown when we're sad, and to smile when we're happy. I rely on the children in my life to remind me to share the gifts of God's bounty with the rest of God's creation; to learn from the other creatures; to listen to wisdom of the wind; and sometimes to stand on our heads in order to see the world from upside down. I am convinced that our children will lead us to this great shalom if we are willing to follow.

Two

It lived and died....

May I & Cohen |ORADOUR-SUR-GLANE, FRANCE, 1997

Why I Do Ashes

The other day I reminded a friend that Ash Wednesday was fast approaching. She quickly responded that she didn't do ashes, and then she told me a funny story. She said that one year, she was walking downtown around noon, and she saw a woman walk by with a big bruise on her forehead. She thought, "Wow! That woman really slammed into something." A moment later, she saw a man walk by with a big bruise on his forehead, and she thought, "What a coincidence. Perhaps, there's a low-hanging branch over the sidewalk." Immediately, she saw yet another person walk by with a bruise on the forehead, and she thought, "What's up with this?" And then she looked up and realized that she was walking by a church, and people were coming out with ashes on their foreheads. It was Ash Wednesday.

A lot of people don't understand Ash Wednesday and "don't do ashes." Frankly, I think Ash Wednesday is one of the most powerful liturgies of the entire church year, and I believe that marking our foreheads with ashes is one of the most powerful rituals in our tradition.

For me Ash Wednesday really means a right beginning. It truly is a chance to start over, to say with absolute certainty and resolute assurance that today is the first day of the rest of our lives. I don't know about you, but I need starting over days and new beginnings. I also need symbols and rituals to mark these rites of passage. Ashes are the powerful symbol of the starting over as we pray, "Create in us a clean heart, O God, and put a new and right spirit within us." They remind us of who we are and to whom we belong.

We have forgotten who we are. We have forgotten our rightful place in the great scheme of things. And who are we? According to the Book of Genesis, we are creatures of the earth—fallen, broken, and disobedient—but nevertheless loved by our creator, the one who made us. In the words of Phyllis Trible, the story of Adam and Eve is "a love story gone awry."[1]

According to one of the creation myths of our tradition, in the beginning, God created *Adam*, literally translated as "the earth creature."[2] In the beginning, *Adam*, the first human being, was created in the image of God. But Adam, a creature without gender, needed a companion, a helper, and a partner in order to carry out the God-given task of stewardship for the earth. So God, innovative and creative, took from Adam a rib and made another earth creature. In creating *Eve*, the second human being, God gave gender to creation. For until there was a woman, there was no man, only a lonely, genderless earth creature.

At first, life in the garden for Adam and Eve was very good. They were naked and yet not ashamed. There was plenty of food to eat and lots of animals to tend. But over time, things went bad.

Both Adam and Eve forgot who they were. In reaching for the forbidden fruit from the tree in the middle of the garden, Adam and Eve were the first human beings to seek their own security, to exploit creation for their own ends, to distort their God-given knowledge, and to abuse their God-given power. They got too big for britches they weren't even wearing. And in forgetting who they were, they wanted to become something they weren't. They wanted to be like God. They wanted to have the knowledge of God. And so they ate from the forbidden tree of good and evil and had to pay the very expensive price for their stolen meal.

The cost of this forbidden fruit was expulsion from the garden. Adam and Eve were forced into the world to learn how to use the knowledge of good and evil, to learn how to live as free will agents of God.

As they left the security of the garden, Adam and Eve began the long journey of remembering who they were—who God intended them and their children to be and to become.

Who are we? According to the Episcopal burial office, "We are dust, and to dust we shall we return."[3] We are still earth creatures. As the psalmist reminds us, we have been made "a little lower than God . . . crowned with glory and honor . . . [and] given dominion over the works of [God's] hands" (Ps. 8:5–6).

According to religious tradition passed down from generation to generation for thousands of years, we human beings are both insignificant in the context of all creation and of utter importance to the God who created it all. On the one hand, we are so small in comparison with the grandeur and splendor of God's creation. We might live seventy, eighty, or at the very most a hundred years, but the stars last for billions of years. As God reminded Job, the ocean waves, the prairie winds, and the daily sunrise and sunset happen whether we're here or not. At the same time, we have been given the unique opportunity and obligation to care for God's precious earth and all that inhabit it. Unlike any other creature, we have the particular vocation of stewardship—the responsibility for taking care of something that ultimately does not belong to us. We are the stewards of God's great vineyard we call planet earth. We have the capacity to destroy our planet or to make it a better place to live. We have the capability to create life or to kill life. We have the ability to form loving relationships and to destroy those relationships. We have the power to build nations and to take them apart overnight. We are no different from Adam and Eve; after all, we are their descendants. And because we are of their bloodline, we too have the knowledge of good and evil, right and wrong.

So what have we forgotten? According to the United Nations Environmental Sabbath Declaration, we have forgotten our place in the great scheme of things. We have forgotten our connection to all the other parts of the created order. We have gotten too big for the britches we now wear. We have forgotten that tuna comes from the ocean and not from a can, that beef was once a brown-eyed cow, and that picking lettuce is back breaking work. We have forgotten that trees existed before roads, that deer and rabbits lived in our backyards before there were houses, and that birds must fly in the sky we pollute. We have lost our place in creation.

St. Francis of Assisi remembered who he was. After returning from the Crusades, through a genuine, near-death, religious experience, Francis opened his eyes and saw for himself how the responsibility of stewardship had become, in his medieval world of wealth, a violent act of domination. St. Francis chose to abandon his privileged place in society and to assume a new place among the poorer creatures of God. In the humility of remembering who he was, St. Francis was able to praise God fully, and in praising God, he was able to serve God's creation.

Our place in creation is one of servanthood. We have the awesome responsibility for serving God by caring for God's earth. This is our particular calling as human beings. As servants of God and God's created order, we have a vocation to care for the earth, to care for the creatures of the earth, and to care for our relationships on the earth. It is a sacred vocation, and we need to remember it.

We need to remember who we are. We need to reconnect ourselves to the unfolding of the cosmos. We need to become reacquainted with the movement of the earth. We need to pay attention to the cycles of life and love.

We need to remember whose we are. We need to seek not our own security but the well-being of the whole created order. We need to stop exploiting the earth for our own ends. We need to reorder our knowledge and stop abusing our power.

We need to remember where we are. We need to remember that much of our land is barren, much of our water is poisoned, and much of our air is polluted. We need to remember that our forests are dying, our creatures are disappearing, and our human neighbors are despairing. Our common habitat, the earth, is suffering in large part from our neglect, abuse, overuse, and exploitation of it.

We need to remember who we are before it's too late. We need to ask forgiveness. We need to ask for the strength to change. We need to believe that we can change. We need to change the way we live on this earth.

How might we remember who we are? We can begin by recalling the majesty of God and the reason why God created us. For remembering who God is and why we were created is the first step in becoming who we're meant to be. And when we become the cocreators that God intends us to be, we will be able to alter the way we live in the world before it is too late.

Ashes mark our frail humanity, our connection to this planet, and our relationship to Adam and Eve, who were symbolically formed out of the dust of the earth and molded out of the clay by God's hands. Ashes are mentioned throughout the Bible. In Genesis, we hear our forefather Abraham say, "I who am but dust and ashes." Tamar, after she was raped, put on ashes in mourning. Job, as his life fell apart, put on the ashes of grief. Almost all of the prophets speak of ashes and dust. It is right and good that we mark our foreheads with ashes to remind ourselves that our God is here—ready, willing, and able to receive our repentance, our anger, and our grief—and to offer in return divine forgiveness, comfort, love, and hope.

If we ever wondered about our mortality, if we ever doubted that we were made of dust and ash, September 11, 2001, took those wonders and doubts away. A friend whose daughter lives in New York City told me that after 9/11 she received a panicked phone call from a distressed young woman. This well-educated college student almost screamed into the phone, "Mom, don't you get it? We're breathing in the ashes of human remains. The dust of lost lives is literally floating through the air."

Yes, there is a lot of dust and ash floating about these days, as there are lots of wounded souls trying to start over. There always have been and always will be. As a people, we need to return to God. We need to make the connection with the pain and brokenness in our own lives, and that of our families, friends, and communities. We also need to get connected with the pain and turmoil of our nation, our world, and the rest of God's creation.

Ash Wednesday is about making the connections with God, our neighbors, and ourselves. Ash Wednesday is about reminding ourselves that everything we do and say has an impact on somebody else.

The prophet Isaiah, writing thousands of years ago, said it well. God calls us to a fast—a fast of making justice, loving kindness, and walking humbly with God (Mic. 6:8, paraphrased). And if we take these steps, "[we] shall be like a watered garden . . . [our] ancient ruins shall be rebuilt, [we] shall raise up the foundations of many generations, and [we] shall be called the repairer of the breach, the restorer of streets" (Isa. 58:11–12). It seems that when we live the fast of God, we feast. What an interesting paradox. The Scriptures actually suggest that when we make the connections, when we live for those less fortunate than ourselves, we ourselves become rich in God-terms. For as Jesus reminds us, "where your treasure is, there your heart will be also" (Matt. 6:21).

On Ash Wednesday, we pray a litany of penitence in which, like on Yom Kippur in Judaism, each petition acknowledges an element of our broken and wounded soul. We confess that we have not loved God, our neighbors, and ourselves with our whole heart, mind, and spirit. We acknowledge that we have sometimes not heard the call to serve. We admit that we have been unfaithful, prideful, hypocritical, and impatient. We disclose that we have been self-indulgent and exploitative. We declare that we have been angry, envious, and dishonest. And guess what—it's true. The truth is that, like the collect for Ash Wednesday, sometime during the past year, we have been sinful and wretched, and we need to be forgiven and made whole. In short, on Ash Wednesday, we plead that God "create and make in us new and contrite hearts."[4]

I'm not sure why people don't like Ash Wednesday and don't do ashes. I for one am grateful for the opportunity to get down on my knees and say—God, please let me start over. Please give me a second chance. And once again, mark my forehead so that I may remember whence I came and where I am going.

The Preacher's Prayer on the Eve of War | BOSTON, MASSACHUSETTS, 2001

The Preacher's Prayer
on the Eve of War

It was the week in 2003 when our nation went to war with Iraq. Everyone was caught up in prewar madness. Even as I planned peace vigils and prayer services at the cathedral, I felt like I was part of the war effort, the pregame show, accepting with fatalism the inevitable while hoping and praying for a miracle. I also felt like the prophet Jeremiah, carrying a heavy load of powerlessness, grief, and predestined doom. I sensed that in a few hours or a few days, Rachel would once again be weeping for her children, who "are no more" (Jer. 31:15). Like the prophet of old, I longed for a "balm in Gilead" (Jer. 8:22) that would restore health and wholeness to our broken world.

Later that evening, as sandstorms in the desert continued to blow and impede the progress of military troops, a priest in our congregation remarked, "Maybe this is the miracle we've been waiting for." Yes maybe, I thought, this is the work of *Ruach*—God's spirit, God's breath, God's wind—preventing this war from becoming reality. My hopefulness rallied for an hour and then plunged as I learned that the first bombs had dropped on Baghdad. No miracle, I'm afraid. Neither sand, nor wind, nor sun, nor dark of night could stop this war from happening before our eyes on television. The war was accompanied by an extraordinary stock market rally.

By Thursday, the day after the first bombs fell, antiwar demonstrators were in the streets, the Iraqi desert was littered with convoys of troops, and downtown Baghdad was in flames. On Friday, as a thousand missiles dropped from the sky, tens of thousands of troops marched through the desert, and hundreds of thousands of antiwar demonstrators took to the streets, our city's religious and civic leaders gathered to pray in the cathedral. The stock market continued to climb in record fashion.

On Saturday, I returned home from a funeral to learn that more bombs had dropped, more people had taken to the streets in demonstrations of protest and support, and more soldiers and news reporters had been killed in battle. Oh, and yes, the polls indicated an increased American endorsement for the war effort. The stock market was closed for the weekend.

And then on Sunday, I had to stand in the pulpit and try to make sense of it all. I was not sure what I should say. I was angry that our nation had initiated this war, but I did not want to blame our military men and women and all those in harm's way. I was angry with my nation's president, but I sure didn't like Saddam Hussein. I was angry that the stock market had applauded this war, but I was relieved that my retirement savings might begin to reclaim some value. I was angry that we were celebrating the capture of oil fields, but I was grateful that they would not be set on fire. I was angry that the media was having a field day with its war coverage, but I found myself glued to the television set. I was angry that antiterrorism measures were eroding our civil liberties, but I did not want to experience another September 11. I was angry that our nation was discounting, maybe even dismantling the legitimacy of the United Nations, but I did not want to seem overreactive. I was angry that preemptive war was becoming the new style of U.S. diplomacy, but I am not an expert in foreign affairs. I was an angry American, but I did not want to appear unpatriotic. In fact, what I was really angry about was that those who did not support this war effort were considered in some circles as naïve and unpatriotic, while those who supported this war were considered knowledgeable and patriotic.

Anger is an emotion that we're usually told not to express. But on the Sunday following the bombing of Iraq, the appointed reading from John's Gospel (2:13–22) was serendipitously about the anger of Jesus and the actions it led him to take. In the gospel account of his life and ministry, Jesus showed us that anger is like fire: it can be used to clear a field and rejuvenate it for the next planting, or it can destroy an entire forest or desert. The fire ignited in Jesus' belly was hot, long-lasting passion for peace and justice fueled by righteousness, grounded in prayer.

In John's gospel account, Jesus arrived in Jerusalem the day before Passover. It was like the day before Christmas in a suburban mall. The streets were bustling with activity: women were picking up last minute items for the Seder meal, children were running around excitedly poking their heads in and out of shops, and men were gathered in small groups talking about politics.

The Temple, located in the center of it all, was equally crazed. What should have been a place of prayer had been turned into a sprawling religious market, roughly the size of two football fields. There were thousands of pilgrims, hundreds of merchants selling animals for sacrifice, and yes, the moneychangers.

In first century Judaism, animal sacrifices were required for Temple worship. A faithful Jew had to make an annual pilgrimage to Jerusalem to offer ritual sacrifices and pay the yearly temple tax. Because the tax could only be paid with shekels, Roman currency had to be exchanged, and the moneychangers demanded a ridiculous surcharge for this essential service. They remind me of check cashing businesses that charge exorbitant interest and fees in poor communities where there are no banks, or predatory lenders that prey on ill-informed borrowers.

As for the Temple market, it wasn't like the café, gift shop, or bookstore on our cathedral campus. These merchants supplied the sheep, oxen, turtledoves, and pigeons certified by the rabbis as "free from blemish" and thus acceptable for sacrifice in the Temple. What began as a service of convenience for out-of-towners evolved into an abusive, expensive, and exploitative practice, which would today have been exposed by Dan Rather on "Sixty Minutes."

On the day before Passover, the Temple was a three-ring circus filled with corruption. No wonder Jesus got angry. Jesus' anger was a kind of righteous anger that came from loss or grief rooted in memory and history. It was the anger that propelled Moses to kill the Egyptian. It was the fiery passion of the burning bush, anger that does not consume itself or those around it. It was the anger of God who saw the people suffering in Egypt and caused the divine voice to say, "Let my people go." Jesus was angry that the Passover, a feast of liberation, had turned into a religiously licensed activity of economic exploitation.

His righteous anger led him to an act of public engagement. He went into the Temple and, with a whip of cords, drove out the animals and their merchants, and then overturned the tables of the moneychangers. As he acted, he yelled with great passion, "Take these things out of here!"

Jesus' action in the Temple was a symbolic one. He knew that within a day or two, business would resume. It would all be back in operation. Like any demonstration, he had not solved the problem; rather, he had simply disrupted the situation, bringing attention to the injustice. Was his action successful? The answer depends on what his intentions were. If the purpose of his public action was to get a reaction, to be recognized, he was successful. In John's Gospel, many came to believe in his power. In Matthew, Mark, and Luke, Jesus was arrested following his temple action. Were the protest marches and demonstrations before and during the war on Iraq successful? Again, the answer depends on the intentions of the protesters. For those who were trying to prevent the war, one could say they were not successful. For those who wanted to make a public witness of their disagreement with the actions of our nation, one might say their intentions were achieved.

Successful or not, Jesus' action, and that of many of the protesters against the war on Iraq, was rooted in prayer. Roland Gittelsohn, a Jewish theologian, once wrote: "By far the most important outcome of prayer is action. Without ensuing action, even the most eloquent prayer is like a stage set for a symphony concert—instruments tuned and in place, music opened to the proper score, audience pleasantly expectant—but the musicians' chairs occupied by men who cannot play."[1] If prayer is talking with God about our deepest thoughts, our anger, hurt, joy, passion, and love, then action is the most important outcome of prayer.

There certainly was a lot of action in the week our nation went to war with Iraq. Bombs were dropped and troops were deployed. Journalists were embedded with military units. Civilians prepared for attack, and humanitarian organizations prepared for the aftermath. Peace activists prepared for the day-after demonstrations, and police departments prepared for local public safety. Politicians prepared remarks for the press, and clergy prepared prayers for the people. It is said that our president and his cabinet are men and women of prayer and that their actions were grounded in prayer. I was not about to argue that point. But what I wanted to say from the pulpit was that the expression of anger in the form of peaceful demonstrations and nonviolent civil disobedience is not the betrayal of patriotism. Rather, it too is the expression of prayer.

One of the great gifts of our country is our freedom to publicly express dissent. As we entered this new phase of international engagement, I prayed a lot. I prayed for military forces and their families. I prayed for the people of Iraq and their neighbors. I prayed for the journalists, relief workers, and all in harm's way. I prayed for the leaders of the nations, especially our president and his cabinet, that they would make wise and just decisions that would lead to peace. I prayed for safety workers protecting us at home. I prayed for immigrants, especially of Arab descent, who were worried about their civil liberties. I prayed for frightened children and worried parents around the world. I also prayed for those who were expressing dissent; I prayed that they would not be called unpatriotic, and that they in turn would not condemn those in military service.

During the week we went to war with Iraq, lots of people asked me, "What do you think Jesus would do or say right now?" How could I be certain what a first-century peasant rabbi would say about this twenty-first century crisis? I believed that the spirit of Christ was in the turmoil with us—on battlefields, in frightened towns and villages, in public squares around the world—calling the world to peace and reconciliation, even in the midst of war and destruction.

And I prayed that somehow we would hear and respond to Christ's call for a just peace before it was too late. Isn't that always the preacher's prayer during a time of war?

The Garbage Tree | PATERSON, NEW JERSEY, 1997

The Garbage
Tree

On a cold, clear day in the middle of March, I decided to go look for God through a camera lens. I started walking around my church neighborhood. Though it was a familiar landscape that I walked almost every day of the week, with my camera in hand, the bleak streets, broken sidewalks, vacant lots, abandoned buildings, and run-down houses looked different to me. I began seeing details that I had not noticed before.

One such previously ignored and unnoticed detail was a tree on Ellison Street. It was big and tall. Buds were just beginning to form on its leafless branches. It leaned a bit toward a rooming house on the south side of the street, then it shot straight up in the air. The branches on the street side had been clipped and removed. The branches overhanging the sidewalk, interspersed with telephone cables and electric wires, were a sight to behold. They were a virtual store of abandoned bicycle tires, toilet paper, a plastic garbage bag, an old mop, and a doll with her face cut off from the rest of her head. The bike tires were looped over branches, and the toilet paper looked like Christmas tree tinsel. The plastic bag contained some undisclosed object and was tied up in a knot. The doll's leg rested on a branch and her upper body and head were literally trapped between three wires. The mop looked like hair, and at first glance, I thought that this inanimate human replica had not only been beheaded but also scalped. Perhaps, this was a tree symbolizing some exotic religious ritual I did not know. Next to the tree, hanging from the wires was a pair of boots, symbolizing a ghetto ritual I understood all too well; they signaled the presence of a local drug dealer.

As I was examining the tree through my camera lens, three young boys approached me. "Rev. Tracey, what are you looking at?" they asked. I pointed to the tree and said, "Isn't that cool?" They looked up at the tree and its paraphernalia; then they looked at each other; and then they looked back at me as if I were crazy. "It's just some dumb old tree that people were throwing junk in," one said. "What's so special about a garbage tree?" another inquired winking at his friends. "And *you know* what those boots are about, don't ya?" the third one said, glancing over at his buddies. "Yes, I know what the boots are about," I responded. "So why are you taking a picture of it?" one asked. "Because I see God in that tree," I answered. "God? I don't see God. Where is he?" "God's everywhere, even in you," I laughed as I pointed to my young friends. "Take a picture of us," another said; and I did. In front of that old tree on a snowy sidewalk, I made photographs of all three boys and sent them on their way so that I could continue to make a portrait of the tree. Without a doubt, I had found God through the lens of my camera that morning.

I quickly took the roll of photos to be developed. When I went to pick up the prints, the camera store attendant said, "Where did you see that tree?" As we reviewed my newly minted photographs, I told her about the tree on Ellison Street and about the boys on the block. I took the opportunity to tell her all about my church and its neighborhood. On Sunday, the camera store woman showed up in the pews and two of the boys showed up at the door.

On my daily walks around the neighborhood I continued to look at the tree, and then back in my office, I reviewed my photographs. As the weeks and months wore on, leaves grew, hiding the garbage. In the autumn, when the wind blew the leaves off the tree, its unusual collection of items began to fall as well. Eventually, the tree was devoid of its interesting and provocative display. But the boots in the wires remained.

My tree on Ellison Street symbolizes many things for me: poverty, despair, abandonment, death, life, humor, creativity, and urban energy. Eventually, I enlarged one of the photographs, and it became the first serious expression of my avocation as a photographer. I gave it different titles. Initially, it was called "Valley of the Dolls," for it reminded me of that drug-ridden world the book and film portrayed. One Lent, as I was working on a sermon about the prophet Ezekiel, I renamed the photograph "Dry Bones," because it reminded me of the awesome power of God to enflesh a skeleton and rebuild a broken world. Finally, one Advent, as I was preparing for my first photography exhibit, I again retitled my now beloved tree. I decided to name it "Jesse's Tree," in honor of both my nephew named Jesse and his spiritual ancestor, the father of David.

Trees are essential for life. They provide food, shelter, water, and even the oxygen we breathe. Many trees also teach us about the cycle of nature as they seem to die in the fall and are reborn in the spring. Others defy the rhythm of the seasons, remaining green all year round. The health of trees is often an indicator about the health and vitality of our environment, and the elimination of ancient forests is one of the leading concerns of the environmental movement. Because of their deep and abiding connection to the earth's energy, ancient people (and some modern folks as well) looked to the tree as a source of divine power with mystical, life-giving virtues.

It is no wonder that trees are central to the story of our faith. The tree of life and the tree of knowledge, both planted in the Garden of Eden, form the boundaries of human existence. God's instructions to Adam, the first human being were "You may freely eat of every tree of the garden; but of the tree of the knowledge of good and evil you shall not eat, for in the day that you eat of it you shall die" (Gen. 2:16–17). Our primal ancestors ate the fruit of the tree of knowledge and were expelled from the garden of ease into the world of toil on the land from which they were formed. And the Eternal One placed an angel and a flaming sword to "guard the way to the tree of life" (Gen. 3:24).

From the beginning of our sacred story, trees have symbolized God's presence. God introduced the divine self to Moses through a burning bush, protected Elijah under a solitary broom tree, used a tree to explain to Job the regenerative power of creation, and appointed a shade plant to teach Jonah about compassion. Consistent with his spiritual forebears, Isaiah employed the metaphor of a tree to predict the coming of God's anointed one. Jeremiah compared those who trust in God to a tree planted by water that does not fear the drought or become barren of fruit. Daniel envisioned a great, strong, tall tree at the center of the earth whose "foliage was beautiful, its fruit abundant, and it provided food for all . . ." (Daniel 4:12).

In the Christian tradition, a shoot from a stump and a branch growing out of roots (Isa. 11:1) came to symbolize the dawn of a new age in the promise of a Messiah and the coming light of Christ. The Jesse tree, as it came to be known, represented Jesus' family tree, connecting Jesus, son of Mary, to the royal line of David and his father Jesse. Over the centuries, the Jesse tree morphed into what we now call the Christmas tree. It amazes me that to this day a fir tree, complete with sparkling lights and colorful decorations, is one of the few mystical, magical, and mythical traditions that has not been destroyed by modernity. But the story of trees does not end with Christmas.

Two trees, a palm tree and a cross tree, bracket the Holy Week before Easter. The first, a royal date palm, tall and green, is a symbol of power and hope. In ancient times, a person broke its branches and used them to signal the route for a king to arrive in royal splendor. Excited crowds waved the palms and placed them on the ground below the king's feet. The second is an anonymous tree—lowly, dry, dead, and leafless—cut up and carved into a cross, an instrument of execution. A person was nailed to this tree: stripped, beaten, and left to die. Birds roost in both trees. In the first, they make their nest. On the second, they find their prey.

The jolting juxtaposition of these two disparate trees is the backdrop to Palm Sunday and Good Friday, respectively. The lush, royal palm tree and the stark, barren cross tree represent the two visible faces of God incarnate—the God of saving power and the God of suffering pain. These two trees also symbolize the complicated and unique truth about humanity—our ability to choose good and evil, life and death. As the Gospel narrative demonstrates, Jesse trees evolve into palm trees and become cross trees when people are seduced by the evil, silent to the pain, indifferent to the suffering, passive to the injustice, and overwhelmed by the oppression that surrounds and confronts us.

The betrayal, arrest, trial, and crucifixion of Jesus was a complicated affair. No one can deny that fear, expediency, apathy, cynicism, indifference, grief, and loyalty are all elements of the passion narrative. Those who ran away at the time of his arrest were fearful, those who tried and sentenced him were politically expedient, those who ignored him on the cross were apathetic, those who challenged him to save himself were cynical, those who played lots for his clothing were indifferent, those who wept by the cross were in grief, and those who buried him were loyal. And it happens over and over again. All you have to do is read the morning newspaper.

For instance, what prompts a state legislature to adopt conceal-and-carry handgun legislation? Is it fear? Too many people are afraid and think that carrying concealed handguns will protect them. Is it political expediency? Too many state legislators do not vote their conscience. Or, is it apathy and ignorance? Those who see the madness in allowing every citizen to carry a concealed handgun don't speak up.

Why did the "War on Terrorism" get so out of control that our political leaders enacted the Patriot Act? Is it because they believed that arresting suspicious-looking people and locking them up without just cause and due process would actually prevent terrorism? Or, was it that too many citizens were overwhelmed, frightened, and afraid to speak their minds because we did not want to be accused of being unpatriotic or naïve?

Why is there so much talk about a Defense of Marriage Act that actually forbids states from legalizing same-sex marriage? Do our politicians really believe that validating gay and lesbian relationships will threaten the institution of heterosexual marriage? Or are we all so afraid of human sexuality, that sacred mystery that nobody really understands, that we believe we can legislate family stability when what we are really enacting is discrimination, exclusion, and bigotry?

This juxtaposition of good and evil is how my garbage tree came to be. A long time ago, a tree was planted along the sidewalk of a new neighborhood in the city. It grew and matured along with the housing stock. As time passed, newer neighborhoods were developed in the suburbs outside the city. Lots of people left the old neighborhood and bought new houses in the suburbs. Some of them held onto their city houses and became landlords. Some landlords took good care of their former homes; but others allowed their properties to deteriorate. As the neighborhood began to change, realtors encouraged people to sell and get out while they could. Many did, and now greedy absentee landlords took over, allowing properties to deteriorate even more. Apathetic building inspectors looked the other way. Expedient politicians accepted campaign contributions from the landlords. Indifferent civic leaders ignored what was happening on the other side of the river. Cynical drug dealers seized the opportunity to take over and set up a profitable business. Fearful homeowners sold to absentee landlords, leaving a handful of determined residents and faithful churches to struggle with neighborhood decline. Eventually, an abandoned house, a broken sidewalk, a pair of shoes hanging over telephone wires, and a tree filled with garbage became the symbols of urban blight.

It's easy to see how Jesse trees evolve into palm trees and become cross trees during the course of a liturgical season. After all, this painful evolution has occurred in our cities, over and over again.

Fortunately, death does not have the last word. Yes, the Jesse tree and the palm tree developed into a cross tree, but on Easter the cross tree was transformed once again into the tree of life. In Glastonbury, England (the place Joseph of Arimathea supposedly settled after Jesus' death), there is a thorn tree that blossoms every spring. Legend has it that Joseph (the one who buried Jesus in his own tomb) took the crown of thorns with him when he set sail for England. Upon arrival, he buried it on the spot where this remarkable tree grows. Fact or fairy tale—who knows? But the truth of this legend is that every spring when the thorn tree blooms, it is a powerful symbol, a reminder, that death does not have the final word. For out of pain and death comes new life. Resurrection is the final word of the Jesus story.

And so, as we make our annual pilgrimage through the story of salvation, from the tree of knowledge in the Garden of Eden, to the burning bush at Sinai, to Jesus' family tree in Bethlehem, across the towns and villages of Galilee, to the royal entrance into Jerusalem, through the city streets, to the cross at Calvary, we are called to remember that, like our spiritual ancestors, we have choices to make at every intersection. We can be indifferent or engaged, apathetic or involved, overwhelmed or determined, expedient or courageous, fearful or hopeful. The truth is that we, ourselves, are capable of moving from Jesse trees to palm trees to cross trees. Fortunately, God says no to evil and death and so transforms cross trees into trees of life. And the story goes on.

Last summer I returned to visit the garbage tree on Ellison Street, and much to my surprise, there was a brand new house smiling on this old tree. Just like the last book of the Bible promised, my tree was flourishing. There in front of my very own eyes was a tree healthy and vital surrounded by new life. This tree once filled with garbage still reminds me that God makes new out of old, whole out of brokenness, and life out of death. It teaches me that with God all things are possible. Maybe I'll now call the garbage tree simply God's tree.

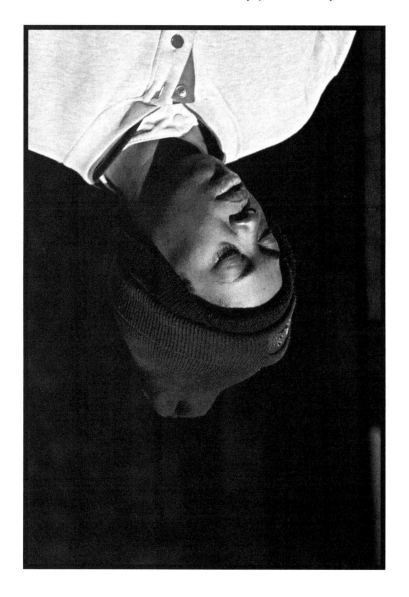

Stolen
Shoes

"They stole my shoes! They stole my shoes! Rev. Tracey, while I was sleeping, they stole my shoes! How could they steal my shoes?" Yvonne barged into the quiet of Good Friday. It was the evening service on a Good Friday that coincided with the first night of Passover. Knowing that many of our Jewish neighbors were sitting at Seder, I had taken the opportunity to mitigate the anti-Semitism of John's passion narrative and preach to my inner city congregation about how the Jews did not kill Jesus.

In she bounded. Dirty, smelly, and hungover; angry, dishevelled, dismayed, and utterly distraught. "They stole my shoes! How could they steal my shoes?" she cried.

Yvonne had lived on the streets for many years. She was what one might call a classic "bag lady." She moved around Paterson, New Jersey, with her shopping cart filled with old clothes, bottles, cans, leftover food, and lots of papers. Yvonne's life was overflowing with our discards wrapped up in old newspapers.

Yvonne was a character, to say the least. Throughout the seasons she usually wore several layers of clothing, and atop her head she always had two or three hats. At the beginning of the month, she had money from her SSI check tucked away in her purse or her sock. By the middle of the month, it was gone: half to liquor and food, and the other half had been given away, borrowed, or stolen. Some days Yvonne was in a good mood, flirting and laughing with everyone she met. Other times, she was withdrawn and pensive, lost in her own thoughts. But sometimes, she was, for lack of a better word, emotionally out-of-control. This was one of those days.

Yvonne didn't care that she had interrupted the Good Friday service. After all, she was having her own Good Friday. While she slept, someone had stolen her shoes. When Yvonne barged in, I stopped preaching. What else was I to do? Her eyes pleaded with me. "How could they steal my shoes?"

I looked at her, and I saw Christ hanging on the cross. Suddenly and painfully, it all made sense. I saw my Lord hanging naked with nails in his feet where shoes should have been. And so I responded, "Yvonne, I don't know how they could steal your shoes, but if you can have a seat and wait till after the service, I'll find you another pair of shoes." As Yvonne sat down barefoot and in tears, I turned to the congregation and said, "You see what I mean. The Jews did not kill Christ. The people who stole Yvonne's shoes—they killed Christ, and every time someone's shoes are stolen, Christ is killed again."

At St. Paul's Church, following the sermon on Good Friday, we had developed our own liturgical tradition. A large wooden cross was carried to the front of the church and placed on the bare altar. A hammer and a box of nails were also set upon the holy table. People were invited to come forward and hammer a nail into the cross. It is a way of acknowledging that we all participate in the crucifixion. Sometimes we crucify others; sometimes we ourselves are crucified; and sometimes we just watch and ignore, or walk away as the crucifixion happens.

That Good Friday, as people filed up the center aisle to hammer their nail into the cross, I glanced over at Yvonne, who was waiting and watching with great patience. When everybody had finished the ritual, Yvonne walked barefoot up to the altar. She stared at the cross for a few moments, and then with a look of sheer determination, she picked up the hammer and nail and started to bang it. As she hammered, tears flowed from her eyes, and, when finished, she simply glared into space and returned to her seat.

After church, Yvonne and I went into the clothing closet and found a relatively new pair of shoes as well as a coat, a sweater, and a skirt. Knowing that it was the end of the month, I gave her some money. She thanked me and walked out into the night. I locked up the church and went to my car. I couldn't help but think how I was going to the safety of my home as Yvonne faced the insecurity of another night on the streets.

Pulling out of the parking lot, I saw Yvonne standing with her shopping cart by the side of the church. I rolled down my window, and asked if she was all right. She smiled and replied: "Yes, I was waiting to make sure you got into your car safely, 'cause you never know what will happen out here." I returned the smile and drove away leaving Yvonne standing in the dark.

Over the years, I have thought a lot about Yvonne's parting words of caution and concern. Did she realize that her loving statement was a deep reflection of ironic wisdom and prophetic insight? We do know what's happening out here. We know all too well what is happening to the Yvonnes of the world. Their shoes are stolen every day, and so are their jobs, their homes, their children, their health, their safety, their dignity, and their freedom. We don't like to admit that we know. It leaves most of us more fortunate than Yvonne with a sense of discomfort and unease. It makes some of us feel guilty and others angry. But if we're honest with ourselves, we all know what is happening.

Yvonne's life was one of hand-me-downs and cast-offs, deteriorated buildings and abandoned factories, littered street corners and garbage-filled vacant lots, alcohol, drugs, violence, and fast food. Her urban environment was poisoned by lead paint, asbestos, industrial toxins, air pollution, noise, rats, roaches, stray dogs, and lack of access to sanitation and clean water. But her world was just a microcosm of the economic injustice and environmental distress confronting many of the poor throughout our global village.

In the United States, the vast majority of landfills and toxic waste sites are located in or near predominately low-income, minority communities. Large numbers of our nation's industrial polluters, abandoned brown fields, and "sick" school buildings are also situated in our inner cities. Access to decent housing and quality food at affordable prices remains a challenge for many inner city residents.

On the international scene, polluted water, deforestation, soil erosion, and toxic contamination are disproportionately higher in poor, developing nations than in comparatively wealthy countries. Approximately one-sixth of the earth's population does not have access to safe water, contributing to the death of about fifteen million children under the age of five every year. Third world agribusiness, encouraged by first world economic demand, eliminates traditional subsistence farming and forces millions of people into overcrowded cities and shantytowns. While many of us in the first world are becoming increasingly overweight, more than one billion people around the globe suffer from hunger. And as we import much of our clothing from third world sweatshops with environmentally unsafe working conditions, we export electronic waste (our high technology trash) to the third world in alarmingly increasing amounts. Every time you and I upgrade our computers, palm pilots and cell phones, the old ones have to go somewhere; and if we're honest with ourselves, they don't end up in our backyards.

Today's urban economic and environmental crisis is an attack against the Yvonnes of the world—the poor and the powerless who live in the shadows and on the edges of the global economy. But it's not the first time in history we've witnessed such a dilemma. According to the prophet Amos, a similar situation was going on in ancient Israel at the pinnacle of its political power and economic prosperity.

In the words of Abraham Heschel, "There was pride, plenty and splendor in the land, elegance in the cities, and might in the palaces. . . . The rich had their summer and winter [homes] adorned with costly ivory and gorgeous couches with damask pillows, on which they reclined at sumptuous feasts. They planted pleasant vineyards, anointed themselves with precious oils . . . and became addicted to wine. At the same time, there was no justice in the land. The poor were afflicted, exploited, and even sold into slavery, and the [politicians] were corrupt."[1] Sounds familiar, doesn't it.

Amos, a shepherd and dresser of sycamore trees, a man living close to the earth in a village near Bethlehem, heard the word of God and spoke to the rich and powerful of the land:

> *Hear this, you that trample on the needy,*
> *and bring to ruin the poor of the land . . .*
> *and practice deceit with false balances,*
> *buying the poor for silver*
> *and the needy for a pair of sandals,*
> *and selling the sweepings of the wheat. (Amos 8:4–6)*

God is not pleased with you, Amos said. In fact, God is angry with you. "Shall not the land tremble on this account and everyone mourn who lives in it" (Amos 8:8). You will be punished, he said. Wait and see. You will experience "the end of it like a bitter day" (Amos 8:10).

Like the prophets before him, Jesus of Nazareth was also enraged by the exploitation of the poor and the affliction of the oppressed. Quoting from the great Isaiah, he announced in his inaugural address, "The Spirit of the Lord . . . has anointed me to bring good news to the poor" (Luke 4:18). And that's what he did.

Jesus was a revolutionary who declared the impoverished, the hungry, the homeless, and the outcast as blessed, and he chose to live as one of them.[2] He was a radical who affirmed children, viewed in the first century as nonpersons with no rights or protection under the laws, and he took them unto himself, even when his band of followers tried to keep them away. He was an activist who healed those who were cast out of society, disregarded because of illness and disease, and then pronounced them clean. He was a rabble-rouser who violated the societal and religious rules of meal companionship and table fellowship by choosing to eat with the unclean and the rejected. He was a reformer who broke religious laws that he deemed were abusive to those whom they were intended to serve, protect, and empower. And then, he had the nerve to take his revolutionary act to Jerusalem, a center of social, economic, religious, and political power. Jesus so threatened the power of the status quo with his radical vision of justice, compassion, mercy, and love for all of God's children and the rest of creation that the religious and political authorities of his day wanted him dead. The Jews did not kill Jesus of Nazareth; the religious and political establishment and all those supposedly innocent bystanders who didn't stop it—they crucified this remarkable agent of God in the world.

Yes, Jesus of Nazareth died once and for all. But the crucifixion of Christ continues to happen, each and every day. For as Jesus told us in the parable of the Great Judgement, "Just as you [do] to one of the least of these who are members of my family, you [do] to me" (Matt. 25:40). The person who stole Yvonne's shoes while she slept killed Jesus. Every time somebody loses her shoes, his coat, or their lives on the streets, the incarnate word of God suffers. When entire families are forced to live on toxic landfills, the holy child born in a stable resides with them. When migrant farm workers get sick from pesticides, the Good Shepherd cries out in pain. When an infant is poisoned by polluted water, the Beloved One of God says, "I thirst." When miners suffocate from black lung disease or inner city children languish from asthma, the author of our salvation breathes his last.

As I said on that Good Friday, and have reiterated every Good Friday since, the Jews did not kill Jesus. The person, company, organization, religious institution, gang, or government that steals shoes, clothing, credit, food, houses, jobs, health care, freedom, safety, security, dignity, and quality of life from any individual, community, or nation kills the spirit of Jesus, over and over again. When little boys and girls become political footballs in budget debates, Christ the Son of God is violated. When innocent people are forced out of their homelands and into refugee camps, Christ the beloved is exiled. Every time a racist or anti-Semitic remark is spoken, Christ our Lord is maligned. Each time an individual is excluded from the gathered community of faith because of his or her sexual orientation, Christ the head of the church is rejected. Whenever and wherever people are oppressed, Christ our Savior suffers and dies. Whether we want to admit it or not, people just like you and me crucified Jesus of Nazareth. And in our fear, doubt, expediency, laziness, denial, greed, and hatred, we crucify Christ—over and over again.

But the truth is that the bad news often precedes the good news. The prophet Isaiah reminds us that we are a people with dirty lips, and we must acknowledge that fact before we can be made clean. Or as they say in some church traditions, you can't get to Easter without passing by Calvary. The good news is that there still is time to change. We have the ability—individually and collectively—to make a difference and not just stand by the cross and watch or walk away. If we're serious, if we're determined, if we organize: we can feed the hungry, clean up the water, reduce air pollution, regulate toxic waste management, redevelop brown fields, remove lead paint and asbestos from old buildings, and use organic pesticides. If we are willing to act with faithful determination, we can change the direction in which we're headed. We can create sustainable jobs with living wages. We can build decent and affordable housing with environmentally responsible construction materials and techniques. We can revitalize our cities and towns and protect our forests and farms. We can provide protection, safety, care, and healing for the most vulnerable and powerless among us. We can honor and respect our neighbors as ourselves. People of faith have done it before; we can do it again. It's not too late, but I fear the time is short. And the systemic and structural change we're talking about requires commitment over the long haul, not just passing interest today.

Myles Horton, the great labor organizer, civil rights activist, and popular educator tells a marvelous story in his autobiography.[3] It's a story about the backlog, the log that is set at the back of the fireplace to keep the fire burning. The owner of a southern plantation once said to his slaves, "This year, the Christmas holiday will last as long as the backlog in the master's house burns." The slaves knew that Christmas meant less work and more food. So they went out into the swamp and found the biggest, greenest tree they could find. They cut it down and soaked it in the swamp water for weeks. Then when Christmas came, they hauled the log into the master's house and got it lit with lots of dry kindling. And you know what, that green, water-soaked backlog burned for a very long time.

This parable suggests that our angst—our grief and anger about the world's situation—needs to burn slowly like a backlog so it won't consume us or give out too soon, but rather that it might last for the long haul. In order that we might stand in solidarity with the least of our brothers and sisters, we need to become a phoenix of love rising out of the ashes of death, roaring our resistance to injustice as we hammer shoes for every foot.

Sounds overwhelming. In many ways it is, but no one person is called to do it all or to do it alone. The wisdom of Talmud, a wellspring of rabbinic learning about the sacred texts of Jesus' Jewish heritage, teaches that while we are not free to desist, we don't have to complete the work of restoration. In Yvonne's concern for me in the parking lot on that Good Friday, she was doing her part in the work of resistance and restoration so that I could do my part. As I drove away from the church, the cross of that Good Friday began to fade in the dark as the phoenix of the coming Easter emerged.

In my office hangs a portrait of Yvonne taken over a dozen years ago when I was a beginning photographer. I was outside playing with a new camera, and Yvonne with her devious smile wheeled her cart up the sidewalk to my side and said, "Rev. Tracey, will you give me some money?" "No," I replied for the hundredth time, knowing by her smile and her breath that she was looking for drink money. "Will you take my picture?" Thinking for a moment about the reality of Yvonne's life, her extraordinary beauty, and our unusual friendship, I responded, "I won't *take* your picture, but I'll pay you to sit as a model so that I might make a portrait of you." Yvonne was delighted. She arranged her several hats, looked into her pocket mirror, sat down on the bench by the front door of her church, leaned her head to the side, closed her eyes, and said, "I'm ready for my portrait."

Three

It rose again....

The Unfinished Story

"But the women were afraid; so they told no one." It is a strange way for Mark to end his Gospel—so strange that later editors added various endings to proclaim the message of resurrection and eternal salvation. However, most biblical scholars agree that the Gospel of Mark ends with this sentence: "But the women were afraid; so they told no one."

Think about the irony of that ending for a moment. Mark's Gospel opens: "the beginning of the good news of Jesus Christ, the Son of God" (Mark 1:1). It then closes: "and they said nothing to anyone, for they were afraid" (Mark 16:8). The Gospel of Mark, the earliest of the narratives adopted as Christian canon, ends with the good news of Jesus' resurrection still unproclaimed because of fear. It's all so incomplete and unfinished. Isn't it odd that some 2,000 years later, we often begin our Easter morning worship shouting what the women on the first Easter morning were afraid to whisper: "Christ is risen!" It is downright scandalous.

Why were the women afraid to speak? What was there to fear by proclaiming the good news of Jesus' resurrection? Were they afraid that it was their imagination? Were they afraid that no one would believe them? Were they afraid that they would get in trouble? Were they afraid that their hopes would be raised and then shattered once more? Were they afraid that nothing would ever be the same again?

I am not certain what these women were afraid of, but whenever I prepare to preach on Easter morning, I find myself filled with fear. I am always reminded of a story about Reinhold Niebuhr, a distinguished professor at the Union Theological Seminary, Niebuhr never accepted preaching invitations for Easter Sunday, and he usually sought out a liturgical church where there wouldn't be much of a sermon. He said he didn't want to listen to some preacher making a fool of himself trying to explain the resurrection. It is easy to make a fool of oneself from the pulpit on Easter. It's hard to find adequate words to express that which is at the core of all that I am and all that I believe.

One year as I began my annual Easter sermon struggle, my best critic said to me, "Whatever you say, don't try to make it seem like all is right with the world because it's not. Tell the truth but give us hope." Right there is reason to avoid the pulpit on Easter morning. In the words of Pontius Pilate, the one who condemned Jesus to death, I ask, "What is truth?" (John 18:38).

The truth is that the world is not all right. In fact, it's a mess. The Middle East is in utter disarray. North Korea hates South Korea and visa versa. AIDS is devastating much of Africa and Asia. The European Union is disgusted over U.S. foreign policy and divided over continental economic policies. Latin America continues to struggle with drug trafficking, political unrest, and financial instability. Our national economy is in trouble, our state budgets are a mess, and our cities are struggling to survive. The environment continues to be threatened and abused. Gun violence is on the rise. Racism, classism, homophobia, and religious prejudice haven't disappeared; some might even say we're taking two steps backwards. People are worried about terrorism. And bad things still happen to good people.

The larger truth is that the world has never been all right. There has never been a time in history without war, famine, disease, poverty, prejudice, and hatred. Bad things have always happened to good people. If it were not so, Moses would not have been sent to tell Pharaoh to let the Hebrew people go. If it were not so, Rachel would not have wept for her children. If it were not so, Ezekiel would not have sat in exile by the River Chebar. If it were not so, John the Baptist would have kept his head. If it were not so, Jesus would not have been crucified. If it were not so, Paul would not have written to the Christians in Rome about "hardship, or distress, or persecution, or famine, or nakedness, or peril, or sword" (Romans 8:35). The world has always been a troubled place, and we humans have always made a mess of things. How's that for truth?

So in the words of suffering Job, "Where then is [our] hope?" (Job 17:15). Our hope is in the truth that God's ways are not our ways; that God is not finished with humanity; and that God will not take "no" for an answer. Our hope is in the truth of Easter. And the truth of Easter is God's "yes" to Jesus, and to the rest of humanity, made real by an empty tomb.

Emptiness—the emptiness associated with death—is what the women discover on the first Easter morning. It's what many a widow or widower has experienced the morning after the funeral. It is what an unemployed person often feels on the first Monday morning that one realizes that one has no place to go to work. It's frequently one of the emotions we feel when the final divorce papers are signed. It's the dark cloak of depression when something inside of us has died. Emptiness is a place associated with loss, separation, and painful endings.

We all know the empty landscape of death and despair. Remember when you first lost your innocence, when you became cynical, when you began to struggle with your faith. Remember when you first lost the hope that life could be happy and that you could be fulfilled. Remember when you first betrayed someone or got betrayed. Remember when lying and cheating became easier than being honest. Remember when addiction took over pleasure. Remember when homelessness, hunger, and violence became so commonplace that it didn't faze you anymore, not even when you walked by a person sleeping on the sidewalk in the bitter cold. Remember when you lost a job and unemployment seemed to go on forever, or when you sat in jail for months that turned into years. Remember when the lovemaking ended, the silent distance began, and a relationship withered away. Remember when your son threw a fist at you, or you threw a glass at the wall. Remember the first time a close friend died or moved away. Remember when you lost a part of your body, or found out that you had a serious illness. Remember when you began to notice your parents getting older, or when you looked in the mirror and saw for the first time that you were growing old. Remember all those things and you will recall the empty landscape of death and despair. We all know the place. In fact, too many of us spend too much of our lives in that awful place.

The resurrection is not something to be reasoned or rationalized. Resurrection is an event beyond time and space. It is a moment that only God can name. And it is an experience born in the depths of death and despair. Resurrection is the place beyond emptiness. It's a place filled with hope, the hope of new beginnings and new possibilities. Resurrection is an opportunity that wants to be fulfilled like a blank piece of paper that invites writing.

When the young man sitting in the tomb spoke to the women, he said "Do not be alarmed; you are looking for Jesus of Nazareth, who was crucified. He has been raised . . . he is going ahead of you to Galilee; there you will see him" (Mark 16:6–7). Why would the Risen Jesus have gone to Galilee: to start over? No, the Risen Christ went to Galilee to start anew. Galilee is a not a place; it's a state of being. The women were being told that in living their ordinary lives, incorporating and even incarnating the teaching of their Lord, they would meet Jesus again, for the first time. The truth of Easter is that Jesus was not just about the past; he's alive in the present, whenever and wherever people choose to follow him and his way. And when and where we choose to follow him and his way, we meet our Risen Lord, see the face of God, and experience the resurrection.

Resurrection is the renewal of an old friendship once broken by a wound of the past. Resurrection is a free man walking out of jail with another chance at life. Resurrection is falling in love for the second time, maybe even with the same person that you fell in love with the first time. Resurrection is the signing of a peace treaty and the rebuilding of a nation after the war. Resurrection is the new house on a city block destroyed by vandalism or the new seedling growing where a massive forest fire raged. Resurrection is getting over an illness and forgetting what it was like to be sick or accepting a chronic or fatal disease and deciding to live with it. Resurrection is a family gathered to dedicate a park bench in memory of their dad who died too young. Resurrection is the extraordinary ability of God's creation to rise again. It is the resiliency of life and the rebirth of love that is incomprehensible, simply beyond explanation.

So why on earth would the women be afraid to share this incredibly good news with their friends, the disciples? I think they were afraid for the same reason that I get afraid of preaching on Easter Sunday. It is an awesome thing to proclaim the truth that God is alive, that God has not abandoned us, and most importantly, that God expects big things of us. Yes, God's love is unconditional; the Bible tells us so. But God expects us to face the brokenness of our world with the gentle boldness of the one who was crucified. As followers of the Risen Christ, we are called and commissioned to resist and renounce evil, to love our neighbors and our enemies, to strive for justice and peace among all people, and to respect the dignity of all creation. When we do this work, we take on the cross of Christ. But when we claim the Christ crucified, we find new life. And in the promise of new life, there is hope. And thank God, that's the truth.

Perhaps Easter is lived in the ordinary places of life: at home, work, or school; as we raise our kids, clean up our neighborhoods, fix up our houses, and care for our elders. Perhaps Easter is lived whenever people feed the hungry, shelter the homeless, clothe the naked, and visit those in need. Perhaps Easter is lived when the fearful are comforted and the lonely are befriended. Perhaps Easter is lived in places where justice is done, love abides, and peace is the goal.

Perhaps Easter never gets finished. Easter is the widow who understands that her husband is gone but knows somehow that he is going to be all right and so is she. Easter is the survivor of a divorce who is able to love again. Easter is the person struggling with alcoholism or drug addiction getting through another day without a drink or a smoke or a needle. Easter is being unemployed and having the energy to look for another job. Easter is working for a just peace wherever there is unjust war. Easter knows that somehow, some way things are going to be all right. Easter is about more than surviving or bringing back the old or that which is dead; it's about living again and living anew.

I like the Gospel of Mark precisely because it has no ending. In fact, Easter has no ending, no neat conclusion. Rather, it calls us all into the future, into new possibilities and new beginnings. As Thomas Merton once said, the Risen Christ "is history with us, walking ahead of us to where we are going."[1] The resurrection—the message of Easter—is not a doctrine; it is an experience lived. Thank God the Easter story is unfinished. We get to write the Easter script as we go.

Welcome
the Risen Christ

The sun shined through the Tiffany stained glass windows, the freshly polished brass glittered, flowers filled the church with their sweet fragrance of spring, a new paschal candle flickered, and the acolytes robed in freshly laundered vestments were on their best behavior. I was in the high pulpit preaching to the largest Easter congregation the church had seen in years. And then it happened.

A scruffy, drunken man was walking up the center aisle. All eyes were on him. The ushers didn't stop him. The parish wardens, wondering what they should do, did nothing. Who was he, what did he want, and why was he interrupting my Easter sermon?

He stopped in front of the altar, looked up at me, and said, "Hey Mama, what's happening?" With a smile of recognition and relief, I replied, "Good morning, Pop. It's Easter, and I'm talking about Jesus. So have a seat." My friend sat down. Since all eyes were still on him, I took the opportunity to introduce this stranger. I told the congregation that this was my friend Bacardi. He was the very first person I met on my first Sunday as rector of St. Paul's Church.

Bacardi and I met on a hot summer morning. I had just concluded the eight o'clock service. Our senior warden had left the church to pick up his wife and father. I was by myself in unfamiliar territory. I didn't know anybody. I couldn't help but notice a noisy group of men sitting on the litter strewn corner drinking, laughing, and gesturing to me. Feeling out of place and all too obvious in my new suit and neatly pressed clergy shirt, I walked across the street into the Lincoln Restaurant and bought a few cups of coffee and a pack of cigarettes. I went back out on the corner and approached the men. I offered them each a smoke and some coffee, and we began to talk.

"Who are you?" they asked. "I'm the pastor of the church," I responded. They laughed, and one said loudly, "A woman preacher in *that* church, no way." But we kept talking, and I invited them to come to church someday.

Over the years, through all the seasons, Bacardi and his friends hung around the church. Sometimes, one or more would come in for food or clothing; every so often stay in our shelter; periodically seek counsel; and once in a while wander into the church to pray or make a confession. And always, these men faithfully watched over the church and its pastor.

So whether or not the freshly scrubbed and well-dressed Sunday congregation knew it, Bacardi and his companions were part of the St. Paul's gathered community. It made perfect sense that, in spite of his dirty clothes, scruffy beard, and drunkenness, Bacardi came to church on Easter morning. What I didn't quite understand was his unusual entrance.

In silence, Bacardi sat through the sermon, stayed for the prayers, and passed the peace. In silence, his listened to the choir anthem, witnessed the communion table being set, and watched the ushers collect the money and present the food basket. All is well that ends well, I thought.

Just as I was about to begin the Eucharistic prayer, Bacardi again interrupted. As I raised my hands and said, "The Lord be with you," Bacardi stood and walked up to the altar. What was he going to do now, I wondered. He smiled at me, put a dollar on the Lord's table, wished me Happy Easter, and then left, this time walking out the side door. Again, nobody said or did anything.

When I introduced Bacardi to the congregation from the pulpit on that Easter morning, a spontaneous round of applause greeted him. I wondered why. Had Jesus himself come among us and our "eyes were kept from recognizing him" (Luke 24:15)? No, I think in some way we did recognize him. Because we practiced Jesus' radical hospitality, the Risen Christ greeted us that Easter morning, and our hands responded with joy.

Life Happens
in the Interruptions

A number of years ago, I was walking around my parish neighborhood in Paterson. Walking with me was a potential donor for our outreach ministry. This well-dressed, polished, foundation executive from Manhattan was asking lots of pointed questions about our community development corporation, and I was trying my best to provide answers that would convince her that we would be a good philanthropic investment. As we wandered down one deteriorated inner city block, I pointed out the vacant lot we were transforming into a garden and the dilapidated building we were converting into affordable housing.

Midway through our walk, we were interrupted by a frantic woman sticking her head out of a second story window in a run-down apartment building, waving a five-dollar bill in her hand and yelling, "Rev. Tracey, Rev. Tracey, wait up! I need to talk to you."

My potential donor was startled, and I was taken aback. At first, I was tempted to respond, "I can't talk now. Come find me later." Instead, I shrugged my shoulder and stopped and waited for the woman to join us on the crumbling sidewalk. She ran up to us, grasping the five-dollar bill and gasping for breath as she spoke in a shrill voice. "Rev. Tracey, you've got to take this money for the church. You'll do something good with it and if I don't give it to you, it will just get me in trouble."

My walking companion had a bewildered and confused look on her usually composed face, but I knew exactly what my neighbor was trying to say. It was her last five dollars and she was ready to buy drugs to feed her habit. For some reason when she saw me she decided to give the money to me instead of to a local drug dealer. I accepted the five-dollar bill, thanked the woman, turned to our potential donor, and said, "Now you can see what good this church does in this neighborhood." My five-dollar donor smiled, and said, "She's absolutely right!"

The woman with the five-dollar bill gave to the church, out of her poverty, all that she had. The foundation, out of its abundant wealth, agreed to generously support our outreach ministry in the community with a fifty-thousand-dollar grant (that's a 10,000 percent return). We leveraged both the five-dollar bill and the fifty-thousand-dollar foundation grant to raise even more money for community development.

I have come to take interruptions for granted in my daily life and ministry. In fact, I actually enjoy interruptions, unless of course, I don't want to be interrupted. And there are many times that I don't want to be interrupted. I don't particularly like being interrupted in the pulpit by a child competing for attention. I don't appreciate it when an unexpected visitor interrupts my over-scheduled morning or the phone rings just as I'm leaving the office, already late for my next appointment. I hate cell phones intruding into an important conversation or meeting, and I don't like it when bad weather interrupts my vacation or an emergency interrupts my weekend. On the other hand, I like being interrupted when I'm procrastinating over a writing project, and I welcome interruptions when I'm bored. Children who interrupt parties and adults who interrupt public events often amuse me. I chuckle when a cell phone rings during the orchestra. And once I accept the inconvenience of it all, I really like blizzards and snow days. One of the best times of my life was when a broken ankle interrupted my life and insisted that I lie in the backyard reading books, talking with friends, and looking at the sky.

I don't think I'm alone in my mixed feelings about interruptions. Most people believe that interruptions are a nuisance. That's why we call them inter-ruptions, a word closely related to eruption. Interruptions break into the normal state of affairs and stop the continuity of events. It is no wonder we're taught as youngsters that it is not polite to interrupt others.

Christ happens in the interruptions. Though I don't always welcome them in the moment, I see interruptions as divine grace waiting to be recognized and received. In fact, I believe that the Risen Christ is always standing in the shadows of life, and every now and again, more often than not, comes out and is made known to us through some person, action, or event, an interruption into the ordinary realm of possibility. We never know when Christ is going to move from the shadows to center stage. It just happens, and when it does, the normalcy and complacency of our lives is interrupted.

Jesus was interrupted repeatedly in the Gospel accounts. A leper interrupted his preaching in the synagogue, and a paralytic barged in through the roof while he was speaking. The crowds interrupted his mealtime, and the frightened disciples disturbed his sleep in a storm. A leader of the synagogue interrupted Jesus' talking with the crowds, and a sick woman intercepted him on his way to heal a dying girl. A Syrophonecian woman interrupted Jesus in a private home, and a rich young ruler stopped him as he was setting out on a journey. Peter interrupted Jesus at prayer, blind Bartimaeus called out to him as he was leaving Jericho, and the woman with the alabaster jar of very costly ointment interrupted his supper. Wherever Jesus traveled, into villages, cities, or farms, people interrupted him, "laid the sick in the marketplaces, and begged him that they might touch even the fringe of his cloak" (Mark 6:56).

Did Jesus reject or refuse all these interruptions? No, Jesus saw the realm of God at hand as an interruption to be welcomed. Jesus was an interrupter himself. He interrupted the ordinary lives of naïve fisherman by inviting them to follow him. He disturbed unclean spirits and demons that were holding innocent people hostage. He intruded upon the profitable career of Levi, the tax collector. He interjected himself into the argument James and John were having about being the greatest. Jesus interrupted the

Sabbath. He interrupted the storm. He interrupted the corruption of the temple. He even interrupted the finality of death through the resurrection. In all these interruptions, Jesus embodied a new definition of human reality: human power as divine power released in the world.

Jesus' ministry was defined by interruption. He didn't always appreciate the interruptions and he wasn't always prepared for the interruptions, but each time Jesus interrupted others or allowed himself to be interrupted by a father's plea for his sick child or a woman's touching his garment, something profound and unexplainable happened. Divine power entered the world and Jesus healed the sick, fed the hungry, and even raised the dead. Somehow when we are interrupted we are changed. Maybe it is our very unpreparedness, the getting caught off-guard that releases the divine power that actually creates a new human reality. Jesus placed his God-given authority at the service of desperate, needy, and interrupting people whose interruptions were opportunities for connecting human lives with divine energy and grace.

I am coming to believe that the incarnation itself was a divine interruption in the earthly realm. As Johann Baptist Metz, a twentieth-century German theologian observed: "Christianity is something that interrupts something. Christian faith insists upon a different, unexpected and unpredictable kind of future, in a situation where the future is assumed to be already familiar and under human control."[1]

Although this idea that Christianity is something that interrupts something is true, it is difficult to follow in our daily life. Most of us don't like to be interrupted when we don't want to be disturbed, and we're taught that it's not polite to interrupt others. The first challenge of gospel living is to make room for interruptions: to look up and stop what we're doing when we hear, "Excuse me, I don't mean to interrupt but . . ." God only knows what wonderful experience is about to happen, or what gifts might be given and received. God only knows how we might be instruments of God's grace, how we might ease someone's pain or share in another's joy, how we might experience life more fully.

The second challenge of gospel living is to be willing to interrupt: to interrupt our neighbor when we need help; to interrupt our neighbor on behalf of another who needs help; to interrupt the status quo when it needs changing; to interrupt acts of hatred, evil, and oppression whenever and wherever they are found.

As I walk around my neighborhood or go about my daily life, I keep my eyes open and ears attuned to the possibility of meeting Christ and glimpsing God in the interruptions. When I am interrupted, before saying, "I'm busy . . ." I do my best to stop and ask myself, am I really too busy, or am I avoiding the possibility of the Divine disturbing or inconveniencing my life? When I need to intrude for the sake of gospel justice, I remind myself that while it might not be polite to interrupt others, who ever said that Christianity is polite? When my finely tuned plans and well-ordered routines are interrupted, I remember that life happens in the interruptions.

Four

It continues to rise...

Mountain
Musings

"Live by the word and keep on walking," says Alice Walker.[1] Our ancestors in faith understood their religion as a road, a path, a journey, and a way of life. Abraham and Sarah responded to the call of God. On the promise of God's word they left their home and set out for a new land. Nomadic customs and traditions provided a framework for ordering their life on the road. The ancient Hebrews, the people of the Exodus, received God's words as a map for their journey, and the Ten Commandments given by God to Moses on the mountain became a guide for governing their daily life in the wilderness and in the land of milk and honey.

And so it was for the first Christians who were known as people of "the Way" (Acts 9:2). These early disciples were "en route" people—men, women, and children living a particular manner of life centered on Jesus their Lord. Our spiritual ancestors were not much into creedal statements, cultic practices, or strict religious rules of conduct. Rather, they were committed to Jesus and were determined to follow his footsteps. Thus, they committed their lives less to a new religion than to a new way of living in the world.

These people of the path experienced something extraordinary in their daily lives when they followed Jesus and his way. His road became their road, his direction became their direction, his cadence became their cadence, his walk became their walk, and his journey became their journey. And his journey included some serious mountain climbing.

On the top of a mountain, Jesus and his closest disciples experienced an incredible revelation of God's truth. Like Moses and Elijah, they encountered God in a new way. They saw life from an entirely new perspective; they were privileged to experience the fullness of God's glory; and when the ultimate question—"Who do you say that I am?"—was asked, they received an answer. But then, these witnesses of divine revelation were called to come down from the mountain into the land of the living. On a wilderness mountaintop, they received "the word" and were told to "keep on walking."

When I was a senior in seminary, I climbed a mountain that literally transformed my faith, changed my life, and permanently altered my walk. As part of a course entitled, "Ministry in the Caribbean," I traveled with a dozen classmates to Puerto Rico. In the middle of our journey we drove east of San Juan into the mountains. We parked our van and started walking up a steep mountain trail. We had to climb for over an hour in the heat of the day, all of us wondering whether this side trip was worth the climb.

At the top of the mountain, we came upon a small community called La Communidad de Villa Sin Miedo, which means in English "village without fear." It was a collection of brightly colored, handmade houses clustered around a beautifully painted pavilion, which had been readied for the Feast of the Epiphany. There were extraordinary gardens, winding paths, lots of kids, and numerous collections of animals. And everyone was smiling. Why would an economically disadvantaged community of people isolated on top of a mountain with no running water, no electricity, and no vehicular access be smiling? The answer was quite simple: La Communidad de Villa Sin Miedo is a resurrection community, a village of new life and new hope.

The people of this settlement had been squatters on land owned by the U.S. Navy. The navy, in the cause of national security, decided to remove the entire community from their homes, but the community resisted. So the navy burned this makeshift village to the ground and tear-gassed the residents to get them off the property.

The beleaguered but determined group found refuge on the grounds of the Episcopal Diocese of Puerto Rico; and eventually with the help of the Episcopal and Methodist Churches, they were able to buy a mountain in the middle of nowhere. And on top of this mountain, they started a new life, *vida sin miedo*, "a life without fear."

On top of that mountain, I got a glimpse of the realm of God. I saw resurrection firsthand, and I sensed without a doubt the presence of the Risen Christ in our midst. When we broke bread together, I experienced what Robert Capon calls "gospel-centered astonishment."[2] I was astonished at the transforming glory of God, so much so that our hosts appeared radiant. As we talked and shared a meal with strangers who treated us with extraordinary hospitality, my seminary classmates and I were ourselves changed, perhaps even transformed. Perhaps we experienced what Martin Luther King intended in those immortal words preached in Memphis on the night before he died: "I've been to the mountain top. I've seen the Promised Land."[3]

When you're on the top of a mountain, there is often a great desire to remain there: to capture the moment and prolong the experience by taking photographs, planting flags, making recordings and building booths—all ways of avoiding the return to reality, sometimes painful reality. We go to the mountains on a personal odyssey; we encounter God and ourselves in a new way, and sometimes we don't want to return to face what we left behind. As Oscar, a character in *The Gospel in Solentiname* said, "Peter had a vision of heaven, and he wanted to stay there, without having to suffer."[4] So it was for our motley crew. We could have stayed on top of that mountain in La Communidad de Villa Sin Miedo for days.

But we could not stay on that mountain forever. As the evening approached, we had to come down and continue on our journey. As we descended, the world looked different. It was clearer and more detailed. Moving slowly, so as not to slip and fall, we were able to look around and see with new eyes the painful reality of our own world. As is often the case in mountain climbing, the coming down, the return to reality was almost too much to bear. It seemed so overwhelming. In the words of singer-songwriter Carol Etzler, "Sometimes I wish my eyes had never been opened."[5] We were a determined lot, and as we continued our journey through the Caribbean, we saw lots of sights we wish we hadn't seen and some we would be ever grateful for seeing, but we left the mountaintop determined to see with gospel eyes.

Over the years, I have climbed numerous mountains. In fact, when I think about it, I find myself going to the mountains whenever I need to refresh my gospel vision. Perhaps this impulse comes from my Appalachian heritage and my intuitive sense of mountain wisdom.

A few years ago, I went to the highlands of Peru, where I traveled with a group of friends by horseback up into the mountains. Accompanied by a guide, we ascended for almost three hours. At the top of a mountain range, miles from any town or village, we met an old woman tending sheep. She was sitting on a rock spinning yarn with a hand-held bobbin. She wore a dirty old skirt with several petticoats layered beneath, three sweaters, and a traditional bola hat. Tied around her shoulders was a blanket serving as a backpack. Her face was tan and hardened from years of sun and wind. Next to her sat a black sheep dog. While he was barking and carrying on, she sat quietly looking at us. I got off my horse, and in my very limited Spanish, I introduced myself. Pointing to the sheep, she said she was a *pastora*, meaning "a shepherd." I laughed and said that I also was a *pastora—una pastora de iglesia*, meaning "a pastor of the church." I asked for permission to make a portrait of her and she agreed. As I positioned my camera, she just kept looking at my friends and me. I made the photograph, we visited a few more minutes, and then she said that she had to leave for her sheep were straying.

As she walked away, I had one of those ah-ha moments of revelation and "gospel-centered astonishment." I learned a new definition for the word "pastor," and I gained new insight into what Jesus meant when he referred to himself as "the Good Shepherd." There, on top of a mountain in the southern hemisphere, I came to a new understanding of my faith story.

When we think about shepherds, we usually envision an idyllic pastoral scene with a young man keeping watch over a flock of sheep. But in my travels, I've observed that real-life shepherds are frequently old women and men, and young children—those who are not strong enough to do other useful work. The so-called "least of our brothers and sisters" are the ones assigned to tend the sheep, the cows, and the llamas. They are relegated to the sometimes grueling, often boring, frequently undervalued, but essential work of watching over the flock, guiding them to food and water, rescuing them from dangerous situations. The shepherd, young or old, chases the stray, finds the lost, tends the sick and lame, helps to birth the young, and buries the dead. Jesus modeled his leadership style on those who are literally "sent out to pasture."

Like twenty years previous in Puerto Rico, I was tempted to stay on top of that mountain in Peru. I wanted to follow the old *pastora*. I wanted to know more about her. I wanted a glimpse into her ancient wisdom. I wanted to talk with her about what she knew of life and faith from her high altitude perspective. The truth is in that moment, I wanted to build a booth up there and stay for a while. But it was time to come down, time to return, for like La Communidad de Villa Sin Miedo, the mountains of Peru were not my home.

My home is in Cleveland, Ohio. It is here that I am called to live and work and have my being—in a gothic cathedral on Euclid Avenue, in a supermarket market on Cedar Road, at a gas station on Carnegie Street, in a hospital on West Twenty-fifth Street, in a restaurant in Tremont, and in an office building on Public Square. It is here, not in the mountains of Puerto Rico or Peru, but in the urban sprawl of the Cuyahoga Valley, along the industrialized banks of Lake Erie—it is here that I am called to "live the word and keep on walking." Yet, from my journeys to the mountains, I have come to know that home is not just a place. It is more than geography. As the theologian Nelle Morton once said: "Home is a movement, a quality of relationship, a state where people seek to be 'their own' and increasingly responsible for the world."[6]

The journey to far-off places, to the top of mountains, reminds us how interconnected the world really is. What happens in the flat lands of Ohio affects the quality of life in the highlands of Peru. Our gas emissions alter weather patterns in the Arctic Circle. Our coffee drinking affects the quality of life in Guatemala. Our sweet tooth benefits from virtual slavery in the sugar plantations of the Dominican Republic. Our plastic bottles eventually find their way to mountain streams around the world. Our ever-increasing demand for gasoline influences complicated Middle Eastern politics. Our naval operations affect fishing in Puerto Rico. And as we learned all too painfully on one September 11, our power and wealth make lots of people angry—angry enough to destroy life.

Jesus' mission was about right relationship. He wanted us to see how interconnected and interdependent we really are. He wanted us to understand that if we are to love God, we must love our neighbors, nearby and far away. When we practice our faith as a road, a path, a journey, a way of life, there will be many mountains to climb, many valleys to explore, and some very confusing cul-de-sacs along the way. I'm learning to not be afraid, to travel light, to say my prayers, to keep my eyes wide open, so that I will be sure to see the glory of God and accept that the journey is my home.

Crossing
the Great Water

I love the Bible stories about God and the sea: Noah and the flood; Moses parting the Red Sea; Jonah and the whale; Paul on the prison ship; and Jesus stilling the storm on the Sea of Galilee. I especially like them because I love the water—oceans, lakes, rivers, and streams. These are the places where I best know the presence, power, and peace of God. I go to the water to get centered when I am confused, to rest when I am tired, and to find excitement when I am bored. On a boat, a beach, or a riverbank, I am able to sit still and think. When I watch the tides, the waves, and the flow of water, I can see, hear, feel, and smell the reality of God.

As a sailor and once-upon-a-time surfer, I have tremendous respect for the power of wind and water. I've learned the hard way to watch the water and listen to the wind and these holy voices will usually tell me what to do. I will never forget the first time I actually caught a wave on my surfboard and rode it all the way into the shore. It was an extraordinary rush of energy and lightness of being as I balanced myself on a floating perch and crouched into a wall of water permeated with sparkling rays of light. And I can still recall the first time alone at the helm of my own little sailboat when I realized that my hand and arm were mere extensions of the tiller, and if I allowed the wind to lead and guide me, sailing would become as natural as walking. As I sat that day on the high side of my boat with the wind beating at my face, I was at one with creation. On the other hand, I have been in some very frightening storms on the water, and I've experienced the chaotic potential of the creation and my own inability to cope with the storm.

When you are in a small boat, storms on the water (especially inland lakes) always seem to be the same. They happen without very much notice. Usually, there is stillness as the barometric pressure changes. Then the sky darkens, the wind kicks up, the waves get big, and the rain begins to pound on everything around you. If it lasts very long, you might find your boat filling up with water. If you have a motor, you can turn it on and push forward, slowly breaking the waves and hoping that you can outrun the storm to the safety of land. If you are in a rowboat or canoe, you can continue to paddle or row, feeling like you are getting nowhere and hoping that you will not tip over. However, if you are in a small sailboat, there is nothing you can do but drop the sail, secure everything, and wait it out.

When you are in a storm on a small boat, you might find yourself out of control in the midst of chaos. It has happened to me on more than one occasion, and it is precisely what happened to the disciples the night that Jesus said to them, "Let us go across to the other side" (Mark 4:35–41). A great storm arose and the disciples, a group of experienced sailors and fishermen, were out of control and scared to death as the rain pounded on them, waves swamped their boat, and the wind tossed them about. But Jesus, the carpenter, an inexperienced boater, was asleep in the stern. He was either a very sound sleeper, incredibly exhausted, or seemingly unconcerned with what was going on around him.

The story tells us that when the disciples roused him saying: "Do you not care if we perish?" Jesus awoke and spoke with authority, rebuking the wind and the waves, "Peace! Be still!" and "the wind ceased and there was a dead calm."

I know from my own experience that lake storms—the Sea of Galilee is really a large, inland lake—frequently end as quickly and as suddenly as they begin. Often without warning, they are over; the winds cease, and the waters subside. Although you know that there will be other storms, each time you live through one and it ends, it is a moment of awe and relief. And when you are in a small sailboat, it can feel like nothing short of a miracle.

Is this the point of the story—that Jesus performed his first nature miracle for the disciples by stopping the storm? Is this Matthew, Mark, and Luke's way of beginning to prove the divinity of Christ: Jesus' dominion over the elements of nature, and his oneness with God, the force behind the storm? Maybe, but I don't think so.

I believe there is a larger, more central message in this story, especially in the Mark and Luke versions. Its significance is found in the first verse of the passage: "Let us go across to the other side" (Mark 4:35, Luke 8:22). Jesus had collected a small band of people to be his disciples, and he wanted to teach them about being in relationship with God and about the realm of God that was at hand. He also wanted to instruct them about living a life of expectant faith, confident that God was always there to be found. He desired that his friends become aware of the abiding presence and saving power of the Holy One so that they would possess the courage and faith to face whatever circumstances life presented.

Thus far, the disciples' journey with Jesus had been filled with learning, wonder, and excitement. At the same time, it had been fairly safe. They were operating in familiar territory, on their home turf. But now, it was time to break camp and move out, to expand their horizons to new places and new people. Like pioneers in every age, Jesus and his band of followers were being called to leave home and travel across the border to the frontiers of a foreign land. It was time to traverse the threshold of recognizable place, jump over the edge, and "go across to the other side."

To understand more deeply what it means to "cross to the other side," I turn to *The I-Ching*, which outlines an ancient Chinese practice of prayer and meditation based on the synchronicity of chance and probability.[1] As a treatise on change, the phrase "crossing the great water" frequently appears in the text and commentary of *The I-Ching*. It signifies moving across the threshold, over the edge, and to that which is unknown and unfamiliar. When we "cross the great water," we often are confronted with our true nature, facing directly into our own fears and moving through them, not around them. In the crossing, the journey, like a submarine, frequently travels down into the deep water toward the confusion, chaos, and doubt of the dark night of our brokenness. Yet in the crossing, we discover our wounds enveloped by the healing waters of compassion. In the midst of our internal doubts and fears, we find in the deep the love that will not disappear or die.

The I-Ching says that the wise one (the one who has faith) "perseveres by crossing the great water."[2] Throughout our lives, we are called to make many crossings. We "go across to the other side" when we accept a new job or move to a new city, start a new school year or attend a new summer camp. We "cross the great water" when we begin a new relationship or get married, have a child or move in with our grown children. In these situations, we are often forced to leave behind familiar surroundings, material possessions, daily routines, personal habits, and sometimes friends and family. We frequently have to test new skills and unearth talents undiscovered, develop new patterns and routines of daily life, acquire new property, accept new family members, and form new friendships. Some people who like change make these transitions with relative ease and enthusiasm; others (such as military families) learn how to make them out of necessity, and some individuals resist such crossings with all their might. Think about Aunt Millie and Uncle Mort who absolutely refuse to sell their big house and rent a more manageable apartment or Sam who won't get on the school bus the morning after the family move.

We make such crossings when we're told we have cancer or some other life-threatening illness, when our spouse wants a divorce, when we've lost a job, if we are imprisoned, or God forbid, shipwrecked or held captive against our will. In these situations, we might not be in control of the circumstances in which we find ourselves, but we are always capable of determining what attitude we will hold in any given set of circumstances. As Viktor Frankl, Holocaust survivor, psychiatrist, and author of *Man's Search for Meaning* observed, humanity is "ultimately self-determining."[3] He concluded from his own experience in Auschwitz that the meaning of life can be found even when one is confronted with a hopeless situation. Frankl said what then matters is to "bear witness to the uniquely human potential at its best, which is to transform a personal tragedy into a triumph. . . When we are no longer able to change a situation, we are challenged to change ourselves."[4] And the decision to change ourselves might just be what saves us in the storm.

Businesses, organizations, and local governments find it extremely difficult to "cross the great water." Corporate mergers and organizational consolidations are often complicated crossings, with many stops and starts along the way. Just think about how many partnerships, mergers, consolidations, and acquisitions fall apart at the last minute, all because the crossing got too rough for its captains and their mates. It seems to be nearly impossible for municipalities, metropolitan governments, and regional organizations to make agreements that could lead to cost savings and equity sharing of local public services.

Our nation was established and built by many crossings of many frontiers. Some of our ancestors crossed the Atlantic in search of religious or political freedom; some came for financial opportunity; and others left home and traveled to this new land for romance, adventure, and escape. But let us not forget those who made the crossing against their will in chains. The awful Atlantic slave voyages, often called "the Middle Passage," established the foundation for a complicated and painful history of race relations in our country. To this day, our African American brothers and sisters still sing of "wading in the water" and "crossing the River Jordan" to remember the long and arduous journey toward freedom, opportunity, and civil liberties. As in the struggle to end apartheid in South Africa, race relations in

America are still a complex crossing that has not yet been completed.

The business of peacemaking always demands the crossing of tumultuous waters. If there is ever to be peace in Iraq, Afghanistan, Israel-Palestine, Liberia, Northern Ireland, Eastern Europe, the former Soviet Union, and yes, in urban America, numerous complicated and sacrificial crossings will be required. And the erection of walls, fences, and gates make such crossings all the more difficult and challenging. Perhaps government leaders should adopt the wisdom of Robert Frost, who in a poem entitled "Mending Wall" asks his neighbor "*Why* do [good fences] make good neighbors?' and then responds:

> *Before I built a wall I'd ask to know*
> *What I was walling in or walling out,*
> *And to whom I was like to give offense.*
> *Something there is that doesn't love a wall,*
> *That wants it down.*[5]

Peaceful crossings often require the removal of fences, walls, and gates. Can you imagine a world without them?

What can one say about the church and its crossings? The leadership of women, the acceptance of gay and lesbian people, the interpretation of scripture, and even changes in worship and sacred architecture are always challenging crossings for the community of faith. I'll never forget the time I moved the altar in my church. You would have thought that I had actually relocated God. Unbeknownst to my enthusiastic and inexperienced pastoral leadership, the power of that change was simply overwhelming for some people.

I suppose the ultimate crossing is death—the death of a loved one, and even more so, our own death. Having sat at many a bedside while someone died, I am convinced that when we die, we cross over to eternal life. We die to our flesh as we are born to the spirit. Perhaps, this is what Jesus meant when he told Nicodemus, "You must be born from above" (John 3:7). Maybe, just maybe, this is why Jesus could promise his companions on the cross that by the end of the day, they would "be with [him] in Paradise" (Luke 23:43). I think the promise of eternal life is why lots of people of faith go to their graves with a prayer of thanksgiving on their lips. Death is the final passage, the last great crossing on the journey of life.

When we "cross the great water," when we "go across to the other side," we can be assured that there will be storms in the midst of the crossing. That's almost inevitable in any significant excursion. The challenge is what we do when we hit the storm. Do we turn back to shore, hoping to avoid its tumult, thus avoiding the danger but missing the crossing? Do we panic and risk tipping over our small boat? Do we try to fight the storm, knowing that we cannot win? Do we give up and jump overboard? Or do we follow the wisdom of Jesus, and with faith in the eternal presence of God create a sense of stillness and peace in the midst of the storm? This is hard to do, but it's possible, and Jesus shows us the way. After all, he is our guide and our teacher, "the pioneer and perfecter of our faith" (Heb. 12:2). Jesus traversed the threshold before us, crossed over the edge, and calls us to do the same.

Jesus believed that every storm would pass, that the crossings would continue, and that God would remain in the midst of it all. That night on the lake, he and his comrades just needed to remain calm and faithful. As *The I-Ching* says, sometimes "the taming power of the small" is stillness in the midst of the storm.[6] If you can't stop the storm, you can in the words of Viktor Frankl, "choose your way through it."[7]

We have a magnet on our refrigerator that offers the same advice:

> *peace.*
> *it does not mean to be in a place*
> *where there is no noise, trouble or hard work.*

it means to be in the midst of those things
and still be calm in your heart. —(source unknown)

Remaining calm in your heart when you're afraid, especially when you're in the middle of a storm on the crossing, is not an easy task. It goes against the grain of our instincts to panic and jump ship. Do you remember poor Jonah, who asked to be thrown overboard and landed in the belly of the whale? Do you recall all those men, women, and children who jumped from the deck of the Titanic because they thought there was no other escape from their stormy fate?

So what is the secret formula of faithful calm in the midst of fearful chaos? For me, the wisdom is found in the words of St. Paul, who said to the people of Athens, "In [God] we live and move and have our being" (Acts 17:28). For many years my mantra when I'm afraid has been this powerful assertion that we are intimately and innately connected to the One who created us. I claimed this affirmation of faith when I was a seminarian in the South Bronx during the early 1980s, working in a neighborhood not so affectionately called Fort Apache.

In the beginning of my time there, I was afraid of those drug-ridden, poverty-entrapped, often-abandoned streets, but then one night I had a dream. I dreamed I was living in a great big, very elastic balloon. If you poked the balloon with a pin, I could feel the pinprick. If you struck it with a knife, I could be hurt. If you hit it hard enough, I could even be killed. But no matter what was done to the balloon, it would not break. The elasticity of the balloon simply would not give out.

Eventually, I came to understand the balloon as the elastic womb of God in whom I live and move and have my being. I came to realize that when I acted out of that protective and loving, albeit contrived, environment, I could be hurt, I could even be killed; but I could not be separated from the inexplicable source of energy that gives me life and will be with me—even to the grave. Through my dream, I came to understand what Paul wrote in his letter to the Romans: "For I am convinced that neither death, nor life, nor angels, nor rulers, nor things present, nor things to come, nor powers, nor height, nor depth, nor anything else in all creation, will be able to separate us from the love of God" (Rom. 8:38–39).

It was the knowledge that we live in the ultimate source of life, love, and creative energy that provided me then and continues to give me now the courage to live without fear. This belief in what I affectionately call "God's red balloon" or "God's lifeboat" enables me to take risks, to venture into foreign territory, and to cross the great waters of my life and ministry. When I remember it, my mantra—"God in whom I live and move and have my being"—calms me when I am anxious, directs me when I am lost and confused, embraces me when I shake with fear, and most importantly, reassures me that I'm never alone, no matter what might come my way.

But sometimes—when the waves get really big, the rain pounds really hard, and the wind blows really strong, when my boat falters in the storm—I, like the disciples on the Sea of Galilee that night, forget that I live and move and have being in God. And when I forget that truth, I no longer remain calm and peaceful. Instead, I panic. Not unlike the disciples, I flail about, screaming for help, often becoming paralyzed and sometimes doing more damage than good. And I cry, "Jesus, do you not care that I am perishing?" But when I quiet down and listen, he is there, he does care, and he helps me steer my boat to safety. Most importantly, he reassures me that I am not alone; rather, that God is not only with me, but also in me.

On that fateful night, when the disciples called out, Jesus was there; he responded to their cries and calmed both his companions and the storm in their midst. This story reminds me that sometimes it actually takes another voice to tell the storm to cease and to remind me to remain at peace. Sometimes that other voice is a friend's, and other times it is a stranger's. It was Harriet Tubman, Sojourner Truth, and other determined abolitionists who risked their own safety to guide many a frightened slave across dangerous waters to freedom in the North. It was the righteous Gentiles who jeopardized both freedom and life to save Jewish men, women, and children during World War II. It is courageous people of faith who risk arrest to transport undocumented persons across the Mexican–United States border and shelter them from the immigration authorities. It is the therapist, the doctor, the lawyer, and the friend who accompany the victim of domestic abuse through the storms of escape and recovery. It is the hospice nurse and the loving spouse who holds the hand and wipes the brow of a dying person as he or she crosses the threshold to eternal life.

When I look back over my journey, especially my most challenging crossings, I now realize how many friends have come to me in the midst of the storms and silenced both my turmoil and the turbulence around me. I remember a particularly difficult time when the world seemed like it was crashing down around me, and I could not discern the trees from the forest; it was raining and blowing too hard, and the sky was too dark. Ready to throw in the towel, I instead accepted the saving grasp of a friend. He stood with me on a darkened street corner, looked me straight in the eye and said, "Tracey, it's not you; it's them; don't drown in their craziness." Suddenly, I heard what other friends had been trying to tell me all along. I was in a storm not of my own making, and I had to stay the course until it passed. But the wind was strong and the waves were so enormous that I simply could not steady myself. My friend did it for me. For that lifeline, I will always be grateful. In that moment, and others like it, my friend became the saving presence of Christ in the stern of my little boat, guiding me through the storm to safe harbor on the other side.

After the storm passed, Jesus asked the disciples: "Why are you afraid; have you still no faith?" (Mark 4:40). He was trying to tell them that they had to accept the storms and have faith that there would always be deliverance of some sort. In fact, I believe that Jesus was saying something most of us know deep in our souls. Life is filled with storms, and these are the times and places we often most powerfully experience the saving power of God.

The lesson of this story is not to avoid the crossing. Nor is it to negate the power of the storm. Rather, it teaches that the wise one relies on the saving power and presence of God—sometimes in the form of friends and strangers—to cross the great water to get to the other side. The tough part is remembering this wisdom in the midst of the storm.

Keep On
Singing

When I lived in Paterson, I would walk around the neighborhood on Sunday afternoons and hear shouts of joy and songs of hope coming from storefront churches. At first, I would get uncomfortable. I didn't trust the emotions that I heard. How could these folks be so joyful and hopeful amid all the poverty and pain around them? I thought their ecstasy was sheer escapism.

Then I heard Bishop Michael Curry talk about African American spirituality. "Why didn't slaves go crazy?" Curry asked. "They had no doctors, no therapists, or social workers. Even families were separated and sold. I believe it was their singing. Spirituals took away their shame, wiped away their tears, and made them part of God's own family."[1] Their singing freed their spirits in the midst of captivity, fed their souls in spite of physical hunger, healed their wounds of abuse, and sustained them in the wilderness of oppression. Some spirituals like "Follow the Drinking Gourd" and "There is a Balm in Gilead" actually guided the slaves to freedom on the Underground Railroad.

The heritage of praising God, the sacred memory of jubilation—even in the midst of oppression, pain, and trial—has been passed down through the generations. Miriam sang after crossing the Red Sea, and Deborah sang after defeating her people's enemy. David praised God in song and his son Solomon sang of love. Isaiah wrote songs of exultation, Jeremiah lamented in song, and Zechariah sang as he rejoiced. Hannah and Mary sang when they learned of their pregnancies. Paul and Silas sang in prison, and the voices of heavens sang a new song of the lamb before the throne of God. Singing has always been a way of embodying Jesus' final prayer: "They may have my joy made complete in themselves" (John 17:13).

What does it mean to have the joy of Christ made complete in ourselves? Does it mean to be happy all of the time? Does it mean never to face hardships or suffer pain and loss? Absolutely not! In his own final prayer, the next words out of Jesus' mouth to God were: "I am not asking you to take them out of the world, but I ask you to protect them from the evil one" (John 17:15). What kind of joy is our Lord talking about, and just who is the evil one?

St. Paul suggests that joy is a fruit of the Spirit, along with love, peace, patience, kindness, generosity, and faithfulness (Gal. 5:22). Joy comes unplanned and unbidden, like finding a treasure hidden in a field, or a pearl of great value, like a net thrown into the sea catching fish of every kind (Matt. 13). The kind of joy that Jesus speaks of comes as a great surprise in the midst of the ordinary and mundane. It is sheer delight even if our longings are not fulfilled. It co-exists with the pain of life, so that the psalmist may proclaim "those who sow in tears reap with shouts of joy" (Ps. 126:5).

According to Marcus Borg, gospel joy is "a sense of connectedness with what is,"[2] with the union of heaven and earth, spirit and flesh. Joy is nothing less than what we feel in the presence of God. In fact, it is probably the most certain sign of God's presence. Joy is an expression of thanksgiving and praise to God.

And who is the evil one? The evil one is that which seeks to take away our joy: it is cynicism, hopelessness, and powerlessness, the feeling that nothing we do makes a difference. The evil one tricks us into believing that there is nothing more than meets the eye, that the present is as good as it gets, so why try to aspire to anything else? The evil one whispers to the inner city youth: "Why bother to get good grades and graduate from high school? You'll never make anything of yourself. No one from your neighborhood ever does." The evil one confronts the environmentalist: "Your efforts to save the earth are foolish. Do you really think you can beat big business at its own game?" The evil one says to the abused spouse: "Why leave him? You'll never make it on your own. And don't you really deserve what you get?" The evil one shouts to the peace activist: "Who are you kidding? You're just naïve and foolhardy. You don't know what you're talking about, and why can't you be more patriotic?" The evil one says to those who believe in the reign and shalom of God: "Give it up. Let it go. Get real." The evil one tempted Jesus in the wilderness and upon failure, promised to return at "an opportune time" (Luke 4:13).

When I first arrived in Cleveland, I experienced God's joy and the evil one competing with each other. I had to attend a mandatory new clergy retreat with the bishop at a conference center just outside Toledo, Ohio. I didn't really want to go. It was a beautiful day, and I wanted to play in the garden. But it was a command performance. So off I went, driving west on the Ohio Turnpike to spend a day and a night with my new colleagues. I decided to get off the turnpike and drive on a smaller highway running along Lake Erie. I opened the windows and began to experience the power of the lake. I could smell the water, I could feel the force of the wind, and I could see the extraordinary beauty of a blue sky and green trees. I imagined that I was sailing. Before I knew it, I was sailing—sailing in the Spirit of God with a joy that I could not explain. As I sailed along in my car, I began to pray. I prayed for those on the prayer list; I prayed for friends and family far away; I prayed for my church, my city, and those entrusted to my care.

Suddenly, I came upon a huge nuclear power plant rising up out of the earth. It was spewing clouds of vapor. I realized that it hadn't been there the last time I was on this road. Its enormous size overwhelmed me. It was something so foreign to the landscape that I almost drove my car off the highway. I thought to myself, how could this thing be here ruining God's earth? But I continued to pray. I prayed for the world, for the earth, and for the power plant.

In the midst of my prayers, driving on a beautiful lakeside road by a nuclear power plant, I realized that I was intimately connected to it all. The connection was not of my own making. It was of God, and I found myself singing. At first I started singing the hymns of my youth and the songs of Taize, and then I found myself moving into the folk songs of the early seventies when last I lived in northern Ohio. Before I knew it, I was singing the protest songs of Phil Ochs, Pete Seeger, and Woody Guthrie, shouting to the power plant at the top of my lungs, in my off-key voice, "You can't destroy the beauty of this lake." When I reached the conference center, an hour late as a result of taking the back roads, I walked in ready to be present to my brothers and sisters. I was able to let go of my previous weekend desires and willing to allow the Spirit to do her thing.

At some point in the weekend, I was asked what I wished for my new congregation and its people. I thought about it for a moment, and then I answered with simple words. May it be a place of joy. May its people be sustained by joy in the midst of our complicated world and sometimes painful lives. May we together "sing to the Lord" and "make a joyful noise to [God] . . . with songs of praise" (Psalm 95:1–2) so that the passerby would stop, listen, and wonder what it's all about, and maybe, just maybe, walk through our doors. In that moment, I realized that I was praying Jesus' prayer.

When the conference ended on Saturday afternoon, I decided to drive into nearby Toledo and see my old neighborhood. In doing so, I returned to a time and place that had not been joyful for me. Rather, it had been a painful and foolish period of my young adulthood. As I drove into the Old West End, I was amazed at how a rundown neighborhood had been transformed over the past two decades. I saw houses that had been rebuilt, parks that had been reclaimed, streets that had been cleaned up, and a huge neighborhood festival in progress. I wandered around the neighborhood looking for familiar landmarks.

I bought a birdhouse from a woman who had restored my old house. I had dinner with new friends from the diocese, and then I walked to my car parked behind a storefront church. As I approached my car in the early evening light, I heard shouts of joy coming from within this little church. This time, I didn't get uncomfortable. I realized that these were brothers and sisters in Christ, celebrating the resurrection of our Lord, testifying to the joy of Christ with songs of praise. I smiled and said a prayer for them. I offered up my own grace. As I drove home with the sun setting at my back, I experienced the joy of Jesus complete in myself.

I have given a lot of thought to the power of singing since that day. We sing to celebrate. We sing as we work. We sing as we hike. We sing to make the world smaller. We sing to drown out voices of hate. And sometimes, we sing to stay warm, awake, and even alive. Singing is one of the most important forms of prayer. In fact, it's said that Augustine believed, "To sing is to pray twice."

Singing is also a powerful act of spiritual and political resistance. The early Christian martyrs sang as they faced lions in the coliseums. African American slaves sang as they toiled in the cotton fields. Striking workers sang outside the mines, mills, and factories.

One of the most powerful resistance songs, "How Can I Keep from Singing," is an old Quaker hymn that dates back to pre–Civil War North Carolina when members of the Religious Society of Friends suffered for their opposition to slavery. Its words spoke volumes about the power of song: "Through all the tumult and the strife, I hear that music ringing. It sounds an echo in my soul. How can I keep from singing?"[3]

Singing has always been a way of mentally surviving the horrors of prison and torture. "The Peat Bog Soldiers," written by some of the first prisoners in the concentration camps, was sung as they marched to work in the forest. The words, "But for us there is no complaining, winter will in time be past," kept their spirits alive as their bodies suffered from abuse and hunger. "One day we shall cry rejoicing: Homeland dear, you're mine at last," became a shout of resistance in the face of hopelessness.[4]

In the movie *The Power of One*, a white boy named Peekay grows up in South Africa during World War II, partly in a prison where Africans are horribly treated. Through his friendship with the African prisoners, Peekay comes to understand the evil of apartheid. In an effort to build solidarity among the prisoners of various tribes, Peekay organizes a concert where the prisoners sing for the oppressive camp administration and their families. Sitting in rows on the ground, the men perform a beautiful acappella concerto of ironic resistance in tribal language that is not understood by their Afrikaner audience. Singing words like, "The guards are stupid," lifted the spirits of the prisoners in ways that were applauded rather than punished by their oppressors.[5]

Perhaps one of the most poignant examples of the power of song during fateful times was the final message of Etty Hillesum. Scribbled on a scrap of paper and thrown from a boxcar as she and her family departed for Auschwitz, this young woman wrote, "We left the camp singing."[6]

Where would we be without singing? I don't know. My beloved Emily says that when she sings, it is the only thing she can think about; for her, singing gives her head a vacation from her cares and worries.

I have always loved to sing, and I am told that I usually sing a little off-key but with lots of gusto. When I was in school, we had to go to daily morning chapel. We all lined up in the hallway by class and height and we were inspected for appropriate skirt lengths, clean blazers, and polished saddle shoes. If any one of these items was not in order (especially the length of our skirts), we were sent home to redress and marked tardy for school. Those who survived dress inspection processed into chapel singing a variety of familiar hymns, our favorite being "For All the Saints."

When it came time for graduation, our class selected this same hymn for the processional. And so, on June 12, 1972, twenty-nine girls marched down the aisle in long white dresses, carrying a dozen red roses, with tears in our eyes and singing at the top of our lungs: "O blest communion, fellowship divine! We feebly struggle, they in glory shine; yet all are one in thee, for all are thine, Alleluia, Alleluia!"[7]

Whenever I hear this hymn, I think about my school days and laugh. There we were a bunch of kids singing about the saints. We didn't know anything about the saints, and we certainly didn't think we were saints, much less behave like saints, and yet, we were drawn to sing about them. Now, when I think back on those days, I realize that we were what novelist Anne Tyler phrased "saints maybe."[8] We were, each and every one of us, potential saints. And only time would tell what would become of us. My classmates and I grew up. We became doctors, lawyers, writers, accountants, teachers, artists, civic volunteers, and ministers. We grew up to be wives, lovers, mothers, and yes, grown-up daughters.

Singing has been an integral piece of my journey as a "saint maybe." I have a song notebook (lyrics and guitar chords) in my own unique chronological order, and every song in my well-worn blue notebook reminds me of a particular phase, relationship, or event in my life. On pages wrinkled with use and yellowed with age, there are lots of folk songs that I sang with my junior high friends in our band, the Checkerboard Squares: Wheat Checks, Rice Checks, Corn Checks, and Dog Chow (that was me). In the middle of the notebook, there are protest songs that I sang during my college years. Throughout the binder, you will find falling-in-love songs and breaking-up-is-hard-to-do songs (one or more for each relationship). There are the mountain songs and bluegrass melodies that I sang as I claimed my maternal roots. These are interspersed with the songs of the women's movement that I learned during graduate school. And needless to say, there are songs for folk masses and youth groups, the music that brought me to the Episcopal Church in high school.

But the most important songs in my notebook comprise what I call My Coming Out Collection. This is the music that helped me claim the deepest, most intimate, and, at times, the most complicated aspect of my life—my sexuality. One might say that I came out to music. I can still remember, as if it were yesterday, sitting at a concert in Cincinnati listening to a singer/song writer named Teresa Edell. She sang a song by Alex Dobkin entitled, "A Woman's Love." I'll never forget the final verse; it interrupted my life as nothing else ever would until seven years later when I met God on Forty-second Street.

> *Because I'm a woman, a way was laid out for me.*
> *I always thought I'd need a man to love.*
> *And while the men I've known were as loving as they could be,*
> *There's no one can match her beauty*
> *It's because she's a woman,*
> *And she feels so much the sweet touch of a woman's love.*[9]

Those words expressed the very essence of my sexual being. In my hearing, Teresa was singing both about me and to me. Sitting in a darkened auditorium, I looked around at all the other women and tried to distance myself from them. "I'm not like them," my head insisted. "Yes, you are," my heart replied. There was a dance following the concert. A woman asked me to dance. "No thanks," I politely responded to her invitation. Then I immediately left the dance hall and went directly home. A friend who had taken me to the event called and scolded me, saying, "What's wrong with you? You're willing to fight for everybody else's freedom, but you run away from your own." At first I protested, and then I gave in and thus began the journey of coming out.

Music has always been an essential aspect of gay and lesbian life. The women's music scene, complete with concerts, collectives, and communes, provided a safe haven for many young lesbians in the seventies and eighties. And the bars with their pounding disco sound offered a place to go and be gay and lesbian. We had our own songs, like the Village People's "YMCA" and Meg Christian's "Ode to a Gym Teacher," but gay men and lesbians also became adept at translating "top 40" lyrics in our heads so that we too could claim their truth about romantic love.

Pride Day parades and rallies have always featured song and dance. I still chuckle when I think about walking down Fifth Avenue with my gay and lesbian Episcopal brothers and sisters singing hymns in the New York City Pride Parade. As we turned the corner at Washington Square singing "Jesus loves me this I know, for the Bible tells me so,"[10] bystanders on the sidewalks and overhanging balconies would hoot and howl, cheering us brave religious folk. We'd end up at St. Luke's Episcopal Church to sing Choral Evensong before heading out to the Hudson River Pier to dance the night away. I think it was our singing that won converts to both of the causes found at the intersection of religious expression and sexual honesty.

For almost three decades, the gay and lesbian movement has sung itself into freedom, tolerance, and acceptance, and now inclusion, welcome, and a tentative embrace. At the consecration of Gene Robinson, the first openly gay bishop in the Episcopal Church, our festival worship ended with the singing of "For All the Saints." As Gene joyfully walked out of the arena, thousands of the faithful sang those words Christians have been singing for centuries. I stood with my clergy colleagues (gay, lesbian, and heterosexual alike), singing at the top of our lungs while watching one of our own joyfully process through the Whittemore Center of the University of New Hampshire campus in Durham, New Hampshire. Though I knew that this road had been paved with lots of pain, sacrifice, and tears, and that there would be more of it to come as the church struggled with the threat of separation and schism, in that sacred moment, I experienced the joy of Jesus. And right then and there, I made a promise to God and myself: no matter what happens, come hell or high water, no storm will shake my inmost calm, for I'm gonna keep on singing, all the days of my life.

Five

It nourishes....

Bread
of Life

Bread is one of the simple joys and basic necessities of life. Almost everyone likes it. Almost everyone needs it. You can get it almost everywhere. And it's a part of almost every meal of every day.

Bread comes in all shapes, sizes, colors, tastes, and textures. There's white bread and brown bread, whole wheat bread and pumpernickel. There's sweet bread and salty bread, garlic bread and rosemary spiced bread. There's flat bread, puffed bread, hollow bread, and unleavened bread. There are round loaves and oblong loaves, rolls, crackers, tostados, and bagels.

You can do all kinds of things with bread. You toast it, bake it, fry it, sandwich it, and butter it. You can cover it with jam, jelly, peanut butter, mayonnaise, ketchup, and mustard. You can even eat it plain. When bread gets stale, there are even more things to do with it. You can make croutons or bread crumbs. You can feed it to the birds or stuff a bird. I've even seen kids use old bread for fishing bait.

Nationalities and ethnic groups are identified by their bread. There's Jewish rye bread, French baguettes, Irish soda bread, Ethiopian flat bread, and even English muffins.

For different people in different settings bread takes on different meanings. To the chemist, it's a formula; to the food wholesaler, it's a commodity; to the artisan baker, it's a form of art. To the wine taster, bread is a cleanser of the pallet; to the farmer, it's a finished product. To the dieter, bread is something to control because it has too many calories; to a starving person, it's the staff of life. For the new household on the block, the gift of bread is a wonderful way to say welcome to the neighborhood. At the family dinner table on Thanksgiving, a warm loaf of homemade bread becomes almost sacred.

Bread is the gift of God's creation and the work of human hands. Throughout the ages, bread has come to mean life, love, nourishment, sustenance, and survival. As commonplace as it seems, bread remains one of the most powerful symbols of humankind.

In the Scriptures, the image of bread is found over and over again as a metaphor for God and for both divine nourishment and punishment. During the Exodus, when the Israelites complained that they were starving in the wilderness, God responded to their cry by sending bread from heaven on a daily basis. When the chosen people rebelled against God, we read of the "bread of adversity" as the metaphor for their punishment. Throughout the world of ancient Israel, bread was used for cultic offerings and priestly sacrifices.

Jesus understood the nature and power of bread. He knew that people needed bread to live, and when folks were hungry, he fed them with bread. He also knew that bread was a gift from God, and so he never failed to give thanks to God before breaking and sharing the bread. It is not surprising that Jesus would choose this ordinary staple of daily life—bread—as a principle metaphor for his own life and ministry. According to the fourth gospel, after feeding a group of hungry people, Jesus said it directly: "I am the bread of life. Whoever comes to me will never be hungry, and whoever believes in me will never be thirsty" (John 6:35).

Jesus took the staple of the human diet, the most basic form of physical nourishment, and claimed it for himself. Not only did he claim the image of bread, but he said that he is the kind of bread that would not spoil, but would endure to eternal life. In essence, he said, "I am that which can nourish you forever."

Jesus did the same thing with water. When he met the Samaritan woman at the well, and she offered him a drink of water, he offered in return the water of life, water that would quench her thirst forever.

Can you imagine standing in the presence of this man, having just filled your belly with bread that he has given you, or water that he poured, and then being told that this is not what you really crave or need? This isn't all there is. There's more that he has to offer you—the bread and water of life, food and drink that will satisfy you forever.

Every time I read or hear these passages, I am baffled and taken aback. What kind of audacity is this? How dare Jesus tell me that he will fill my every need, especially when I am struggling for food and drink for my hungry children.

For most of my ministry, I have served in churches that had food pantries or feeding programs for the hungry. In Paterson, we operated a food pantry that sometimes served over 1,200 people every month, and at Trinity Cathedral in Cleveland, we have a lunch ministry that feeds upwards of two hundred people every Sunday. A lot of time, energy, money, space, and people power goes into the acquisition, storage, cooking, and delivery of food.

Some of my fondest memories of my time at St. Paul's Church were getting up early in the morning with a small but faithful group of volunteers and going to a local food reclamation center to spend hours rummaging through bins of rejected canned goods, finding all kinds of surprises. I'll never forget my friend Bill, a retired food broker, waking up the guys in our transitional housing program to go to the food bank. We had to get there early to get the best of the rejects. Arriving with coffee and donuts in hand, Bill could never understand why his formerly homeless buddies were not excited to get up before dawn and go picking in the warehouses of plenty.

One of the challenges of running modern-day, faith-based food pantries, shelters, and soup kitchens has to do with evangelism. As a pastor, I hold the strong opinion that no church has the right to insist that the hungry pray before receiving bread or that the homeless listen to a sermon before getting a warm bed. I've always said, "Let them know we are Christians by our actions, not our words."

One day, a woman came to apply for a position in our food pantry. During the course of the interview, she told me that she was a born-again Christian and then asked me about prayer and Bible study in the food pantry. I explained our policy, and said that we did have a midweek Bible study, but it was voluntary and not tied to our social services. She looked at me with a shocked expression and then leaning forward in her chair, she raised her voice and righteously said, "How can you call yourself a faith-based hunger program and not feed the people what they're really hungry for, salvation in the name of Jesus?" Once again, I explained our policy and philosophy, but she wasn't satisfied. She stood up, ended the interview, and stomped out, exclaiming, "May the Lord have mercy on your soul."

I was stunned as she walked out the door. I felt like I literally had been attacked in the name of Christ. At first, I was angry: how dare she confront me in this manner? But then I thought, what would Jesus do? Frankly, that used to be my first question when confronted with a really difficult situation or challenge to my integrity. What would Jesus do? WWJD are the call letters to a Christian radio station in New Jersey. WWJD are beads on bracelets worn by teenagers in love with the Lord. WWJD are letters found on bumper stickers driving down the highway. WWJD—What would Jesus do?

Perhaps Jesus would have had a pithy remark or question for the woman before she walked out the door. Maybe he would have simply dusted off his feet after she left. Possibly he would have turned to his colleagues and shaken his head saying, "She just doesn't get it."

As for me, I'm not Jesus. I don't live in Jesus' day. I don't walk in Jesus' shoes. I follow his way, his truth, his life, his teaching, and his example two thousand years later. I realize that the question is not

"What would Jesus do?" The question is "What would Jesus have me do?" I am not called to replicate the ministry of Jesus. Rather, I am called and commissioned to continue the ministry of Jesus (and the prophets before him) in my time and place. I am called to be faithful in the context of my twenty-first century North American urban setting, just as Jesus was faithful in the context of his first-century rural Palestinian setting. And it's mighty complicated.

I'll never forget the young man who interrupted worship one Sunday in Paterson. There were a dozen or so people gathered for the quiet early morning communion service. I was in the middle of my sermon when he entered through the parking lot door. At first, I saw him standing at the door, looking in with curiosity. He was wearing a black, overstuffed, down jacket; a black ski cap pulled down low over his head; baggy, black trousers, and worn-out gym shoes. He had a boom box under one arm, and his other hand was in his coat pocket. I glanced at him and, with my arm, signaled for him to come in and have a seat. After a moment of hesitation, he slowly and deliberately walked down the center aisle to the middle of the sanctuary. He turned to the right, and walked over toward the chapel where we were gathered. He abruptly stopped and stood in the aisle at the back of the chapel section. I again welcomed him with a cheerful, "Good morning. Have a seat." He didn't move. Instead he just stood there staring at me. I must confess that I was getting a little nervous. Once again, I said, "Welcome to St. Paul's Church. We'd be delighted if you would join us for worship. Please, have a seat." He remained standing and staring with his left arm wrapped around the boom box and his right hand in his pocket.

Now, I was getting really nervous, and I could tell that my small congregation was beginning to get anxious. I actually thought to myself, "I hope this guy isn't carrying a gun in that pocket." For the first time in my two decades of urban ministry, I feared for my safety.

Not knowing what else to do, I looked directly into the eyes of this young man and asked, "What do you want?" He looked directly into my eyes and responded, "I am hungry. I want something to eat." I must admit that I was relieved, and I'm sure it showed on my face. He's hungry, I thought. He wants food. What a simple request to make of God's church. The sermon that never really began ended. I looked at the bewildered congregation and said, "Why don't you continue with the prayers of the people, and I'll go get our guest some food."

Relieved and scared to death, they nodded in unison. I turned to our young visitor and said, "Come with me, and I'll get you something to eat." We went into the men's shelter and I asked one of the residents to make our guest some breakfast. I returned to celebrate the Lord's Supper with the congregation. After worship had ended, one member of the congregation said, "I didn't think it was going to turn out this way." We all knew what she meant. After saying goodbye to the congregation, I walked into the men's shelter. Our visitor had eaten breakfast and was gone. I guess he had been hungry. Thank God the church was there to feed him.

Hunger and thirst are not always physical deprivations. As Jesus pointed out in the Sermon on the Mount, sometimes we who have full bellies are spiritually hungry. I have encountered children with all the toys and clothes they could possible want and yet they hunger for affection and attention. I have known wealthy but burned out corporate executives who hunger for time off and who believe there must be something more than worldly success. I've talked with successful writers and artists who have discovered the dry well and thirst for inspiration. I have met single men and women who hunger for a lasting relationship, and I have listened to middle-aged adults longing for a relationship with God.

One Saturday evening, I overheard a woman articulate her spiritual hunger outside a movie theater in an affluent suburb where physical hunger was never known. I had just seen *Regarding Henry*, a 1991 film about an arrogant corporate lawyer who gets shot in the head, develops amnesia and experiences a change of heart as he gradually recovers his memory. Two women were walking out of the theater in front of me. One remarked to the other, "I wish someone would shoot my husband in the head." She didn't really want to see him hurt; rather, she wanted to see him have a change of heart. She hungered for a kinder and gentler husband.

Giving away loaves of bread, bottles of milk, and canned goods is not the answer to hunger. Handouts are not the end all and be all. As Jesus learned during his wilderness journey, human beings cannot live by bread alone. Yet, as Jesus taught his disciples through the feeding of the multitudes, the answer is also not found in preaching to someone with an empty belly. Ultimately, no meal of fine wine and gourmet food can satisfy a hungry spirit; and no amount of money, security, popularity, or power can fill an empty soul.

WWJD—What would Jesus do? I think the answer is really very simple. Feed the hunger and quench the thirst. If a man asks for something to eat, give him bread. If a woman is thirsty, offer her something to drink. If a child is lonely or frightened, befriend and protect him. If a teenager is angry and confused, engage her emotions and her intellect. When a community is starved for knowledge, teach the elders to read. When people hunger for freedom, provide bread for the journey. If a nation thirsts for justice, let it flow like a mighty stream of water. The vocation of the church is to feed the body, enliven the spirit, engage the mind, and heal the soul. One without the others just doesn't make sense. For it is in the bread, the water, the friendship, the love, the engagement, the knowledge, the freedom, and the justice, that we see glimpses of God.

Sally's Feast

Sally was a member of my friend's church.[1] She was a widow, but society would call her a bag lady. She had a room where she slept, but she carried all her possessions with her in shopping bags; her most valuable belongings were pinned into the lining of the many coats she wore. According to my friend, Sally was quite a character, and it took some time for the other parishioners to consider her a member of the congregation.

On communion Sundays the altar guild at Sally's church always feared that there would not be enough bread to feed the people in the pews. But every communion Sunday there were leftovers that were placed on the back of the altar.

One communion Sunday, a rather extraordinary thing happened. As the minister was preparing to lead the post-communion prayer, Sally, with all her shopping bags in hand, walked up to the altar, genuflected, crossed herself, and proceeded to put the leftover communion bread in a shopping bag. When she finished, she crossed herself again, walked back to her seat, and told the minister he could continue with the service. As you can imagine, the congregation was aghast. How could she do such a thing? How could she interrupt the sacred moment and take the sacred bread off the sacred table?

After church, the minister spoke to Sally and asked her what she intended to do with the bread. She looked at him and simply said, "Eat it." The altar guild was perplexed. What should they do? The minister said: "On communion Sundays, let's put the remaining bread in a bag and give it to Sally." He

159

then asked Sally if she would wait for the bread, and she agreed. And for five years, on every communion Sunday, the loaf was passed among the people, the leftovers were placed in a bag, and after the service the bag was given to Sally.

When Sally died, she left written instructions that the church was her family and that they should bury her. Over seventy-five parishioners attended her funeral. Sally, the eccentric shopping bag widow who appalled the congregation with her post-communion behavior, had become a member of the community, and out of her poverty she had given new meaning to the words "Holy Communion."

Holy Communion, or Holy Eucharist as it is often called, is a central act in Christian worship. In the Episcopal Church, we say it's a sacrament, a visible sign of God's invisible grace. It's based on Jesus' Last Supper with his friends. The accounts of this event in the Synoptic Gospels tell us, "While they were eating, he took a loaf of bread, and after blessing it he broke it, gave it to them, and said, 'Take; this is my body'" (Mark 14:22). He did likewise with a cup of wine, saying, "This is my blood" (Mark 14:24). Jesus emptied himself, shared what he had to give, and it was enough.

The sharing of what one has to give, especially out of one's poverty, and having enough to go around with leftovers is gospel truth. Jesus demonstrated this truth over and over again in his teaching, his healing, and his feeding of the people whom he met along the way. There is no better example of this than when Jesus fed the multitudes with five loaves of bread and two fish. He took what he had at his disposal, gave thanks to God for it, divided it among the gathered community, and there was enough to go around. In fact, like on communion Sundays in my friend's church, there were twelve full baskets left over after the people had eaten and were satisfied (John 6:13; Matt. 14:20; Mark 6:43; Luke 9:17).

The miracle of the loaves and fish, the feeding of the multitudes, is sort of like a potluck, only much more. The teachings are the same: the combination of community, sharing, and faith lead us from scarcity to abundance.

Potlucks can only take place in community. They cannot happen alone. Hence, Jesus teaches us that Christianity does not happen privately. It is a community event. Certainly, there will be private moments in a Christian life, but for the most part, our Christian living takes place in the context of community. The community takes what we have to offer, and the community gives back what we need.

Potlucks are about sharing with the community. I believe the miracle of the story of the loaves and fish is that by his own behavior, Jesus taught people to share what they had and there was enough. Think about how he orchestrated the feast: he told the crowd to sit down on the grass; then, he took what he had (actually what the disciples had), looked up to heaven, blessed it, broke it, and gave it to the disciples to distribute among the people.

Now here's what I think happened. When the disciples started distributing what they had to offer, others came forward and began to share what they had to offer. Surely some of the five thousand people gathered had brought a basket or sack of food with them; some of them were probably on the way to and from market. Others probably had a piece of bread, cheese, or fruit in their satchel. Back in those days, few people left home for a day in the countryside without something to eat. It would have been foolish to leave home so unprepared. After all, there were no McDonald's on the highway. So when Jesus lifted up his food to God and instructed the disciples to pass it among the crowd, people came forward with their offerings. "I have a little cheese, I have a loaf of bread, I have an orange, I have a bit of wine—let me share this as well." And when the people shared, there was enough to eat, and there were leftovers.

Potlucks are about having the faith that when we share what we have with the community, there will be enough. It is the simple principle of living into abundance rather than scarcity: a radical, deviant, and creative concept that if lived out to its fullest potential could change both individual and collective life on this earth. I think the story of the bread and the fish summarizes one of the most essential teachings of our Lord. That's why it is recorded in all four gospels!

When Jesus lifted the bread and fish up to ask for God's blessing upon it, he was taking a risk. He didn't know for certain if there would be enough. He had faith that there would be enough. And that's what it's all about: faith that there will be enough: enough food for the meal, enough money for the budget, enough time for the project, enough of whatever we need to fulfill our obligations to ourselves, our families, our communities, our world, and our God.

And here's the kicker, the punch line. Jesus always asks more of us than we think we have to give: more love to offer when we're running on empty; more tests to take when we're exhausted; more mouths to feed when the pantry shelves are bare; more bills to pay when the checking account is empty. Bring to me what you have, Jesus says. Bring me your skills and weaknesses; your strengths and fears; your burdens, challenges, and responsibilities; your hopes, dreams, and convictions; your past, present, and future. Bring it all to me and I will make you adequate.

This is a really difficult lesson for most of us middle-class people. The more we have, the more protective we become of it. But the poor seem to understand it; perhaps, that's why Jesus called them blessed. They have much to teach to those of us who possess much.

Think about the widow whom Elijah approached for food during the famine (1 Kings 17:1–16). When the prophet came upon her, she was gathering sticks to build a fire and prepare a final loaf of bread for her and her son before they died of starvation. The widow had nothing left to lose so she shared what she had, and then God did a miracle by providing enough grain to get through the famine. The widow did not know what the outcome of her actions would be when she was first approached by Elijah. There were no guarantees. The widow at Zarephath simply shared what she had, believing the prophet's promise that there would be enough.

Think about the widow who gave all that she had to the temple treasury. She donated two small copper coins worth about a penny, and Jesus affirmed her when he said, "For all of them have contributed out of their abundance; but she out of her poverty has put in everything she had, all she had to live on" (Mark 12:44). She did not know whence her next meal would come or who would pay for medicine if she got sick. She simply trusted that she would have enough.

When I was in seminary, I met a widow who gave to me of her poverty. I met her when I was volunteering at a women's shelter next to the Port Authority Bus Terminal in New York City. The Dwelling Place was run by an incredible group of Franciscan nuns. My job was to serve dinner and staff the clothing closet. I had mixed motives for my volunteerism. I liked and admired the sisters, and I was learning about ministry from them. I enjoyed hanging out with the women guests, and I was learning about life from them. And if the truth were told, I really appreciated getting a decent meal three nights a week.

A fascinating crowd of women came to the Dwelling Place, and over time, I got to know some of them quite well. There was one woman in particular that I remember. She was a widow from Eastern Europe. She lived in a room in the Garment District. She said she had been a princess who lost her title when the Communists invaded her country. She also claimed to be a fashion designer who just needed a break. I got pretty involved with this woman; I still remember calling my brother Tom (then a fashion buyer with Neiman Marcus) and arranging for him to visit her "studio" and examine her designer dresses. He actually tried to help her but she never followed through.

One day in the dead of winter this same woman showed up at the Dwelling Place with a beautiful navy blue, cashmere coat. She walked up to me and said, "I brought this for you." I looked at the coat for a few moments and politely responded, "I can't take this coat. You should keep it and wear it yourself." With a look that pierced through my eyes to my heart, she said, "No, it's for you. If you're going to be a minister, you need to wear a decent coat." I looked down at my dirty, orange ski jacket and nodded. I accepted that coat and wore it through seminary and well into my first job. In fact, I wore it and patched it until it was literally threadbare—far too worn out to go to any clothing closet or thrift store.

The widow who fed Elijah during the famine, the widow who placed the two coins in the temple treasury, the widow who gave me a coat, and the widow who ate the communion leftovers are incredibly powerful examples of faith, radical faith that God will really provide enough to meet our needs and the needs of those with whom we share.

Jesus understood the essence of giving and living out of one's poverty. In his singling out the poor widow, he was saying to us: "People ask who Christ is. I am not a mighty and triumphant warrior like David. No, I am a poor widow willing to give everything I have, my whole living to my God. That is who I am."

Genuine giving comes from our poverty, not from our wealth. It happens because we realize that there is nothing else to do but to give. When we give out of our poverty, we become rich and blessed, not by the world's standards, but in God's eyes. Thank God for all the widows who teach us this most precious lesson. May we do right by them and ensure that they have adequate food, clothing, housing, and, yes, health care.

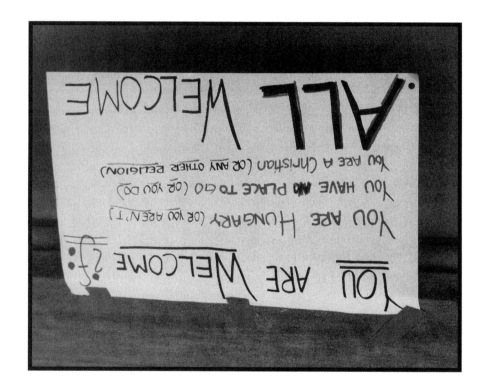

Beckoned
to the Banquet

He drew a circle that shut me out—
Heretic, rebel, a thing to flout.
But Love and I had the wit to win:
We drew a circle that took him in!
—Edwin Markham[1]

This is my favorite poem. It's what I would like carved on my gravestone. It speaks to my own sense of Christian discipleship, about what it means to be a follower of Jesus—a Christ bearer.

I understand God to be the power of love: extravagant, indiscriminate, abundant, unconditional, all-inclusive love. I understand Jesus to be the embodiment of that love; he loved extravagantly, indiscriminately, abundantly, and unconditionally. I understand discipleship to be a commitment to extravagantly, indiscriminately, abundantly, and unconditionally demonstrate that love. I don't do it very well, but I know it's what I'm supposed to do if I call myself a Christian disciple—a follower of Jesus of Nazareth.

Jesus said to the Pharisees as they criticized him for eating with tax collectors and sinners: "Go and learn what this means, 'I desire mercy, not sacrifice.' For I have come to call not the righteous but sinners" (Matt. 9:13).

The Greek word for "mercy" is a parallel to the Hebrew word *hesed*, which means God's faithful and merciful love. According to Jesus, to offer mercy means to extend faithful and merciful love. When Jesus told the folks to "go and learn what this means," he was referring to a phrase used by both the prophets

Micah and Hosea.[2] God said to the people of Israel, "I don't want sacrifices made in the temple; I want *hesed*—that is, I want your faithful and merciful love." What God demands and expects of God's people is nothing more and nothing less than what God offers from the divine self: extravagant, indiscriminate, undeserved, abundant, unconditional, all-inclusive love.

This love offered by God, demanded and expected of God's people, was embodied in the person of Jesus. Jesus loved everybody: tax collectors and tax payers; pious women and those of ill repute; high-ranking military officers and conscientious objectors; the deaf, the blind, and the lame. Jesus loved the brilliant scholar and the village idiot; the ruthless merchant and the honest farmer; the exalted governor and the common thief; the rich and the poor; the oppressed and the oppressor; the clean and the unclean; the religious and the nonreligious. Jesus also loved the birds of the air, the animals of the land, and the fish of the sea. He loved the flowers, the grass, the trees, the water, the sun, the moon, and the stars. Jesus loved the earth, the sky, the universe, and the rest of God's creation.

Nobody and nothing was exempt from his love. Jesus loved people whom nobody else could love. He loved folks who couldn't love each other. He loved individuals who couldn't love themselves. He loved those who had looked for love in all the wrong places. He even loved those who tried to destroy him. "Abba, forgive them" were words of unconditional love. Jesus' love was not about respectability; it was about acceptance. Thus he said, "I have come to call not the righteous, but sinners" (Matt. 9:13).

Jesus' table fellowship—those with whom he ate—was symbolic of his all-inclusive love. Everybody was invited to Jesus' dinner table, and if you showed up you couldn't be certain who else would be in attendance. Sometimes, it would be quite a surprise. Imagine receiving an invitation to the party of a very important person—the most popular kid in school or the most prestigious family in town. You probably would feel that you were special because you were invited to this particular party. You probably would be flattered, because you thought you weren't very important, maybe even a nobody. You accepted the invitation without hesitation. You made all of your preparations. You got to the party and guess who else was there: all the other nobodies in town; plus all the losers; all the outcasts; and all the despised and despicable folks around town. How would you then feel?

This is exactly what happened when Jesus gave a dinner party, and the disciples often had a hard time with his guest list. But whenever they challenged his enlarging of the circle or sought to protect him from unseemly folks, he rebuked them. "Blessed are the poor and the hungry," said Jesus; not blessed are the respectable, the successful, and the self-confident. Our Lord was one hundred percent clear—there are no insiders and no outsiders in God's domain. All are included because it is the very nature of God's love to include us all.

Now we come to the church—the beloved body of Christ. Does the church really love in the inclusive spirit of Jesus? We have made so many rules about who's in and who's out; what kind of love is acceptable and what's not; who's invited to the party; and who can or cannot feast at the table. In my humble opinion, the continuing debate over homosexuality, women in the ordained ministry, salvation, and an open invitation to the communion table—as well as the ongoing struggle with race and class in the church—is an utter embarrassment. It misses the whole point of Jesus' ministry. Frankly, I can understand why so many spiritual seekers avoid church; it's not always very spiritual; nor does it really welcome seekers. We've forgotten, or maybe like some of the disciples, we never really learned the essential message of Jesus—radical, abundant, indiscriminate, extravagant, inclusive love.

I learned about *hesed* from a most unlikely group of people—commercial sex workers, commonly known as prostitutes. For twelve years I worked with the prostitutes in Paterson, New Jersey. The corner in front of my former church was, in fact, one of the principal prostitution hot spots in all of northern

New Jersey. The women who worked on my corner came in all shapes and sizes: as young as fourteen and as old as forty; black, white, Hispanic, and Asian; poor, middle-class, and even wealthy; single and married; with and without children—often with children in the child welfare system. Many were HIV positive; some had full-blown AIDS, tuberculosis, hepatitis, and other infectious diseases. Many had outstanding bench warrants and other legal difficulties. Some were gregarious and outgoing; others were shy and reserved. Some were very smart and some had serious learning disabilities; a few were mentally ill, and most were depressed. Almost all were addicted to one drug or another.

In the beginning of my tenure, I would pass these women standing on the corner as I pulled in and out of the church parking lot. They were doing their work and I was doing mine. I would wave; some would wave back; and others would glare at me or avoid my eyes. Then I started walking by them and saying hello. Again, some would respond; others would stare at me or avoid my eyes altogether. At block club meetings, I would listen to neighborhood residents complain about the prostitutes. At clergy meetings, I would listen to my colleagues condemn them. As for me, I was both fascinated and repelled by these women. They were some of the smartest, funniest, most ingenious, courageous, capable, and messed up women I've ever met. I was simultaneously fearful and curious about them. I both admired and pitied them. It was a come-close, go-away sort of reaction: a combination of voyeurism and honest concern.

One day, it dawned on me; these women of "ill repute" were part of my geographic parish. While I worked inside the church, they worked on the outside. In the words of folk singer Leonard Cohen, "They were the sisters of mercy."[3] They ministered to the lonely and the broken-hearted, just like me. In fact, pushing the metaphor a little, they ran a church of sorts on the same block. More importantly, just like me, these women were made in the image of God. They were my neighbors, my sisters in Christ, a part of God's beloved creation; and I needed to get to know them. So I did.

At first, I shared cigarettes, coffee, and donuts with them. We had cautious and casual conversation about the weather, the block, and neighborhood gossip. Then over meals at McDonald's, we began to talk about our work: my ministry to redevelop this inner city church and their activities on the street. I would ask questions about their work, and they answered honestly, sometimes saying, "Rev. Tracey, you don't really want to know about that, do you?" As time passed, we became more than acquaintances but not yet real friends. When they would disappear for a while, I would worry. But they would reappear saying they had been locked up or were drying out. It was a vicious cycle of addiction, sex, and arrest. Eventually, I raised some money and hired an outreach worker; and now, a dozen years later, my former parish runs a large outreach program that works with hundreds of commercial sex workers throughout the region.

Over the years, I became a pastor and priest for some of these women. But one thing never really changed; they never felt very comfortable in church. Once in a while, they would come in and sit quietly in the sanctuary by themselves. If they were getting sober and staying off of the streets, they might even attend a church service. But if they were active—working on the streets—they felt unworthy being in the house of God. They might arrange to have a child baptized, and then wouldn't show up for the baptism. They would say, "Rev. Tracey, I'm going to surprise you one day and join the church," but it rarely happened.

Then one year, just around the time of World AIDS Day, two women died of the virus and another woman was murdered on the street—her battered body discovered by the railroad tracks. I was invited to the women's support group to talk about their loss. Well, did we ever talk about God, sin, sex, forgiveness, humility, addiction, healing, abandonment, rejection, and anger. You name it; we talked about it. At the end of the meeting I said to them, "Would you like to do something in church for your friends who died?" They said yes, and I suggested that World AIDS Day might be an appropriate time. They agreed and we began working on a service. We picked hymns to sing, wrote prayers, and decided who would do what. It was unlike any liturgical planning process that I had learned in seminary.

The long-anticipated night arrived, and people began showing up. The lights in the church were dimmed and candles were lit. The youth group was there, as were others from the congregation and community. After all, it was World AIDS Day. Frankly, I didn't know if the women were going to show up, and I was ready to cover, just in case. But at exactly 7:30, when the service was about to begin, the women came in en masse. They took their places in the front of the church and led the worship. They said the prayers and read the lessons. And some of them decided to testify—to share their stories with the attentive and inquisitive congregation. One by one they stood up at the lectern and told their tales of life on the streets as a prostitute. Then they lit candles for sisters who had died: killed by the virus, killed by the drugs, killed by the pimps and johns—dead from a life of prostitution, addiction, disease, poverty, and despair. We sang a closing hymn, I said a blessing, and we all had cookies and punch. It was a remarkable evening. There was a healing on that night and the outcast women of the street were the healers.

When the morning light came, some of my "sisters of mercy" were back on the corner again. What happened—I don't know for certain. What I do know was that God gave a party and we all showed up. And life was different, at least for a moment.

Several months later, during Holy Week, the congregation was gathered for our traditional Liberation Feast on Maundy Thursday. A number of the women from the streets joined us that evening, and when it came time for the foot washing, they presented themselves at the basin. After supper, the gathered community walked through the parking lot of the church on the way into our sanctuary to strip the altar for Good Friday. Several people paused to look up at the night sky hoping to see Haley's Comet. As we gazed upward, a flock of wild geese, the ancient cultic symbol for the Holy Spirit, flew in perfect formation over our urban parking lot right under Haley's Comet. It was an extraordinary sight. We all stood there looking up the sky, and one of the sisters of mercy quietly whispered to me, "God is good," and I responded, "All the time."

Six

It challenges...

The Demolition
Contractor

Have you ever noticed how a seemingly simple home improvement project can quickly become a complicated renovation undertaking? When I moved to Cleveland, I bought a ninety-year-old house. I decided to make a few cosmetic improvements. They were simple projects, or so I thought. Before I knew it, I was replacing steam pipes, plumbing, electrical wiring, and even the steel beams holding up the house. I was moving walls, repairing ceilings, replacing windows, and restoring floors. And everywhere I looked, there was dust, dirt, and disarray.

The same thing happened at my new job. Less than twenty-four hours after the moving van pulled out of my driveway, I received the architectural plans for the redevelopment of the cathedral and diocesan headquarters. At first, it was going to be a simple project—remodel some offices and classrooms and put a fresh coat of paint on the buildings. But as the building committee met, they began to realize how much work these century-old buildings really needed. They also began to consider issues of environmental stewardship and the demands of the information age. Then they started to dream and envision the possibilities for mission and ministry, and by the time I arrived as the cathedral's new dean, a seemingly simple church renovation project had become an exciting but major restoration and construction effort.

As I sat in what felt like a war zone, with contractors invading both my home and work, I took comfort in C. S. Lewis's wonderful metaphor of the "living house."

> *Imagine yourself as a living house. God comes in to rebuild that house. At first, perhaps, you can understand what He is doing. He is getting the drains right and stopping the leaks in the roof and so on: you knew that those jobs needed doing and so you are not surprised. But presently He starts knocking the house about in a way that hurts abominably and does not seem to make sense. What on earth is He up to? The explanation is that He is building quite a different house from the one you thought of—throwing out a new wing here, putting on an extra floor there, running up towers, making courtyards. You thought you were going to be made into a decent little cottage: but He is building a palace. He intends to come and live in it Himself.[1]*

It was as if John the Baptist had entered my life shouting: "Prepare the way of the Lord. Cut down the trees that don't bear good fruit. Burn the chaff left on the threshing room floor. Clear the stones away. Get ready to build God a mansion."

If Jesus is a carpenter, then John the Baptist is a demolition contractor. John the Baptist is like my friend, the explosives guy. He uses dynamite to explode bedrock so that roads and highways can be built through mountains. Some people might think this is a politically incorrect vocation. But let us remember the words of the prophet Isaiah: "I will go before you and level the mountains, I will break in pieces the doors of bronze and cut through the bars of iron" (Isa. 45:2). Sometimes, explosives are required to level the ground and break down the doors and cut through the bars.

John the Baptist is the contractor hired for the first stage of reconstruction, the tearing down so that the new can be built up. Living as an ascetic in the desert—eating odd food, wearing odd clothing, speaking in an odd manner, John knew that the Lord was coming. How or when or where he didn't know. He just knew that people had to get prepared for the great restorer of humanity, the repairer of the breach, and the salvation of God.

John the Baptist teaches a harsh lesson: a cleansing of the hands, feet, arms, and legs, the whole body. John's baptism is the treatment you received as a kid for talking dirty and your father washed out your mouth with soap expecting that you would clean up your language. It is what happened when you played in poison ivy and your mother scrubbed your skin raw hoping that some day you would learn to recognize the three-leafed plant. John's baptism is intended to clean away the old so that the new can be born. And that cleansing can be pretty abrasive and harsh.

John's baptism has nothing to do with feeling sorry for yourself. It is not about listing all the things you wish you had not done in your life and feeling bad about them. Nor is it wishing you were a better person, keeping track of your faults as if God could be persuaded to overlook them. John's baptism is not an intellectual exercise described in a self-help book.

John's baptism is like a cold shower, a bright light, or an assaulting alarm clock. John's baptism is what happens when you encounter poverty for the first time and realize how much wealth you really possess. John's baptism is how you feel when you finally get in touch with your own racism, classism, or homophobia. John's baptism is waking up in the middle of the night with a lump, a pain, or a fever and beginning to rethink your life. John's baptism is when you've got it all together—your career established, your savings plan begun, your daily routine in order—and someone else (an elderly parent, a sick friend, a child in trouble) comes into your life and turns it upside down. John's baptism is what happens when a city wakes up to the reality that its last major steel company has gone belly up and thousands of people are unemployed. John's baptism is turning the radio on one morning to find that two hijacked commercial jetliners have destroyed the World Trade Center, killing thousands, and thrusting the world into a "War on Terrorism."

The baptism of John is a rude awakening, often a painful crisis, and almost always a profound wake-up call that tears our individual or corporate life apart so that it may be put back together again. This baptism of awakening often leads us to frightening places where we realize that we are not immortal, invincible, God-only wise, but rather, that our mortal lives are fragile and foolish.

Repentance is the turnaround that follows the crash; the rebuilding that follows the demolition, the construction that follows the destruction. Repentance is a change of course, a change of heart, mind, and life. Repentance is too busy redeeming the present to apologize for the past. It is a matter of being grasped by God, of being picked up and put down so that everything looks and feels different.

With repentance, you lose your old bearings and you get new ones. And when it happens—however it happens—life is changed and can never return to where it was. Repentance is deliberately and intentionally changing the pattern of one's life and creating a new one.

The remarkable thing about repentance is that we get to do it over and over again. Repentance is not something that happens only once. Life is full of turning points, both large and small. If we miss the opportunity for repentance the first time, if we become complacent and forget, if we choose to ignore the invitation, or if we just can't help ourselves, the opportunity for repentance will present itself again and again. That's the way life is.

When the demolition is complete and the dust settles, when the storm has passed and the waters recede, when the illness is over and you start to feel like you might live, when the old pattern is really gone and you're looking at a blank palette, then, and only then, can the reconstruction begin in earnest; then, and only then, can new life and new patterns appear. This process of regeneration, reconstruction, re-creation, doesn't happen overnight. It takes time.

Sometimes after the demolition, as we try to figure out how to reconstruct our lives, there's nothing we can do but wait, watch, wonder, and hope. We can join a fitness class. We can read a self-help book. We can update our resume. We can prepare a budget. We can give away more money. We can change our lifestyle. But after all the actions are taken, we still have to wait and watch for the new life to emerge. We can plant the seeds, but we have to allow them time to germinate, sprout, and bloom. We can paint the room, but we have to give it time to dry. We can start to exercise, but our muscles need time to develop. We can begin the treatment, but we have wait for it to work.

And in the meantime, the time being, the time between time, as we wait for something good to happen, as we watch for the new beginnings, as we wonder what the future holds, as we hope for rebirth, we are called to trust and to believe that we are "a living house." We are called to trust and believe that God is rebuilding our house, that instead of being made into a decent little cottage, God is constructing a palace better than anything we ever dreamed or imagined is possible.

The good news of John the Baptist is not the demolition that he demands. Rather it's the new construction that he promises. This good news is true in our individual, institutional, and global life. I have certainly learned this truth as a city planner, nonprofit executive, parish priest, and cathedral dean at the dawn of a new millennium. Ours is an age of deconstruction and reconstruction. The institutions we've known, loved, and respected have been challenged to the depth of their cores. Prior to entering seminary, I served two such organizations: the Girl Scouts and the United Way.

In our parents' and grandparents' generations, scouting was at the heart of American life. Many families expected that boys and girls would join the scouting movement and rise up through the ranks of Brownies and Cub Scouts to Senior and Eagle Scouts. I can still remember the excitement of getting my first scout uniform, the awesomeness of being inducted, the hard work of earning service badges, and the pride of wearing them proudly on my sash. To this day, I can remember my brother and I dressed in our scout uniforms attending our first minor league baseball game. In the long run, the Girl Scouts didn't really take with me. I got bored with making carrot cake and wanted to learn rope tying instead.

But back then, rope tying was not part of the Girl Scout curriculum, so I dropped out.

In 1981 I returned to scouting as development director for Patriots' Trail Girl Scout Council in Boston, Massachusetts. This time I had the privilege of overseeing one of the world's largest cookie sales and raising money for tents, latrines, and inner city program development. It was hard work persuading donors of the relevance of the Girl Scouts in the 1980s. As more women entered the workplace and feminism became an acceptable word in polite company, we had to convince the public at large that the Girl Scouts were still capable of preparing young women for the future. In a campaign to recruit men as scout leaders, we developed slogans like "You don't have to be someone's mother to be a leader among girls"; and in campaigns to recruit professional women we proclaimed, "The Girl Scouts are looking for a few good women." Where were the simple days when we could go off to troop meetings, parades, and camp without needing to reinvent ourselves and our beloved organization?

I experienced much of the same need for reinvention when I worked for the United Way right after graduate school. Once the most creative and forward thinking civic arm for raising and allocating voluntary funds, United Way has recently come under scrutiny, attack, and suspicion. The official gatekeepers of civic philanthropy are constantly being challenged about inclusivity, diversity, competency, honesty, and fairness. No longer are employees willing to say, "Of course, I'll give a percentage of my pay to the United Way, just because my company expects me to do so." Now people insist on opening the doors of workplace giving to a much wider variety of nonprofit organizations. This trend is good for other community organizations previously excluded from workplace giving, but tough on the United Way. As the voluntary, not-for-profit sector matures, will those familiar organizational names established in the last century survive this period of institutional deconstruction and be reconstructed into something new and better? Or, will they go by the wayside into what Parker Palmer once coined, "the golden age of memory?"[2]

When I entered seminary to prepare for the ordained ministry, I clearly chose to affiliate with another institution under attack by the winds of time. Without a doubt, the church of my generation is an institution in various states of demolition and reconstruction. The turmoil the church is experiencing over modernity, diversity, inclusivity, and justice is symbolic and symptomatic of the tumultuous nature of modern society. As one older parishioner remarked to me at a recent coffee hour, "I used to associate church with the words 'stability' and 'permanence.' Now I use the words 'chaos' and 'change.'" Urban cathedrals, perhaps even more than rural and suburban congregations, are a paradox of change in the midst of permanence and chaos within the walls of stability.

At the 2000 North American Cathedral Deans Conference, the Rev. Dr. James P. Wind, president of the Alban Institute, suggested in his keynote address that the cathedral is "a feudal invention" and "an emblem of a lost world," thus an anachronism in post-modern American life. In the Middle Ages, when the great cathedrals of England and northern Europe were built, the cathedral and the city lived in "an intimate and symbiotic relationship." "Our challenge," he said, "is to take an institution created for that world and somehow make it fit a very different one." He suggested that we think of our cathedrals as "reconstruction sites . . . where people find resources with which to reconstruct their lives . . . places where people reconstruct—or build anew—relationships with God."[3]

As I listented to Dr. Wind I was reminded of the late Bishop Paul Moore's thoughts about the unfinished Cathedral of St. John the Divine in New York City. I recall Bishop Moore saying that in his opinion, this great cathedral, located in the heart and soul of urbanity, should never be finished until God's justice is realized and there is a house not just for God, but a house for every one of God's people.

Here I was a new cathedral dean, sitting in a conference of deans, thinking lofty thoughts about the role of the urban cathedral as I stood on the precipice of a major construction project. We were about to redevelop the entire campus of Trinity Cathedral and the Diocese of Ohio. What would the future bring? I wondered.

On Ash Wednesday, at the beginning of our annual Lenten journey, we vacated our parking lot and the Parish House (which included the meeting rooms, offices, kitchen, pantry, even the restrooms) to make way for the construction. At the annual congregational meeting preceding the move, I presented our situation as an adventure saying, "For the next six months, we will learn how to live in this Cathedral in an unfamiliar and somewhat primitive manner, much like our ancestors did a century ago. This phase of construction will force us to simplify our community life so as to become a one-room church for a while. Frankly, I'm thinking of it as an Exodus: a journey into the wilderness with the bare necessities—maybe just the sacred texts, the holy table, and the tent of meeting."[4]

Over the course of our eighteen-month construction project, Trinity Cathedral was transformed into a chaordic organization. The word, chaordic, means "the blend of chaos and order." According to its proponent Dee Hock, a chaordic organization releases what people have in the depth of their being— the passion they have—and the integrity they bring to the attempt.[5]

This word aptly describes what is still going on at Trinity Cathedral and what I think is called for in the life of a cathedral. We have become a place of organized chaos where people are released at the depth of their being to realize the passion they have and the integrity they bring to it. Through the very bricks and mortar of the cathedral, as well as the preaching, teaching, art and music, outreach, healing, and community that happens in a cathedral, individuals are invited to discover their true vocation: "the place where their deep gladness and the world's deep hunger intersects."[6]

In the middle of Trinity Commons, our newly redeveloped campus, there are two piazzas connecting three late-nineteenth- and early-twentieth-century buildings with the great gothic cathedral. The outside piazza faces the main street of Cleveland. It functions as an outdoor dining area for our café and as sacred public space in the midst of a bustling, urban, university neighborhood. The indoor piazza is a large circular building complete with fountains, symbolizing the river of life, the waters of faith, and the various bodies of water that flow through our city and region.

This piazza design was influenced by the teachings of St. Angela de Merici (1474–1540), a third-order Franciscan and founder of the Ursulines, a religious community of women. In her writings, St. Angela, a native of Northern Italy, instructed her sisters to "be like a piazza."[7] She told them to be open, gracious, hospitable, and engaged in the world—to stand with Christ in the crossroads of life.

A few summers ago, I had the opportunity to travel to Desenzano, Italy, the birthplace of St. Angela. I sat in her piazza, I prayed in her church, and I visited the site of her vision, an ancient olive grove outside of the city.

The olive grove was a peaceful, serene, and well-ordered orchard with wise old trees silently worshipping the midday sun. I imagined Angela taking a break from her harvest chores to sit quietly in the middle of the grove, listening to the wisdom of the trees, and experiencing her call to ministry. Angela's hometown piazza was just the opposite of the olive grove. It was (and still is) a lively community square surrounded by restaurants and shops, the church, and a few civic buildings. I could envision Angela visiting with neighbors and sojourners as she drew water from the town's five-hundred-year-old well. It's not surprising that Angela had a sense of order and stability in her life as well as an energetic faith lived at the crossroads of a piazza.

Each Italian piazza is different in ethos, but all have essentially the same characteristics. They are open, inviting, restful, refreshing, and yet filled with energy. They are found in the center of the city and tucked away in neighborhoods. They are centers of commerce, culture, charity, conversation, and collaboration. Since most Italian piazzas are located in front of a church, they also are places of worship, protest, and celebration. You never know what will happen in a piazza; rather, you should always expect the unexpected, the serendipitous, and the spontaneous.

Being steadfast like an ancient olive grove and hospitable like a piazza were the centerpieces of St. Angela's spirituality. And that was and remains our hope for Trinity Commons. We have constructed and are continuing to create a piazza to express the unity, integrity, and community of God, the city, the church, and the rest of creation.

Trinity Commons is becoming everything we hoped it would it be. University students and faculty are crossing the street for the first time in decades; people are gathering to work on peace, justice, and environmental issues; artists are creating in the light and shadows of the great cathedral; and seekers are discovering this sacred public space. When I walk in the café, I frequently see people reading our cathedral newsletter as they drink a cup of coffee. When I go in the cathedral, I often watch pilgrims silently walking the ancient way of the labyrinth. When I wander into the bookstore, I see children and adults of all ages and backgrounds browsing the shelves filled with books about religion and spirituality. When I enter the Ten Thousand Villages gift shop, I hear conversations about fair trade. When I gaze into the gallery I see exhibits that explore the intersection between art, religion, and daily life. In a short time, Trinity Commons has become a place of culture, charity, commerce, conversation, and collaboration—a commons for making connections and a holy place to worship, celebrate, create, and hang out.

The building of Trinity Commons, as the campus is called, has deeply transformed the mission, ministry, and daily life of the cathedral. Once a dark and dreary place, we have been overcome by light. And the light attracts new people. And with new people come new ideas. And with new ideas, there's even more change. It's a never-ending, ever-expanding, open circle, a chaordic institution, otherwise known as God's construction zone.

A few years ago, following a lecture called "Why Christianity Must Change or Die,"[8] I was seated at a dinner table next to the speaker, my former bishop of seventeen years, John Shelby Spong. In the midst of a lively table conversation, I said to Jack, "You frustrate me. You spent all these years serving the church, leading worship, proclaiming the creeds, and now, when you're retired, you are dismantling, demolishing, and deconstructing them—leaving the rest of us with a mess to clean up." Jack, in his ineffable and unflappable way, immediately responded, "Tracey, you're absolutely correct. So what are you going to do about it?"

Long after that evening, it dawned on me that I have followed several deconstructionists in my career, Jack Spong being the most recent, influential, and notable. On more than one occasion, I have found myself picking up where they left off, and beginning a new phase of building. As usual Jack was correct in his response to my frustration. If my mentor's vocation has been about deconstruction, then mine is about reconstruction. I, and other leaders of my generation, are called to reconstruct our sacred and civic life for a new millennium.

Some of our beloved, familiar, and trusted institutions, structures, tools, and ideas will be left by the wayside; others will be demolished and rebuilt; and new ones will be constructed. Yet, the noble foundations of our common life will remain, buried deep within the soil and close to the surface. On those solid foundations, we will continue to build. Other foundations made of toxins or rotting material will have to be carried away so that the new may be built on clean ground. The demolition of this age will be painful for some, a relief to others, and wrenching for most of us.

Annie Dillard was correct in her observation that while seated in a church pew, "We should all be wearing crash helmets."[9] For God is not finished with us. God is still building a great mansion, a living house, a home for all of creation, a body worthy of the divine residence. We just have to remember that construction usually involves both carpenters and demolition contractors.

Love
for the Fallen Flower

Every time I bemoan the ninety-degree heat and humidity of summer in the city, I am reminded of a 1999 *New York Times* op-ed entitled "Indifferent to a Planet in Pain." The author, environmentalist Bill McKibben, argues that we are as blind to the threats that plague our environment as our parents' generation was blind to the wrongs of segregation.

> *They were good people of good conscience; so why had the inertia ruled for so long. . . . It took the emotional shock of seeing police dogs rip up the flesh of protesters for white people to really understand the day-to-day corrosiveness of Jim Crow. We need that same gut understanding of the environmental situation if we are to take the giant steps we must take soon.*[1]

Roger Gottlieb, in *The Spirituality of Resistance*, likens our avoidance and denial of the environmental crisis to the world's early response to the Holocaust. He says this ecocide is "so vast and yet so diffuse that it is extremely hard to hold it in our minds."[2] But the crisis is very real. In the words of the prophet Hosea, "The land mourns and all who live in it languish; together with the wild animals and the birds of the air, even the fish of the sea are perishing" (Hosea 4:3). We are living in a world in pain.

The environmental sentinels have been crying in the wilderness for sometime now. Rachel Carson, a marine biologist with the U.S. Fish and Wildlife Service, was one of the first ecological prophets to speak out. In her landmark book entitled *Silent Spring*, Carson warned of the dangers of pesticides. It's ironically funny, in a scary way, that when *Silent Spring* was first published in 1962, I and my playmates were busy riding our bicycles behind the mosquito-killing DDT truck as it passed with its death-dealing

191

fog through our middle-class suburban neighborhood. We imagined that we were riding into the clouds of heaven; little did we know that were breathing in toxic pollutants.

Not surprisingly, both *Silent Spring* and Rachel Carson met considerable resistance from U.S. industrial leaders. Major chemical companies tried to suppress the book's printing and distribution, and when excerpts appeared in *The New Yorker*, a chorus of voices immediately accused Carson of being hysterical and extremist.

But Rachel Carson did not give up. She continued to plead with passion, devotion, and scientific knowledge for her beloved environment. At the end of her life, she wrote: "The beauty of the living world I was trying to save has always been uppermost in my mind—that, and anger at the senseless, brutish things that were being done. I have felt bound by a solemn obligation to do what I could—if I didn't at least try I could never be happy again in nature. But now I can believe I have at least helped a little."[3] Rachel Carson's determination to speak the truth and work for environmental justice over the long haul was an act of deep heart love for the earth and its inhabitants.

In the scriptures, we are instructed to practice our faith by loving deeply from the heart. This commendation comes from the First Letter of Peter (1:22). It is a message that was presumably written by an early Christian leader, sometime around the last part of the first century and circulated among Christian communities in Asia Minor. Addressed to new converts, "strangers in a strange land," as they were called, this ancient letter was considered a guide to early Christian living. Two thousand years later, the First Letter of Peter is still relevant and applicable, especially as it speaks to our environmental dilemma.

The text beckons the reader to be attentive to all the paradox, ambiguity, and mystery of the Christian story, and then to get busy with faithful living in the world. At its core, the message invites us to be an Easter people, empowered and enlivened by the resurrection—the good news that death is not ultimate—so that we might do as Jesus did and taught, love one another—deeply from the heart.

What does it mean to love deeply from the heart? In essence, it means to live with compassion, seek and serve God in all creation, love your neighbor as yourself. But in our present age, we have to broaden our definition of neighbor to include nonhumans.[4] Our neighbors on earth are also the soil, air, and water; forests, deserts, mountains, and glaciers; horses, cats, monkeys, and birds; grass, trees, and plants. There are at least ten million, maybe as many as one hundred million species on earth. All of them—the bright rumped attila, the slime mold, the long-necked agra, mating termites, common dog fleas, and even the fallen flower—are our neighbors in the eyes of God.[5] As the Quakers teach, to love deeply from the heart is to see God in all creation and every living creature, both great and small.

Loving deeply from the heart is not easy. It demands self-sacrifice, a generous spirit, and an expansive embrace. It implies the risk of abandoning the baggage and letting go of old habits that become burdens on the journey. Deep heart love for the earth can be as simple as leaving one's car in the garage and riding the bus to work. It is as complex but as essential as converting the automobile industry from gasoline-dependent combustion engines to ecologically sound cars in order to reduce greenhouse emissions and depletion of fossil fuel reserves. If all the members of all the faith communities in our nation made a commitment to make their next purchase of an automobile a hybrid car, think of the impact that would have on the U.S. auto industry. They would respond to market demand by producing more ecologically sound cars.

Loving deeply from the heart is to speak and act with compassion from the heart, asking ourselves along the way: is what I'm about to do or say helpful, useful, and truthful? Is it kind? Does it help those around me grow in grace? Does it respect the dignity and sacredness of the other? Does it seek to build up or to break down the common good? Deep heart love for the earth asks us to stop spraying our lawns with pesticides and to learn to cohabit with the insects that also enjoy our gardens. It also summons us to support agricultural and industrial policies that protect the earth's biodiversity from extinction.

Loving deeply from the heart is intentional. It loves in spite of us, even if it hurts. It means getting beyond what holds us back. Deep heart love makes peace with the earth and those who inhabit it. It invites us to walk in the woods, paying attention to the sound, the smell, and the touch of trees. It bids us to recycle our newspapers and purchase recycled paper products. But it doesn't stop there. Deep heart love for the earth challenges the expedient, profitable, but shortsighted practices of the logging industry and all its tangential markets that destroy tens of thousands of square miles of primary forest every year.[6]

Loving deeply from the heart is about forgiveness. It remembers Jesus' words from the cross, "God forgive them, for they know not what they do." It forgives previous generations for lead paint, asbestos, and PCB's. Loving deeply from the heart also means praying, "God forgive us even though we know what we do." Deep heart love forgives our nation's refusal to meet the Kyoto protocol and our own generation's urban sprawl littered with gas guzzling motor vehicles. At the same time, it calls us to active repentance, to a change in both attitude and behavior. It says, "don't just sit there feeling bad; do something to make it better." When we are able to forgive and be forgiven, we make room for repentance, and repentance leads to a different kind of living.

Loving deeply from the heart gets beyond that which keeps us stuck in the past and prevents us from moving into the future, such as thinking that natural resources are infinite and defining "use as our primary relationship with the planet."[7] For people of faith, deep heart love for the earth calls us to reexamine our age-old creation theology of human dominion over the earth and its nonhuman inhabitants. Deep heart love for earth encourages us to reimagine a new cosmic theology, a new story of God, humans, and the rest of creation.

Sabbath is an intentional part of the cosmic theology, and Holy Scripture tells us that on the seventh day God rested and admired the creation. As a part of the divine plan, God instructed the creatures to keep a rhythm of activity and rest. Sabbath has been understood from the beginning to be an essential element of life's pattern. While much of nature and her creatures understand and honor Sabbath time by resting from labor and play, we humans (particularly in the developed nations of world and most especially in the United States) have decided that this aspect of divine wisdom can be dismissed and negated, and we now live in a twenty-four-seven world of work, shopping, entertainment, and travel.

Perhaps, if we were intentional about Sabbath time, we might begin to hear the new story longing to be told. If we closed down the shopping malls, mega stores, and entertainment centers for just one day out of each week, think how much energy we might save and fuel emissions we might reduce. Since our religious faiths can't agree on a common Sabbath day, maybe for the sake of the earth and all its creatures (including humans), the new cosmic story is calling for a new Sabbath day. How does Wednesday sound to you?

Deep heart love is a call to conversion. In the words of that old Shaker hymn, it is "to bow and to bend . . . to turn and to turn . . . till we . . . come round right."[8] When it comes to the environment, most of us need a wake-up call to conversion. The environment that sustains life on earth in is pain. Some would say it's dying but still resilient. Our neighbors—the whales, the bats, the cranes, the oceans, the sky, and the land—are hurting and wounded, and human actions are largely responsible. Time is short, and the problems are profound. Radical change is long overdue. Theologian Jürgen Moltmann says: "What we call the environmental crisis is not merely a crisis in the natural environment of human beings. It is nothing less than a crisis in human beings themselves."[9]

Frankly, I'm having a hard time with my environmental wake-up call. It is a rude but necessary interruption of God's grace. I'm struggling with this conversion to an intentional, ecospirituality, because I know it will require an essential change in how I live and work in this world. My friend Bob Morris says that as we pass through the change of heart required for radical spiritual awakening and conversion to action, we must begin with a period of grief. "The ancients called it 'the sorrow that leads to repentance'."[10] Like many of you, I have just begun to love the earth deeply from my heart, but I have vowed to heed the voice of God's prophets and not be "indifferent to a planet in pain."

As I walk outside carrying the kitchen garbage to our compost pile, I pause to listen to the birds, smell the flowers, watch the squirrels, admire the insects, touch the trees, and look up at the sky. In doing so, I recall the wisdom of Teilhard de Chardin: "The future of the earth is in our hands. Let us then, for the love of our Creator and of the universe, throw ourselves fearlessly into the crucible of the world of tomorrow. . . . The task before us now, if we would not perish, is to build the earth."[11] And then I recommit myself to the work of honoring and healing the earth and all of its creatures, both great and small. May we all do the same for the love of the fallen flower.

The Trinity
of Love

Glimpses of God are usually momentary and fleeting. Noah got a rainbow and Moses stood before a burning bush. Abraham encountered God in a vision, and his son Isaac was rescued by divine power on the altar of sacrifice. Elijah heard the Eternal Voice in the sound of silence while God answered Job out of a whirlwind. Isaiah fell prostrate on the ground as a cleansing coal from the eternal flame burned his lips, and Ezekiel saw the Holy One in visions by the River Chebar. One Mary encountered God through an angel, and another Mary met the Risen Christ in the garden. Peter saw him walk on water, Thomas touched his wounds, and his light blinded Paul. I encountered the fullness of God in a McDonald's restaurant. Go figure.

Those who have wrestled with God in the night walk with a limp for a reason. They always lose the match and carry the scars of battle. I am no different. A few years ago, I had my neck rebuilt, and I am now the proud owner of a few donated bones, a titanium plate, and four screws. I sometimes joke that I ruined my neck while wrestling with God. Actually, I think it might be the truth of the matter.

My wrestling with God got fierce during my first semester at seminary. I went to New York City to grapple with the question of faith and explore an unrelenting pull to ordained ministry. At the Union Theological Seminary, as I pursued my studies with vigor, I found myself entering into dialogue and debate with "the doctrine, discipline and worship of the Episcopal Church."[1] I decided that I had better sort this out, as I would be obliged to declare loyalty to its authority and teaching if I ever became ordained. I carefully read the Christian theologians assigned by my professors and tried to find meaning and validity in their thinking. I dug deep into the Bible, examined the lessons of church history, and started thinking about ethics from a religious perspective. I began to test my vocation of ordained ministry as a seminarian in a local parish. But perhaps most importantly, I started facing directly into my own darkness and confusion as I worked with a spiritual director and prayed for clarity, guidance, and a deeper knowledge of God.

By the end of my first semester, I had talked and rationalized my way out of pursuing ordination. I was convinced that my theology was inconsistent with the church's orthodoxy, that my reasons for seeking ordination were not valid, and that I was not "called" by God to be a priest. My reasoning was grounded especially in the fact that I could not claim the traditional words of the Trinity—"Father, Son, and Holy Spirit"—to express my understanding of God. For me, those words were not adequate. I knew them by heart, I could say them by rote, and sometimes they offered a poetic sense of the familiar. Yet, they simply did not express my understanding of the Divine.

In the first place, my God was much more than a father. My God was also a mother, a friend, a lover, a power, an action, a source of energy, a way of being—something bigger and more encompassing than my childhood image of God as an old man with a long white beard sitting high up in the clouds. My God was a whirlwind and a gentle breeze, a blazing fire and a speck of light, a roaring sea and a babbling brook, a loud thunderbolt and a still small voice, a symphony orchestra and an unspeakable truth.

I also had a hard time with Jesus as God's only son. While I could affirm that the historical Jesus of Nazareth was a male human being, it was difficult for me to proclaim the words of the Nicene Creed, "He came down from heaven . . . and was made man." I simply did not believe that God among us could only be expressed in the male gender. What about me; wasn't I am made in the image of God?

But gender was not the only issue in my struggle with the doctrine of the Trinity. I had often wondered if Jesus was God's son, how could God be in unity? How could God be one? And how could Jesus be both human and divine? It just did not make sense. And if that wasn't enough, I could not get a firm grasp on the concept of the Holy Spirit. Was the Holy Spirit the spirit of God, the spirit of Jesus, or the spirit of their relationship?

Finally, I had difficulties with the notion that it was only through Jesus and the Holy Spirit that I could be in relationship with God. Did this mean that I couldn't talk directly, on my own with God? Did I really need a mediator and advocate before the Almighty?

By the end of my first semester in seminary, I was ready to discard the Trinity (or at least the Trinitarian formula of Father, Son, and Holy Spirit) from my spiritual framework. I did not want to limit God in these ways or put the Eternal One in a box. And I wanted to focus more on how God relates to us rather than how God relates to Godself. At the same time, I knew with greater clarity and certainty that I was "called" to a vocation of ministry—serving, giving, teaching, acting, and living life sacramentally (that is making the invisible grace of God visibly known in my life) and helping others to do the same. In short, I was ready to reenter the secular world as a committed layperson, an active and intentional agent of God in the world.

And then something happened that I could not negate or ignore. As I mentioned in the introduction of this book, one afternoon in late January I met God on Forty-second Street in New York City, and I had a conversation with the Holy One that changed everything. It was the interruption of a lifetime, and it came from the unexpected edge, that place on the margin where earth and heaven intersect, where the profane and the sacred collide, where the conscious ego gives way to unconscious self. There weren't lightning rods, thunderbolts, whirlwinds, or burning bushes. There was no arrow, ladder, broom tree, star, winged creature, chariot of fire, or even writing on the wall. I was not in the belly of a whale, shipwrecked on an island, standing on a mountaintop, chained in prison, or hiding in a cave. I was sitting in a restaurant in the middle of New York City, and I heard a voice and felt a presence that simply would not let me go.

I was Jesus on earth, but I'm still God, and I'm here with you now." That's what the voice answered in response to my question, "Am I talking with God or Jesus?" "Is there a difference?" asked the voice. "There has to be," I replied. "Why?" inquired the voice. "Because my theology will fall apart if Jesus is still present." I had so carefully sorted out my theology about God and Jesus. In fact, I did have God in a box of my own making, and I didn't want to mess it up. But I couldn't deny what I had just heard with my very own ears, *"I was Jesus on earth, but I'm still God, and I'm here with you now."* And though it exploded and destroyed all my fancy theories about God and Jesus, it made sense. I had to say "yes." I had to own it. I did not feel like I had much choice in the matter but to affirm and incorporate these words into my life.

For a few weeks, I kept silent about my conversation with God. I didn't know how to talk about it, and I didn't want anyone to think I was crazy. After all, I had heard a voice, and we all know what they say about people who hear voices. Then, in prayer, I had an ah-ha moment. I realized that the voice had articulated a clear, concise, and definitive statement about the Trinity. I now had the words to make my own confession of faith in a God who somehow was three-in-one, one-in-three, an undivided unity.

The voice said, "I was Jesus on earth." I remember thinking that the voice must have said, "I was with Jesus on earth." But that was not the case. I looked at my notes. I wrote this conversation down verbatim: *T* for Tracey and *G* for God. I still have the original text. The voice claimed to be Jesus on earth, and I had to accept that reality. The voice didn't say, that Jesus was God's only son. It simply said that Jesus somehow embodied or incorporated the very essence, the fullest expression of God in the world.

Then the voice said, "I am still God." Well, how could I argue with that statement? I asked, "Can I see you?" The voice responded, "No, not now." "When?" There was no answer. To this day, I still recall my embarrassment for asking to see the Holy One. I must not have paid attention to God's admonishment when Moses asked the same on Mt. Sinai, "You cannot see my face; for no one shall see me and live" (Ex. 33:20). The voice did not define its essence. The voice simply spoke the ancient Hebrew name for the Holy One, "I am," and in the silence of my heart I replied, and "You will be who you will be."

Then the voice said the words that have reassured people of faith down through the ages, "I am here with you now." The power and presence of the Divine that had been with Jesus on earth was still God and here with me now. For twenty years, almost half my lifetime, I have pondered these words. I have prayed, written, and preached about them, but I cannot fully comprehend them.

My struggle with the doctrine of the Trinity can be compared to those of generations before me. As I've learned over the last two decades, most descriptions of the Trinity—Father, Son, and Holy Spirit; Creator, Redeemer, and Sustainer; God unbegotten, God incarnate, and God among us—emerge from the living experience of ordinary women and men trying to respond in human words to God's activity in the world and to the event of Jesus Christ.

To put it in an Anglican or Episcopal perspective, the statements about the Trinity emanate from a dynamic tension between scripture, tradition, reason, and, with John Wesley, I would add experience.[2] They result from attempts by people like you and me to synthesize—to bring together—what we read in the Bible, what we have learned from believers who came before us, and what we understand from our own experience and our ability to reason. The Trinity, for which my cathedral is named, is our human attempt to talk about God and the ways in which God is made known to us.

I still can't explain it, but I describe the Trinity best by metaphor. The Trinity is like three strands of hair making one long braid. It is like three light bulbs in a hallway making one light. It is like steam, ice, and water, different aspects of the same substance. The Trinity is like three images from different angles in a mirror, yet one person. My friend and colleague Lucinda Laird once explained it this way:

> *We rash fools peer into the dark glass and try to see, and even to name, God. We see creation and the love behind it, and we see wisdom, and we see the Christ who dared to be known simply as Jesus from Nazareth, and we hear the murmured endearment: Daddy. We see the father running down the road to greet the loser of a son; we feel the breath of the universe filling us; and we seem cradled in a mother's arms and born on eagle wings. We are drunk on the Spirit that gives life to the bones, and we know . . . the Lover who weeps with us, and the Friend who gives away life for us. Hope smashes in when we've perfected our cynicism, life reappears when we've cuddled up to death, and love makes the world go round. We encounter God. It sure as blazes involves more than two men and a bird, but if the Doctrine of the Trinity is the best that can come out our poor befuddled mouths for the moment, so be it. Amazing that we can talk at all.*[3]

According to tradition, St. Augustine, one of the greatest Christian theologians in all of history, finally concluded that human beings are not capable of fully understanding the mystery of the Trinity. In medieval and renaissance art, there are numerous depictions of the life of St. Augustine. One famous scene is entitled *The Parable of the Holy Trinity.*[4] It is based on an episode described in an apocryphal letter by Cyril of Jerusalem. The story goes that one day, after spending many sleepless nights trying to comprehend the mystery of the Trinity, St. Augustine was walking along the sea in an effort to clear his mind. He came upon a young boy playing on the beach. The boy had dug a hole in the sand, and was attempting, with a spoon, to fill the hole with water. After watching the boy for a few minutes, St. Augustine asked, "Young man, what is it you are trying to accomplish?" The boy looked up and responded, "I'm going to fill this hole with water." St. Augustine laughingly said, "That is an impossible task." The boy stopped, looked the saint straight in the eye, and replied, "I have a far better chance of emptying the oceans of the world into this tiny hole, than you have of completely understanding the mystery of the Holy Trinity." According to legend, the boy then vanished, leaving St. Augustine alone on the beach. This great theologian believed that he had been visited by an angel, and realized that he had reached the limits of his comprehension of God.

As I think back on my conversation at McDonald's, I realize that, like St. Augustine, I had reached the limits of my understanding of the mystery of the Holy Trinity. But I came away from it knowing that this divine interruption would change my life forever and that I would never be alone in this world. I also came to believe that it was God's voice of love I had heard. The voice was love; it came to love, and it teaches us how to love. John Lennon once said, "All you need is love, love, love."[5] As I grow older the voice has become a Trinity of Love: love unbegotten, love incarnate, and love among us now. The Trinity of Love is my way of understanding the mystery we call God, and I don't need to comprehend its meaning any further. I just need to return the love to God and my neighbor with all my heart and soul and might.[6] I have concluded that the doctrine of the Trinity is a set of imperfect words to articulate our imperfect human understanding of our most perfect and loving God who was Jesus on earth, still God, and here with us now.

Confessions of an Anglophool Winemaker! | DESENZANO, ITALY, 2002

Confessions of an
Evangelical Universalist

In my early days at Trinity Cathedral, I learned that some parishioners were taken aback by my boldness to proclaim the good news of Jesus Christ. In fact, at coffee hour one Sunday a person exclaimed with utter surprise, "I didn't realize we had called an evangelical to be our Dean." So let me offer a confession of identity. I want to come out as an evangelical. And in doing so, I want to reclaim that word, "evangelical," which among many progressive people of faith has a bad connotation.

I am an evangelist about many things. I like to share good news: the good news about the Ohio State Fair, the Ohio State football team, and other attractions of the Buckeye State. I enjoy spreading the good word about Lake Erie and the Cuyahoga River, neither of which is dead or on fire. Both are great natural assets of water, wind, and potential for sustainable economic development and revitalization of Northeast Ohio. I get a kick from turning people on to the Rock and Roll Hall of Fame (particularly the album juke boxes), the Cleveland Museum of Art, the Cleveland Orchestra, the West Side Market, Trinity Cathedral (of course), and other cool places in Cleveland. When I lived there I especially loved reporting good news about Paterson and the South Bronx, places that many people liked to avoid. I think I inherited this trait of civic enthusiasm from my mother. In fact, when my cousin Jon was president of the Greater Columbus Chamber of Commerce, he used to joke about putting Mom on his payroll as the city's official cheerleader.

I like spreading the good news about the book I just read, the movie I saw last night, the new restaurant I discovered, the latest cause to which I'm committed, and the political candidate I have chosen to support. I also like to share the good news about God's unconditional love for all God's people. As the daughter of a traveling salesman, I come naturally and genetically to this vocation. Malcolm Gladwell, author of *The Tipping Point*, might call me a connector, a maven, or a salesperson who is contagious in spreading word-of-mouth epidemics.[1] More importantly, I believe that it is my vocation as a baptized Christian to proclaim in both word and action the all-inclusive love of God made known in Jesus of Nazareth.

God has given us an incredibly wonderful gift in Jesus, and I want to share that gift with everyone I meet. As Bishop Desmond Tutu has said so many times, in the life, death, and resurrection of Jesus Christ, we have proof that "goodness and laughter and peace and compassion and gentleness and forgiveness and reconciliation will have the last word and prevail over their ghastly counterparts."[2] This is such good news that it has to be proclaimed. To keep it to myself is unthinkable. It's like not giving a starving man food or a drowning woman a lifeline.

I have found life and love in the way of this Jesus. I am part of this Christian vine, and I want to invite others to join me. At the same time, I am a universalist. While I come to the knowledge of God's love and truth through the life, death, and resurrection of Jesus of Nazareth, I believe that there are other paths to God's eternal realm of love. How could there not be multiple paths to the Eternal One who is beyond words, symbols, definitions, and human comprehension? As God said to Job out of the whirlwind, "Who is this that darkens counsel by words without knowledge?" (Job 38:2). All of us see with imperfect vision through the mirror dimly and the glass darkly. Who are we to insist upon ultimate knowledge of the Divine Mystery we call God?

In a world that insists upon religious labels, I decided to call myself an "Evangelical Universalist

Episcopalian"—EUE for short. In doing so, I claim to be a part of a particular vine in God's immensely complex vineyard.

A few years ago, as I cleared a patch of ivy and pachysandra for a backyard garden, it became clear to me why Jesus used the image of the vine in his farewell discourse in John's gospel. Vines are amazing forms of vegetation. They are intertwined, and they spread out. They can easily be transplanted because they grow just about everywhere. Vines are also very difficult to kill. If you don't believe me, try to get rid of the ivy in your yard. Just when you think you've conquered the vine, it reappears—first with a little leaf, and then before you know it, it's back in full power.

And so it is with divine love. Just when you think you've killed it, it reemerges. You can kill the body, but you can't kill the spirit of divine love. Survivors of the Holocaust help me remember this truth. Nelson Mandela, Desmond Tutu, and other leaders in South Africa reinforce this conviction. Martin Luther King Jr. and Dietrich Bonhoeffer died for this truth. The parents of Abner Louema, Matthew Shephard, and other victims of violence who seek reconciliation rather than revenge are sages of the wisdom that says you simply can't kill the Spirit of God's love. Thus, the truth of resurrection is proclaimed; the power of God's love has the final word. And this indeed is good news—news worthy of being shared to the ends of the earth.

Jesus instructs his friends and followers to abide in his love. The word "abide" comes from the Greek verb *menein*, which may be translated as "stay," "dwell," or "remain." The metaphor of "I am the vine and you are the branches" symbolizes and expresses the dynamics of a believer's relationship to Jesus. We are called to stay, dwell, remain, or abide in the love expressed by our crucified and risen Savior.

Through staying, dwelling, remaining, and abiding in this love, we are able to bear the fruit of mutual love—love for one another and love for the world around us. In bearing this fruit, we demonstrate that we are becoming disciples of Jesus. By abiding in this love and by sharing this love, we carry on the mission of Jesus. And by sharing the mission of Jesus, the world may come to know the good news of God through Jesus without necessarily accepting Jesus as the only way to God.

The challenge of a life of Christian witness in an interreligious world is how to abide in Christ and share the love of God with those who belong to different vines in the vineyard. A number of years ago, I cochaired a task force on interfaith relations for the Episcopal Church in northern New Jersey (known as the Diocese of Newark). Starting with St. Paul's bold proclamation, "Woe to me if I do not proclaim the gospel" (1 Cor. 9:16), we concurred that the issue was not if, but how we should bear witness to our life-giving faith in a pluralistic age. After months of story-telling, conversation, discussion, and debate, we concluded that in a world where God's grace "blows where it will," Christians are called to treat members of other faiths as neighbors with valid beliefs; to recognize the Divine Image in all people; to avoid using language and behavior that intimidates, oppresses, vilifies, demonizes, and distorts the other; and to affirm the presence and sovereign action of God among all faith communities.[3]

In a pluralistic world, Christians are called and commissioned to say, "Come, see, and taste the fruit my vines have produced," rather than "Your vines are barren or your fruit is sour." Life in the global village allows us to appreciate many different varieties of fruit from God's garden and still claim the one that best feeds and nourishes us.

I think Jesus came to earth to help us understand that God is the vinedresser in the vineyard in which we all grow. God knows how to tend, water, fertilize, weed, and even prune. Our responsibility is to be in right relationship with the owner of the vineyard. Right relationship calls us to relinquish ultimate control of the vineyard and share its abundance with all its inhabitants.

In the bestseller *Life of Pi*, author Yann Martel writes a story about a remarkable young boy in India who explored and practiced Hinduism, Christianity, and Islam. When his parents and clergy discovered Pi's pluralistic and universal religious leanings, they insisted, "he can't be a Hindu, a Christian, and a Muslim. It's impossible. He must choose." But Pi responded, "Bapu Gandhi said, 'All religions are true.' I just want to love God." And that was, according to Pi, his "introduction to interfaith dialogue."[4] What Pi learned in his prayers to Christ, his offerings to Lord Krishna, and study of the Koran was that "Hindus, in their capacity for love, are indeed hairless Christians, just as Muslims, in the way they see God in everything, are bearded Hindus, and Christians, in their devotion to God, are hat-wearing Muslims."[5] I believe that Pi discovered the universal truth about religion: the paths to salvation are numerous, but the saving God is one and the same, just known to humanity by many names and faces.

To move beyond the perils of exclusivity and to live into the true discipleship of embracing love, we have to prune away the vines that strangle good will and common survival. One of the branches that needs to be clipped and pruned is scripture verses and creedal statements that exclude and negate certain of God's beloved children. For example, let us consider the familiar and comforting words from the fourteenth chapter of John's Gospel often read at funerals. "I am the way, and the truth, and the life. *No one* comes to the Father except through me" (John 14:6). The first half of the verse, which proclaims that Jesus is the way to our God, assures us that we have a place in the divine eternity, comforts us that everything will be all right, and is precisely what many of us need to hear at the graveside of a loved one. However, I will no longer read in public worship the second half of that beloved passage of scripture: "*No one* comes to the Father except through me." In a post-Holocaust world of continued cultic, ethnic, racial, and religious hatred, divisiveness, judgement, and destruction, I cannot proclaim in public worship these exclusive Christocentric words. How can we as followers of Jesus allow ourselves to assert either publicly or even privately that we believe Christ to be the only way to the Holy One, to

God the Father/Mother/Creator of us all?

John's fourteenth chapter is part of a longer unit (chapters 14–16) known by scholars as "The Farewell Discourse." Its setting is a second-story room in Jerusalem on the eve of Jesus' crucifixion. Its content includes the last supper, the washing of feet, Jesus' prediction of his own passion, and Peter's denial. In the text, Jesus tells his faithful disciples that he is about to be executed. Understandably, they are confused, unsettled, agitated, disturbed, and troubled. Thus Jesus seeks to reassure his friends and followers that, in spite of what has happened and is about to happen, everything will be all right. "Do not let your hearts be troubled. Believe in God, believe also in me" (John 14:1). That's good, sound, and loving advice from any leader on the eve of a major conflict.

There is something else going on here, however. The text does not stand alone out of time and space. What we have, as is often the case in John's Gospel, is a "two-level drama."[6] When we read the Fourth Gospel, we recall both the story of Jesus and the story of the rejected Jewish community or sect that believed in Christ approximately sixty to seventy years after his death. At that time, in about 85 C.E., in the Jewish community of the city of Jamnia, the rabbinical council had just issued an edict that Christians, known as Nazarenes, were no longer welcome in the synagogue, the center of Jewish life. A variety of political and religious motives led to this proclamation, not the least of which was that Judaism and the Jewish people were seeking a more stable relationship with the powerful and oppressive Roman Empire. At the same time, the Jesus movement was separating itself from the synagogue.

The Johannine community, a group of Jewish Christians, found themselves to be a synagogue people without a synagogue, a church people without a church. Out of their own context, they recorded a unique Gospel account that reflects their separation and isolation from Judaism as well as their faith in the Risen Christ. Therefore, much of John's Gospel has references to Jewish customs, traditions, holidays, and symbols, as well as references to "the Jews" who did not believe. In short, through the Fourth Gospel we see in the making a polemic response to an ethnic cleansing and a religious feud. Thus, the language of this Gospel is a reflection of an in-house fight, and the battle lines in John's text are drawn clean and clear. John says in no uncertain terms, if you've been thrown out of the synagogue or if you've found yourself excluded from the family table because of your faith in Jesus, you have made the right choice. "I am the way, the truth, and the life . . ." is written in the context of rejected Judaism.

The way, the truth, and the life are all Torah (or Hebrew Scripture) images. In the final book of the Torah, it is written, "I call heaven and earth to witness against you today that I have set before you life and death, blessings and curses. Choose life so that you and your descendants may live, loving the Lord your God, obeying him, and holding fast to him; for that means life to you . . ."(Deut. 30: 19–20). According to the Johannine community, belief in Jesus—obedience to his way—was parallel to the Deuteronomic commands of obedience to the Torah, the law of Judaism. Over and against Jewish tradition, Jesus had become for these outcasts and exiled Jewish Christians the *only* way to God's saving grace. This theme of us-against-them is central to the entire Gospel of John, and comes to its dramatic climax in Jesus' final discourse. In the Fourth Gospel, more than in any other Gospel, believers are called to choose—one way or another. There is no middle ground, *via media*, no grey, just black and white.

So what does this mean for Christians today? How does this familiar and comforting text, "I am the way, the truth, and the life," inform our faith in a world haunted by the memories of the Holocaust, the pogroms, and the Crusades? How is this passage of Holy Scripture to be applied in a world filled with the horrors of ethnic cleansing, religious persecution, racially motivated civil wars, and rampant fundamentalism? How can any one make exclusive scripture or creedal claims in a world that uses religious "truth" as a rationale for terrorism, murder, bombing, imprisonment, death squads, concentration camps, hate speech, and war?

If we say we believe that theology is a reflection of our faith, then what do we do with the exclusive claims in John's Gospel? Do we dare insist that Jesus is the only way to God? Do we dare proclaim that salvation does not exist outside the realm of Jesus Christ? Do we dare set aside all other religious traditions and argue for the truth of only one?

No, I say. Absolutely not! Jesus is my way, my truth, and my life. I come to God through Jesus of Nazareth. I claim Jesus as my Lord and Savior. But, in a post-Holocaust world, where millions of faithful Jews died because they did not follow the way of Jesus, I cannot insist that Jesus is the only way. In a world of ethnic and religious warfare, I cannot proclaim Jesus to be the only door to God. In a world of violence and killing because people are different, I cannot make exclusive claims about divine truth.

While we can hardly toss this verse (and others like it) out of the Bible, we can certainly argue for a consideration that these claims of religious exclusivity are not consistent with the rest of the Jesus message of loving one's neighbor as oneself; nor are they compatible with Jesus' deep respect and affirmation of his own Jewish heritage and faith. This verse, and others like it, is based on the circumstances of fear, hurt, and exclusion, which was the reality of those who recorded the Fourth Gospel. Do we throw them out? No, but we need to remember our history and the whole of the story. It is not a verdict about the validity of all religions for all time; rather it is pastoral exhortation to a particular group of people in a particular historical context. Do we proclaim them as Gospel truth, as good news for all people? Yes and no. While I believe we should be compelled to excise from public worship the second half of the verse, "No one comes to the Father except through me," I think we can claim with earnest conviction the first half of the verse, "I am the way, the truth and the life." In his book *Reading the Bible Again for the First Time*, Marcus Borg recalls a sermon he once heard preached by a Hindu professor in Christian seminary. "The text for the day included 'the one way' passage, and about it he said, 'this verse is absolutely true—Jesus is the only way.' But he went on to say, 'And that way—of dying to an old way of being and being born into a new way of being—is known in all of the religions of the world.' The way of Jesus is a universal way, known to millions who have never heard of Jesus."[7]

When it comes to matters of faith certitude, I would rather stand with St. Paul who wrote in his first letter to the Christian church in Corinth: "For now we see in a mirror, dimly, but then we will see face to face. Now I know only in part; then I will know fully, even as I have been fully known" (1 Cor. 13:12). In this mixed up world of violent divisiveness, I would rather claim openness to religious understanding and take my chances on judgement before God at the gates of heaven. I am more willing to risk divine judgement at the end of my life than to participate in religious arrogance that says, "My way is the only way."

We can make faith claims without being exclusive. We can proclaim with integrity that Jesus is our way, our truth, and our life. We can say that we come to God through Jesus. We can say that Jesus is in God and God is in Jesus. We can say that in God's house, there is a dwelling place for us because Jesus has gone there and will show us the way. We can say all of these things and more to be true without insisting that we have the only truth.

As Christians, as followers of the Jesus way, we need to show moral leadership in this confused and embattled world. And that moral leadership is to work for unity while honoring diversity. That moral leadership proclaims that the truth lies in the grey—not the black or the white. And that the God who created us, redeems us, and sustains us is far greater, larger, and more incomprehensible that any one faith tradition can claim.

While abiding in God's eternal love, each of us needs to ask the question: which vines should be removed from the garden of my life and my community? What are the ancillary vines that drain nutrients and energy away from the building up of God's realm? What are the poisonous vines that keep my kindred and me from dwelling in and spreading the love of God?

As we walk through life, we are called to be workers and stewards in God's vineyard. Daily we need to ask God, the owner of the vineyard, to help us see what needs to be pruned, to give us the courage to pick up the sickle and to grant us the wisdom to cut away that which insists on its own way as the only way. This is a venture we call the agriculture of faith, otherwise known as discipleship. And in being God's farmers, we can build up the body of Christ and reclaim with boldness the word "evangelical" while respecting the word "universal." For me, that is a ditch worth digging, and if necessary, a ditch worth dying in, knowing that on the other side is the promise of life eternal in God's garden of life.

Seven

And it calls out...

All-ee All-ee In Free | GUATEMALA, 1999

All-ee All-ee in Free

I had an amazing experience of the holy in our cathedral one Shrove Tuesday. We were in the midst of our annual Mardi Gras Party. Pancakes and sausages were being served in Cathedral Hall to the accompaniment of a Dixieland jazz band. Our youngest children were making peace banners and paper doves on the stage, and adults were engaged in lively conversations around the room. Walking through the promenade, a parishioner asked me: "Is it all right for kids to be in the cathedral tonight?" I agreed to investigate the situation.

I walked into the cathedral and found a group of young boys playing hide-and-go-seek in the shadows of the nave. The pulpit was the countdown location, and the altar was decorated with Mardi Gras beads and paper peace doves had been designated home base. Off to the side was an adult stewarding the labyrinth by candlelight. I sat down next to her, stared at the labyrinth, glanced at the altar, looked at the boys and said: "Now, this is my idea of church." She readily agreed, and before I knew it, I was up on my feet to join the game of hide-and-go-seek. A younger boy squealed with delight: "Look, Dean Tracey is playing. Get her." The next thing I knew I was "it"—up in the pulpit counting to ten and then running through the darkened nave chasing after an eleven-year-old child. A few minutes later, parents began to wander in looking for their children. One by one, they too joined the game. Eventually, there were about ten kids and five adults playing hide-and-go-seek in the holy space on the night before Ash Wednesday.

Later that evening, I remembered the game's homecoming cry, "All-ee, All-ee in free" or as some remember it, "All-ee, All-ee oxen free." In older English it was "All ye, all ye out in free," which is to say, "All who are out come in for free." I thought to myself, this is what the church is all about. In fact, this is the good news of the gospel. All who, for whatever reason, find themselves on the outside, on the margins, on the edge are invited to come in to a place of safety, a place called home—for free.

About the Photographs

The Question | NEW YORK CITY, 1999

For a number of years I met with a spiritual director on West Twentieth Street. Every week, as I walked to the subway, I would pass a fortuneteller's shop. Sometimes I wanted to ask her the questions I couldn't seem to answer.

The Light of Darkness | CLEVELAND, OHIO, 2003

As I was struggling with this chapter about light and dark, I spent many night hours sitting in my darkened dining room staring at the Advent wreath. One evening I realized that for me the candles of Advent represent not a light in the darkness, but the light of darkness itself.

Just Another Homeless Family | PATERSON, NEW JERSEY, 1998

This is Lisa and Ivan. I made their portrait downtown one day when we were all watching the filming of *The Preacher's Wife*. They both died a few years ago. I still miss their smiles.

A Baptism to Remember | CLEVELAND, OHIO, 2001

Once in a while, I carry a small camera in the pocket of my vestments. I made this photograph during the Christmas Eve Pageant at Trinity Cathedral. Mary, Joseph, and the baby Jesus are all young members of our congregation. They are not homeless, but they do remind me of the utter wonder and amazement in the Christmas story, that with the birth of every child, the human potential is born again.

A Child Shall Lead Us | PERU, 2001

He was leading a rag-tag band of a dozen boys in the school yard. Pausing to catch a breath, he looked up at me. I signaled with my camera, "May I make a photo of you?" With a big grin, he nodded and then proceeded to blow his own horn. Children are not only our future; they are our here-and-now.

Why I Do Ashes | ORADOUR-SUR-GLANE, FRANCE, 1997

Among German crimes of the Second World War, the Nazi massacre of 642 men, women, and children at Oradour-sur-Glane on June 10, 1944, is notorious. On that Saturday afternoon, SS troops encircled the town. Soldiers marched the men to one building, lined them up, and shot them. They then locked the women and children in the church, shot them, and set the building and the rest of the town on fire. In 1946 the French State established the entire ruins of the village as a war memorial. This is the sign at the entrance to the village.

The Preacher's Prayer on the Eve of War | BOSTON, MASSACHUSETTS, 2001

"The Partisans" by Andrezei Pitnyski is a tribute to guerrilla freedom fighters everywhere. This war memorial in the Public Garden of Boston Commons symbolizes the horrors of war. Will we ever learn?

The Garbage Tree | PATERSON, NEW JERSEY, 1997

Once in a while I would walk around my parish neighborhood with camera in hand, seeking signs of

God's spirit in our midst. On a cold winter morning, I looked up and saw this tree. It was filled with the remnants of life: a broken doll, an old mop, a bicycle wheel, and a shoe. It reminds me that God makes new out of old, whole out of brokenness, and life out of death. The Garbage Tree reminds me of the gospel and teaches me that with God all things are possible.

Stolen Shoes | PATERSON, NEW JERSEY, 1997

This is my friend Yvonne. She carried all her possessions in a shopping cart and lived on the streets for many years. We were very close. I saw Christ through her, but I could never convince her to move indoors.

The Unfinished Story | LOUISVILLE, KENTUCKY, 2000

Have you ever seen a gravestone that said it all? I did one winter morning, and I said a prayer for a woman I never knew. Thank God her gravestone was not the end of the story.

Welcome the Risen Christ | PATERSON, NEW JERSEY, 1998

This is Bacardi. Actually, that's his street name. His real name is Alfredo. He was the first person I met on my first Sunday in Paterson. We became friends and watched out for each other. I made this photo as we watched the filming of *The Preacher's Wife* on a winter day in downtown Paterson.

Life Happens in the Interruptions | NEW HOPE, PENNSYLVANIA, 1998

This friendly creature was looking at me from atop a wall in New Hope. He reminds me that life happens in the most unexpected times and places.

Mountain Musings | PERU, 2001

On top of a mountain almost 14,000 feet above sea level, I met an old woman tending her flock and spinning her yarn. She knew her sheep by name and they came when she called them. Now I have a greater appreciation for Jesus' metaphor of the Good Shepherd.

Crossing the Great Water | PUNTA CANA, DOMINICAN REPUBLIC, 2002

Every time I see small fishing boats, I try to imagine what it must have been like on the Sea of Galilee that night when Jesus calmed the storm.

Keep on Singing | AMERICAN BEACH, FLORIDA, 1999

American Beach is a historic landmark on Amelia Island, Florida. It was one of the first African American beach resorts in the segregated south. Ma Vynee Betsch or "Beach Lady," as I knew her, lives in a trailer on American Beach. She keeps local history and customs alive in her informal museum. She has very long hair adorned with solidarity buttons. After a lengthy conversation about her community, Ms. Betsch allowed me to make her portrait. In her face, I could see Miriam dancing at the Red Sea with timbres and lyre.

Bread of Life | PATERSON, NEW JERSEY, 1998

At St. Paul's Church, we assembled hundreds of Thanksgiving baskets in the sanctuary. We lined them up on the pews, and gave them out to our neighbors over the next few days before Thanksgiving. The large grocery bags filled with food sat in over half of the pews. One year we taped photographs of people's faces on the bags to remind us that those receiving Thanksgiving baskets were also part of our gathered community.

Sally's Feast | PARIS, FRANCE, 1998

I spent two summers in France. On each visit to Paris, I met the same beggar woman outside of Notre Dame Cathedral. Both times, she was wearing the identical coat, shoes, and hat. I gave her some money and she gave me permission to make her portrait. She reminded me of Jesus' story about the widow's mite.

Beckoned to the Banquet | CLEVELAND, OHIO, 2003

Every year the Cleveland Marathon starts and finishes in front of Trinity Cathedral. We always provide hospitality and refreshment to the runners and onlookers. One year I arrived to find a ten-foot fence erected in front of our doors. Our diocesan youth made banners to say that in spite of the fence, all are welcome in this house of God.

The Demolition Contractor | PATERSON, NEW JERSEY, 1997

I came upon this abandoned crack house. A woman of faith from the suburbs had decided to convert it into an urban retreat center. As I looked out the window at the street below, I was reminded of E. M. Forster's novel *A Room with a View*. While I know this wasn't the view he had in mind, I saw hope, promise, and love in the midst of poverty and despair.

Love for the Fallen Flower | BADLANDS, SOUTH DAKOTA, 1999

I went on a photography workshop to study the human figure in landscape. Our first assignment was to photograph each other's feet and hands. One participant took off her shoes and presented this tattoo. I saw the same flower in the landscape, and it reminded me of the synchronicity of God's creation.

The Trinity of Love | PARAMUS, NEW JERSEY, 1997

This was the first photograph I ever made with a 35 mm camera. It was Valentine's Day, and I went walking through a local park. The three strong branches of this tree with a heart in the middle remind me of the loving community of God.

Confessions of an Evangelical Universalist | DESENZANO, ITALY, 2002

Sometimes graffiti is good. These words were painted on the sidewalk in front of a village church. "I love you Jesus" says it all.

All-ee All-ee in Free | GUATEMALA, 1999

Sometimes we when we try to hide, others seek us out and call us in—for free.

Notes

Preface

1. Christopher Logue, "Come to the Edge," *Selected Poems* (London: Faber and Faber Ltd., 1996), 64.

2. Ina Hughes, "Doing Theology through Personal Narrative," *The Witness Magazine* (December 2000), 9.

Introduction

1. Iggeres Haramban, *A Letter for the Ages* (Brooklyn: Mesorah Publications, 1989), 47.

2. Gordon Lightfoot, "Sit Down Young Stranger" (Burbank, Calif.: Warner Brothers, 1970).

3. This reference comes from the Book of Thomas the Contender 138.7–19, Nag Hammadi Library 189 as quoted by Elaine Pagels in *Beyond Belief: The Secret Gospel of Thomas* (New York: Random House, 2003), 57.

4. Harold Kushner, *When Bad Things Happen to Good People* (New York: Avon Books, 1981).

5. St. Paul's Episcopal Church Mission Statement.

6. Holly Near, "It Could Have Been Me" (Ukiah, Calif.: Redwood Records, 1974).

7. William Stringfellow, "Living Humanly in the Midst of Death," *A Keeper of the Word: Selected Writings of William Stringfellow*," edited by Bill Wylie Kellerman (Grand Rapids: William B. Eerdmans, 1994), 345.

8. Ibid.

The Light of Darkness

1. St. John of the Cross, *Dark Night of the Soul*, translated and edited with an introduction by E.

Allison Peers (New York: Image Books, Doubleday, 1990).

2. Criogono do Jesus, *The Life of St. John of the Cross*, translated by Kathleen Pond (London: Longmans, Green & Company, 1958), 105.

3. John Kirvan, *God Hunger: Discovering the Mystic in All of Us* (Notre Dame, Ind.: Sorin Books, 1999), 96.

4. Edith Hamilton, *Mythology*, Mentor Book ed. (New York: Little, Brown, 1942), 63–64.

5. Ibid., 47.

6. John Milton, *Paradise Lost*, Book VII, verse 210 (www.literature.org).

7. This quotation is intentionally taken from the King James Version of the Bible. In the New Revised Standard Version, John 1:5 says, "The light shines in the darkness, and the darkness did not overcome it."

8. *A New Zealand Prayer Book—He Karakia Mihinare o Aotearoa* (Rotorua, New Zealand: Church of the Province of New Zealand, 1989), 184. Used with permission.

9. "This Little Light of Mine," African American spiritual.

A Baptism to Remember

1. Martin Luther, "The Pagan Servitude of the Church," *Martin Luther: Selections from His Writing*, edited by John Dillenberger (Garden City, N.J.: Anchor Books, 1961), 291–314.

2. *Amistad*, produced and directed by Steven Spielberg, 1997.

3. James Agee, *Let Us Now Praise Famous Men* (Boston: Houghton Mifflin, 1939), 255.

A Child Shall Lead Us

1. I have paraphrased this story entitled "What the Wind Said to Thajir," by Martin Bell, *The Way of the Wolf: The Gospel in New Images* (New York: Ballatine Books, 1968), 23–28.

2. Edmund McDonald, *Presbyterian Outlook* (Richmond, Va.: The Presbyterian Outlook Foundation). The magazine's publisher is unable to ascertain the date of publication.

3. *The State of America's Children Yearbook 1997* (Washington, D.C.: The Children's Defense

Fund).

4. Frederick Buechner, *Wishful Thinking* (New York: Harper & Row, 1973), 13.

Why I Do Ashes

1. Phyllis Trible, *God and the Rhetoric of Sexuality* (Philadelphia: Fortress Press, 1978), 72.

2. There are two creation myths in the Book of Genesis: Genesis 1:1–2:3 and Genesis 2:4–3:24. I am referring to the second account.

3. *The Book of Common Prayer of the Episcopal Church, U.S.A.* (New York: Church Hymnal Corporation/Seabury, 1979), 499.

4. *Book of Common Prayer*, 264.

The Preacher's Prayer on the Eve of War

1. Roland Gittelsohn, *Man's Best Hope* (New York: Random House, 1961), 23.

Stolen Shoes

1. Abraham J. Heschel, *The Prophets* (New York: Harper Torchbooks, 1962), 278.

2. I am indebted to John Dominic Crossan, Marcus Borg, and Robert McAfee Brown for much of my thinking about the revolutionary politics of Jesus.

3. Myles Horton, *The Long Haul* (New York: Doubleday, 1990), 80–81.

The Unfinished Story

1. Thomas Merton, *He Is Risen* (Wiles, Ill.: Argus Communications, 1984), 5.

Life Happens in the Interruptions

1. I read these words from Johann Baptist Metz in Alan Revering's "God Bless America: Patriotism and Political Theology," unpublished paper given at the American Academy of Religion Annual

Meeting, Toronto (Nov. 25, 2002).

Mountain Musings

1. Alice Walker, *Living by the Word: Selected Writings 1973–1987* (San Diego: Harcourt Brace Jovanovich, 1988), 2.

2. Robert Farrar Capon, *The Astonished Heart: Reclaiming the Good News from the Lost-and-Found of Church History* (Grand Rapids: William B. Eerdmans, 1996), 14.

3. Martin Luther King Jr., "I See the Promised Land," *A Testament of Hope: The Essential Writings of Martin Luther King Jr.*, edited by James M. Washington (San Francisco: Harper & Row, 1986), 286.

4. Ernesto Cardenal, *The Gospel in Solentiname*, vol. 3 (Maryknoll, N.Y.: Orbis Books, 1979), 5.

5. Carol Etzler, "Sometimes I Wish" (Bridgeport, Vt.: Sisters Unlimited, 1974).

6. Nelle Morton, *The Journey Is Home* (Boston: Beacon Press, 1991), xix.

Crossing the Great Water

1. When I consult *The I-Ching* or *Book of Changes*, I generally use the Richard Wilhem translation from the Chinese into German, rendered into English by Cary F. Baynes (Princeton, N.J.: Princeton University Press, 1977).

2. Ibid., 75.

3. Viktor E. Fankl, *Man's Search for Meaning* (New York: Simon and Schuster, 1959), 136.

4. Ibid., 67.

5. Robert Frost, "Mending Wall," *The Norton Anthology of Poetry*, 3rd ed. (New York: W.W. Norton, 1983), 908.

6. *The I-Ching*, 103.

7. Frankl, *Man's Searching for Meaning*, 65.

Keep on Singing

1. The Rt. Rev. Michael Curry, Bishop of the Episcopal Diocese of North Carolina, unpublished sermon printed with permission.

2. Marcus Borg, unpublished address at Trinity Cathedral, Cleveland, Ohio, 2002.

3. "How Can I Keep From Singing," Words by Robert Lowry (1826–1899).

4. "The Peatbog Soldiers," *Die Moorsoldaten*, words reprinted in the songbook *Winds of the People*, 88. Originally published in *Songs of Work and Protest*, edited by Edith Fowke and Joe Glazer (Dover Press, 1973).

5. "Southland Concerto," traditional, arranged by J. Clegg (HRBV Music, ASCAP) from the movie *The Power of One*, original score by Hans Zimmer.

6. Etty Hillesum, *An Interrupted Life: The Diaries and Letters of Etty Hillesum 1941–43* (New York: Pantheon Books, 1981), xiii.

7. "For All the Saints," words by William Walsham How (1823–1897).

8. Anne Tyler, *Saint Maybe* (New York: Ballantine Books, 1991).

9. "A Woman's Love," arranged by Alex Dobkin and Kay Gardner (Preston Hollow, N.Y.: Lavender Jane, 1975).

10. "Jesus loves me," words and music by Anna Warner (1820–1915).

Sally's Feast

1. The Rev. Douglas Fromm, minister at the Upper Ridgewood Community Church in Ridgewood, New Jersey, told me this story. I have changed the woman's name to Sally.

Beckoned to the Banquet

1. Edwin Markham, "Outwitted," *Bartlett's Familiar Quotations*, 15th Ed. (Boston: Little, Brown and Company, 1980), 671.

2. In this passage, Jesus is actually quoting Hosea 6:6, but the same phrase is found in Micah 6:8.

3. "The Sisters of Mercy," Copyright © Leonard Cohen and Sony/ATV Music Publishing Canada Company.

The Demolition Contractor

1. C. S. Lewis, *Mere Christianity*, book IV, chapter 9 (New York: Touchstone, Simon and Schuster, 1996) 176.

2. Parker Palmer, *The Company of Strangers: Christians and the Renewal of America's Public Life* (New York: Crossroad Publishing, 1983), 62.

3. Quotes are taken from personal notes. However, Dr. Wind's entire address, *Place of Nostalgia or Place of Reconstruction? The Role of Cathedrals in the New Millennium*, has been published in the *Proceedings of the 2000 North American Cathedral Deans 199 Conference* (The Washington National Cathedral, Washington, D.C.).

4. Tracey Lind, *2001 Annual Dean's Address*, Trinity Cathedral, Cleveland, Ohio.

5. Dee Hock, *The Chaordic Organization* (San Francisco: Berrett-Koehler, 1999).

6. Frederick Buechner, *Wishful Thinking: A Theological ABC* (New York: Harper & Row, 1973), 95.

7. "Be like a piazza" was first explained to me by Geri Hable, a former Urusline nun.

8. *Why Christianity Must Change or Die* is John Shelby Spong's book published in 1998 by HarperSanFrancisco.

9. Annie Dillard, *Teaching a Stone to Talk: Expeditions and Encounters* (New York: Harper and Row, 1982), 40.

Love for the Fallen Flower

1. Bill McKibbon, "Indifferent to a Planet in Pain," *New York Times*, 4 September 1989.

2. Robert S. Gottlieb, *A Spirituality of Resistance: Finding a Peaceful Heart and Protecting the Earth* (New York: Crossroad Publishing, 1999), 29.

3. Rachel Carson, letter to a friend in 1962. Quoted in numerous web sites and articles about Rachel Carson including, "Time's 100 Most Important People in the Century" (www.time.com/time/time100/scientist/profile/carson03.html).

4. For a more extensive treatment of the subject of broadening our definition of neighbor, see Carolyn Irish's article, "Who Is My Neighbor" in *The Witness*, October 1998, 21.

5. David Bank, "All Species Great and Small," *The Wall Street Journal*, 22 January 2002.

6. Natural Resources Defense Council web site (www.wrdc.org).

7. Thomas Berry, *The Great Work: Our Way into the Future*, (New York: Bell Tower, 1999), x–xi.

8. "Simple Gifts," eighteenth-century Shaker song.

9. Jürgen Moltmann, *God in Creation: A New Theology of Creation and the Spirit of God*. This quotation was found in "One God, Family, Earth: the Episcopal Church's Guide for Environmental Education."

10. Robert Corin Morris, "What is blocking us?" *Living with the Earth: The Interweave Journal* (Fall 1996): 11.

11. Pierre Teilhard de Chardin. This quotation was found in "One God, Family, Earth: The Episcopal Church's Guide for Environmental Education."

The Trinity of Love

1. *The Book of Common Prayer of the Episcopal Church, U.S.A.* (New York: Church Hymnal Corporation/Seabury, 1979), 526.

2. According to Urban Holmes in *What is Anglicanism?* (Harrisburg, Pa.: Morehouse Publishing, 2002), 11: "[Richard] Hooker (1554–1600) articulates for Anglicanism its answer to the question of what is our authority. Our authority is the association of Scripture, tradition and reason. Subsequent commentators have spoken of this as a 'three-legged stool' . . . The threefold nature of authority—Scripture, tradition and reason—is not original with Hooker; but sixteenth century Anglicanism felt no compulsion to make claims of originality, since it conceived of itself as the continuing Catholic Church in England." In the eighteenth century, John Wesley added experience to the nature of authority.

3. Lucinda Laird, Rector of St. Mathew's Episcopal Church, Louisville, Kentucky, unpublished sermon.

4. *The Parable of the Holy Trinity* can be viewed at the Church of Saint Augustine, San Gimignano, Italy, in the Apsidal chapel, scene 12, south wall. Another rendition can also be viewed at the Metropolitan Museum of Art, New York City, The Cloisters Collection (61.199). Here is it presented

as a fifteenth-century Flemish painting under the title of *Scenes from the Life of Augustine*, Master of St. Augustine, circa 1691.

5. "All You Need is Love," John Lennon and Paul McCartney © 1967.

6. Deuteronomy 6:5 and Luke 10:27.

Confessions of an Evangelical Universalist

1. Malcolm Gladwell, *The Tipping Point: How Little Things Can Make a Big Difference* (Boston: Little, Brown, 2000), 14.

2. Desmond Tutu, *No Future without Forgiveness* (New York: Doubleday, 1999), 267.

3. The Task Force on Christian Mission in an Interreligious World, Robert Morris and Tracey Lind, cochairs, *A Handbook for Interreligious Relations: Report to the 122nd Diocesan Convention of the Diocese of Newark* (Newark, N.J.: Diocese of Newark, 1997).

4. Yann Martel, *Life of Pi* (Orlando: Harcourt Books, 2001), 69–70.

5. Ibid., 50.

6. My understanding about the "two-level drama" in John's Gospel and the Johannine community of faith are informed by my seminary studies with Raymond E. Brown. For further review of Professor Brown's scholarship on the subject, please see *The Anchor Bible Commentary on John* (Garden City, N.Y.: Doubleday, 1966), and *The Community of the Beloved Disciple* (New York: Paulist Press, 1979).

7. Marcus J. Borg, *Reading the Bible Again for the First Time: Taking the Bible Seriously but Not Literally* (San Francisco: HarperSanFrancisco, 2001), 216.